Business Administration

KNOWLEDGE-BASED DESIGN SYSTEMS

R. D. COYNE, M. A. ROSENMAN, A. D. RADFORD,
M. BALACHANDRAN, and J. S. GERO

University of Sydney

 Addison-Wesley Publishing Company

Reading, Massachusetts ● Menlo Park, California ● New York
Don Mills, Ontario ● Wokingham, England ● Amsterdam ● Bonn
Sydney ● Singapore ● Tokyo ● Madrid ● San Juan

This book is in the Addison-Wesley Teknowledge Series in Knowledge Engineering.
Frederick Hayes-Roth, Consulting Editor

The programs and applications presented in this book have been included for their instructional value. They have been tested with care, but are not guaranteed for any particular purpose. The publisher does not offer any warranties or representations, nor does it accept any liabilities with respect to the programs or applications.

Many of the designations used by manufacturers and sellers to designate their products are claimed as trademarks. Where those designations appear in this book and Addison-Wesley was aware of a trademark claim, the designations have been printed in caps or initial caps.

Library of Congress Cataloging-in-Publication Data
Knowledge-based design systems/R. D. Coyne ... [et al.].
     p. cm.
    Includes bibliographies and index.
    ISBN 0-201-10381-8
    1. Engineering design—Data processing. 2. Architectural design—Data processing. 3. Computer-aided design. 4. Expert systems (Computer science) I. Coyne, R. D. (Richard D.)
TA174.K59   1989
620'.00425'0285–dc19   89-335
                             CIP

ISBN 0-201-10381-8

ABCDEFGHIJ-DO-943210

Preface

Design is a continuing fundamental human activity. While we all have carried out some design activities ourselves, we use professional designers who are skilled experts whenever significant designs are to be executed. Designers are change agents within a society; their goal is to improve the human condition, in all its aspects, through physical change.

Design as a separate expert activity is very old. We know, for example, the name of the architect who designed Queen Hapshepsut's tomb some 3,500 years ago. Even earlier, Hammurabi, King of Babylon, had enacted a building code that both recognized design and made it dangerous:

> If a builder (designer) has built (designed) a house for a man and his work is not strong, and if the house he has built (designed) falls in and kills the householder, that builder (designer) shall be slain.

Many of the apparently modern ideas in design are not new. Much of knowledge-based design systems is based on systems theory, which can be traced to Plato's *Republic*, in which he used the concept of a system to produce the political structure of a state. In *Ten Books of Architecture*, by the Roman architect and engineer Vitruvius, we find descriptions of both declarative and procedural design processes:

> For in all their [the ancient Greeks] works they proceeded on definite principles of fitness and in ways derived from the truth of Nature. Thus, they reached perfec-

tion, approving only those things which if challenged, can be explained on grounds of the truth. Hence, from the sources which have been described they established and left us rules of symmetry and proportion for each order.

And,

Let the front of a Doric temple, at the place where the columns are put up, be divided, if it is to be tetrastyle, into twenty-seven parts, if hexastyle, into forty-two. One of these parts will be the module; and this module once fixed, all the parts of the work are adjusted by means of calculation based on it.

The notion of design as goal seeking can be traced to the 18th century. Only since the 1960s, however, has the idea of externalizing design from human designers and constructing executable design systems been explored. More recently, the formalization, representation, and manipulation of knowledge in computers has made it possible to construct knowledge-based design systems. Such systems have the potential to produce both fundamental changes in design and better designs.

This book describes the bases, approaches, techniques, and implementations of knowledge-based design systems, and advocates and develops new directions in design systems generally. A formal model of design coupled with the notion of prototypes provides a coherent framework for all that follows and is a platform on which a comprehension of knowledge-based design rests. The book is divided into three parts. Part I, Design, examines and describes design and design processes, providing the context for the remainder of the book. Part II, Representation and Reasoning, explores the kinds of knowledge involved in design and the tools and techniques available for representing and controlling this knowledge. It examines the attributes of design that must be described and the ways in which knowledge-based methods are capable of describing and controlling them. Part III, Knowledge-Based Design, presents in detail the fundamentals of the interpretation of designs, including the role of expert systems in interpreting existing designs, before describing how to produce designs within a knowledge-based environment. This part includes a detailed examination of design processes from the perspective of how to control these processes. Within each of these processes, the place and role of knowledge is presented and examples of knowledge-based design systems given. Finally, we examine central areas of human design and demonstrate what current knowledge-based design systems are capable of doing now and in the future.

The aim of the book is to present the state-of-the-art in knowledge-based design systems. Although primarily intended for graduate students, practitioners, and researchers in the design professions, researchers in knowledge engineering will find it useful. Only a modicum of background in knowledge engineering is necessary since the book concentrates on concepts, applications (through examples), and issues. Readers with a strong artificial intelligence background will find novel ideas here. Examples are used as vehicles for the ideas involved, and a wide variety of design problems are presented. These can be explored further through exercises presented at the end of each chapter. References are provided to texts that elaborate the individual techniques.

Readers of this book should gain an insight into the conceptual basis, application, and potential of knowledge-based design systems and the knowledge necessary to formulate their own design problems for solution using such systems.

Acknowledgments

This book evolved from research and teaching carried out primarily in the Design Computing Unit of the Department of Architectural and Design Science at the University of Sydney, at which this research formed the basis of an extensive teaching program. The ideas presented here have been used in courses taught by the various authors at the University of California—Los Angeles, University of California—Berkeley, University of Technology, Sydney, and Carnegie Mellon University, in addition to the University of Sydney. The students in those courses helped us refine many of these ideas, as did a series of graduate students and visitors to the Unit.

Support for this work has come from numerous grants from the Australian Research Grants Scheme, the Australian Research Council, the Australian Computer Research Board, the University Research Grant, the National Science Foundation (United States), and the Science and Engineering Research Council (United Kingdom).

Fay Sudweeks, the administrative secretary of the Design Computing Unit, converted the text into TEX format files and worked on many of the figures. Her energetic assistance is, as before, much appreciated.

The enthusiastic response to the early manuscript of Rick Hayes-Roth, the series editor, spurred us on. We would also like to acknowledge the various reviewers of the manuscript.

Contents

Design

Design is a fundamental, purposeful human activity. Designers are change agents within a society whose goals are to improve the human condition through physical change. Part I examines design and design processes, providing the context for the remainder of the book.

Chapter 1 introduces the notions involved in formal design and design systems and then describes knowledge in design. Chapter 2 delves into more detail by describing models of design. The design process is characterized as a search for a description that matches some intended meaning within a space of designs defined by generative knowledge. This model provides the framework for knowledge-based design systems. The concept of prototypes and their role in design is introduced.

Design is shown to be modelable by reasoning processes. The three models of reasoning—deductive, abductive, and inductive—are placed within a design context. The notion of reasoning beyond mere calculation is postulated as fundamental to the development of knowledge-based design systems. A transformation model of design that utilizes reasoning is presented.

CHAPTER 1

Introduction to Knowledge-Based Design

Design, which vitally affects the safety and well-being of human lives, covers a wide range of behaviors within a large number of disciplines—graphic art, product design, the many facets of engineering, industrial process, and architecture, to name but a few. The products of these design disciplines perform in varied ways. Some products move or carry loads, or function in a physical sense. Others, such as highways and buildings, are by nature inert. Some artifacts must operate in a manner we might regard as purely functional, while others exist simply to convey meaning.

Issues in design include functionality, meaning, expression, and aesthetics. The study of design is made difficult when we consider that design covers activities that exhibit varying degrees of sophistication. Some of these activities essentially involve nothing more than selecting among alternatives offered in a catalog, finding a suitable configuration for a small number of well-understood components, or slightly modifying a pattern. Design activity may involve adjusting dimensions of some well-established form or producing new forms from an entirely plastic medium. It may also range from the subtle adaptation of well-established prototypes, such as dimensioning a ship's hull, to the creation of something entirely novel, such as the first airplane, which may itself be a prototype for subsequent modification.

In design we utilize bodies of knowledge and certain operational tools. Through the mid-20th century accumulated knowledge was primarily set out in books, and the primary operational tools were drawing implements, slide rules, and calculating machines. In the 1970s, computers began to be used in design. The computer provides a single medium both for representing knowledge and for carrying out the operations of design. We can expect this landmark to produce major changes in the way design is carried out. Especially, developments in design knowledge and design tools are expected to lead to direct links between the two—to *knowledge-based* and *computer-aided* design systems.

In this book we approach the field of knowledge-based design as experienced designers who seek a better way to use computers to assist design activity. We seek an understanding of design that fosters the effective employment of computer systems in the design process. This contrasts with some other legitimate motivations such as better methods of teaching design skills or an understanding of the role of designers and their products in society.

In our endeavor we can appeal to a substantial history of research into design theory and methods. The origin of the design methods school is attributed to the researchers of the 1960s who demonstrated the benefit in assuming that design processes are essentially computable. Computer-aided design became possible when design was reduced to the process of computing

Figure 1.1   An image of a desired state.

values within well-defined problem statements. This emphasis on *numerical computation* limited the applicability of the methods. The current state of design thinking, however, assumes that design activity involves not merely computation but *reasoning with knowledge.* The emphasis now is in finding methods of appropriating and rendering operable the knowledge available to designers.

We use the term "knowledge-based" as a temporary expedient, just as it was once common to talk of "digital computers". The term "digital" was later dropped from common speech as the distinction between analog and digital computers became less important. (The same applies to "motor" cars.) "Knowledge-based" is simply a transitional term, emphasizing obvious differences between this view of design and what has gone before.

In this chapter we first summarize some important assumptions about the nature of design and consider the importance of models. This provides background for the discussion of some influential models of the design process that take their impetus from ideas about science, method, problem solving, logic, and language. We then consider the transition to a general knowledge-based model of design and assert that other models are essentially subsumed in a knowledge-based view. In so doing we address something of the nature of knowledge (in the sense in which the term is frequently used in current computer-aided design research and practice) and prepare for a detailed discussion of a knowledge-based model of computer-aided design in the next chapter.

1.1   The Nature of Design

We start with a basic assumption about design. Design is a *purposeful* activity. We have in mind some idea of a *desired state* (Figure 1.1). Design involves a conscious effort to arrive at a state of affairs in which certain characteristics are evident. The realization that a sharpened piece of wood picked up from the ground can be used to penetrate other objects does not constitute an act of design. The realization that if a blunt piece of wood is sharpened it can be used to penetrate other objects, however, constitutes a rudimentary design act. The latter process is essentially purposeful, while the former is opportunistic.

Design is also the process of originating systems and predicting how these systems will fulfill objectives. When we design we have an end in mind, but this end does not constitute the final artifact. The design is merely a description of the artifact from which predictions of performance can be made. A design is therefore an abstraction, providing a description of an artifact that can be interpreted by some other agent for purposes of manufacture or construction.

A design can therefore consist of a set of instructions or some other abstraction suitable for interpretation and realization as a physical artifact.

More specifically, design has been usefully described as a goal-directed activity that involves the making of decisions (Archer, 1969). We can translate this definition into the process of producing a set of descriptions of an artifact that satisfy a set of given *performance requirements* and *constraints*. Design begins with the realization of a need—a dissatisfaction with existing conditions—and a realization that some course of action must be taken. A goal is set, the purpose of which is to attain a state of satisfaction in response to the need. The ability to set goals and plan for their attainment is a powerful and essential human capability. We can even strive for goals that cannot be realized, such as the pot of gold at the rainbow's end.

Even granting this assumption about the primacy of goals in design, we are at only the edge of understanding what design is. For a start, we often find it difficult to specify the nature of the desired state. Goals themselves are the products of processes that are little understood. A strong determinant of the goals we set is our system of values—the frameworks for our models of reality, our appreciation of what is worth striving for. These vary from culture to culture and between individuals. And individuals are not averse to changing their value systems to suit changes in situation. Even within a particular culture and given a particular goal with which most would agree—such as the "public good"—it is unlikely that everyone would interpret that goal in quite the same way. Different designers would set different objectives. Given an agreement on objectives, fewer would agree on the way to reach these objectives.

Furthermore, goals appear to be a product of what is possible. Goals rarely behave as autonomous entities, each of which can be realized independently. Goals interact. We often cannot know what is possible until we have undertaken some kind of exploration. Design sometimes appears as an activity of exploration, in which partial responses lead to the redefinition of goals (Bijl, 1987).

Design also appears to be ill-structured in the sense that there is no straightforward process to be followed. A designer may start with goals and work toward the design of objects that will satisfy these goals, or may begin with the objects and check whether the properties of these objects match those of the requirements. This may involve a "top-down" approach, a "bottom-up" approach, or a combination of approaches. To solve a problem of any size and complexity, many approaches will be attempted, perhaps in parallel. There also appears to be no fixed starting point. A designer may follow a certain line or may attack the problem at different points.

Of course, goals are not the only means of deciding on the form of an artifact. Designs often appear to be a response to some invisible system of conventions as much as to stated objectives. Some conventions, often referred to as style, may be peculiar to individual designers, saving them from a great deal

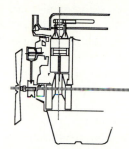

Figure 1.2 Two contrasting artifacts. It can be said that style played a major part in the design of both.

of reinvention. But design style is in evidence as much in the way components are fitted together in a car engine as in the outside form of the body of the car. We must understand how to account for this phenomenon in computer-aided design (Figure 1.2).

1.2   Models of Design

How we address the nature of design in computer-aided design depends on the models we employ to describe design. In this section we consider the role of models, also reviewing some influential models that assist in bringing design and computer systems together.

Problems arise in attempting to formalize design. Compared with many other disciplines, our understanding of design is at a fledgling stage. In his book *The Sciences of the Artificial*, Simon (1969) argued that such a thing as the "science of design" is possible, if not yet realized. If we accept that some day we will be able to talk in terms of well-established theories and practices, then we must also concede that the study of design is currently at a "pre-science" phase. This is the terminology of Kuhn (1970), who argues that a science must pass through several phases before it constitutes a formal discipline. Mature science is that state of a discipline in which there is a coherent tradition of scientific research and practice, embodying law, theory, application, and instrumentation.

If we seek this kind of maturity, we can take consolation from the pattern set by other disciplines that have blossomed into full-fledged sciences. There are two major approaches to increasing understanding in a particular discipline: case studies and models. The *case-study* approach was prevalent in such foundling subjects as early psychology, prior to the establishment of any appropriate experimental method. Case studies are particularly prevalent in discussions of architectural and engineering design. Some design phenomenon, such as the design of a particular building, is observed and analyzed, perhaps in an attempt to extract general principles. In the absence of a clear theory of what is going on, there is a heavy reliance on interpretation. This approach is also prevalent in computer-aided design research, where it is common first to devise a computer application on the basis of uncertain principles, and then assess the value of the operation in retrospect.

The second approach is a reliance on *models*. Here we consider models as limited abstractions of particular phenomena, less ambitious than theories. Whereas theories attempt to explain observed phenomena and predict behaviors that are somehow connected, models are content with explanation and prediction within a subset of connected phenomena. In physics there exist a wave model and a particle model for the behavior of light. On its own, neither model gives a wholly satisfactory explanation of the behavior of light. With the advent of quantum mechanics, however, satisfactory explanations are to be found that effectively subsume the wave and particle models of light. In design, therefore, we must be content with models that account for "design behavior." They enable us to explain and predict certain phenomena, but we may have to posit more than one model, and these models may appear contradictory. In this book both approaches—the use of case studies and the exploration of design models—are in evidence, though we favor the latter.

One of the problems with the current state of design understanding, of course, is that we are not even sure if we are on the road to a mature science. We cannot know for certain until we get there. Some say that we cannot inherit the credibility of science by claiming that our study is a pre-science. In this case we can argue for models of design only from a pragmatic point of view. The argument goes something like this. Design is a human activity that eludes formal description. We wish to use computers to assist in the design process. Computers operate only with formal, repeatable, and rigorously defined processes. If we require a computer system to simulate some kind of design behavior then we must formulate models—imperfect as they may be— of the design process. The models may have very little value as descriptions of human processes. They serve our purpose, however, if they result in useful systems that fit into a designer's mode of operation.

We should not assume that there is a neat definition of what constitutes science. If there is general agreement within the scientific community about the theories of quantum physics, there is rather less agreement about science itself (Chalmers, 1982). The study of science is *not*, in itself, a science. From the field of design research, Cross, Naughton, and Walker (1981) have suggested that, since there is not yet any uniform understanding of science, it is not prudent to look to a science of design. They believe that design belongs to the realm of technology, or applied science. But we need not achieve a science of design before our models are useful to us.

Various views of design find their origins within the disciplines of science, problem solving, logic, and language theory. For our purposes it is sufficient to call upon these disciplines to provide analogies. In some cases it is helpful to discuss the development of a design in the same way that we talk about the development of theories in science. In other cases it is useful to treat design as a phenomenon explainable in terms of logic. In still others a designed object may be treated as a statement in some complicated, multidimensional language system. Whether or not design *is* any of these things is an interesting issue, but one that need not be resolved before tackling the pragmatics of creating useful computer systems.

Before we summarize different models of design, it is worth reflecting on the relationship between science and design. We can make *predictions* in science by the application of theories. The process of prediction is relatively straightforward, belonging to the well-understood process of deduction. If we know where Mars is positioned at the moment, and if we have at our disposal theories—in the guise of formulas—for describing the motions of planets, then with little difficulty we should be able to predict where Mars will be in six months time. The interesting question for science, however, is how we arrive at the theories in the first place. We can regard the process by which theories are derived in science as similar to the process by which we produce a design.

There have been various developments in thinking about how theories are formed, but no uniform model. A candidate model is that of the naive empiricist—that scientists observe phenomena; with enough observations, the theories emerge by a process of induction. Thus, Kepler observed and recorded data on the motion of the planets, which enabled him to produce the theory that planets move in elliptical orbits around the sun (Figure 1.3). The implication is that theories emerge from observations with the same certainty that predictions emerge from theories.

Popper (1972) maintains that there is no clear logical link from a set of observations to a theory. Furthermore, the empirical method presupposes unprejudiced observations, although scientific observations appear to be influenced by theories. (Some of these arguments are summarized elegantly by Medewar, 1964, in a paper entitled "Is the scientific paper a fraud?") This

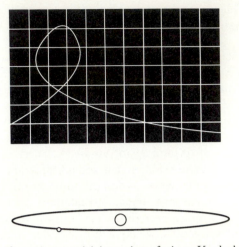

Figure 1.3 From the naive empiricist point of view, Kepler's observation of the motion of the planets led inevitably to his theory that planets move in elliptical orbits around the sun.

rejection of the empirical method has carried over into design. As in science there is no clear logical jump from a set of design requirements to a design (Hillier, Musgrove, and O'Sullivan, 1972), and "there is no legitimate sense in which deductive logic can prescribe physical form for us" (Alexander, 1964, p. 8).

The "empirical view" of science has been generally displaced by a problem-solving model advanced by Popper (1972). Rather than starting from observations, science begins with a problem—a phenomenon to be explained. A hypothesis or conjecture is generated by whatever means are available. The generative procedure is not very important, although the hypothesis must contain within it the opportunity to be refuted. The conjecture pervades within the scientific community until it is refuted since hypotheses can only be disproved, never actually proved valid. Theories are therefore "on hold" until they are found lacking, which creates another problem. A further conjecture is proposed that pervades until it is refuted, and so on. The process is one of trial and error, analogous to natural selection. According to Popper, this process removes the generation of hypotheses or theories from the realm of rational processes. It all depends on what is made of a hypothesis as to the role of reason. Although one can assume the right frame of mind for having ideas, the process itself is outside logic. The success of science therefore appears to depend on "luck, ingenuity and the purely deductive rules of critical argument" (Popper, 1972, p. 53). Whether one is dealing with small problems or major scientific theories, Popper contends, the process is universal.

This view of problem solving as a generate-and-test cycle has received wide acceptance in design research. (Popper is not the only source of this idea.) A design conjecture is generated, and when it is found unsatisfactory a new design conjecture is generated.

These ideas are valuable in discussing design. We should, however, guard against drawing too close an analogy between science and design. How helpful is it to consider design as a process resembling the development of scientific theories? Do designs bear any resemblance to theories or hypotheses? Theories are embedded in principles, rules, and formulas. Designers appear to utilize and to be constrained by rules, but the product of their endeavors is primarily an artifact, the form of which is dictated by those rules. Design and science therefore appear to be about different things. Simon (1969) makes the distinction that science is concerned with natural things, how they are and how they work, while design is concerned with how things ought to be. Natural science looks at the state of things and attempts to postulate hypotheses that explain their state. Design looks at the results that are required and attempts to predict the states of things necessary to achieve those results. That is, science generalizes, whereas design specializes.

Science attempts to formulate knowledge by deriving relationships between observed phenomena. Design, on the other hand, begins with intentions and uses available knowledge to arrive at an entity possessing attributes that will meet the original intentions. The role of design is to produce form—or more correctly, a description of form—by using knowledge to transform a formless description into a definite, specific description.

Moreover, design is a pragmatic discipline, concerned with providing a solution within the capacity of the knowledge available to the designer. This design may not be "correct" or "ideal" and may represent a compromise, but it will meet the given intentions to some degree.

In this book we explore what can be loosely called a knowledge-based model of design that accounts for various design behaviors—various ways we are accustomed to talk about design, as well as ways of formulating computer-aided design systems. The model we follow is a continuation of a substantial tradition of thinking about design. It is useful, therefore, to consider some different views of the design process and the different philosophies behind them, and examine how they can be subsumed within a knowledge-based model.

The models we describe in the following discussion may overlap in various ways. In some cases the models are simply different ways of describing the same phenomena. We may have to embrace several models to gain a satisfactory picture of design. These models are important—we will draw on them in our subsequent discussion of knowledge-based design systems.

1.2.1 Method and Design

Beginning in the 1960s, serious attempts have been made to improve the results of design by encouraging designers to follow certain procedures or methods. These methods constitute techniques or expediencies to produce results

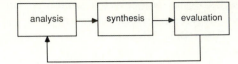

Figure 1.4   The three design phases of analysis, synthesis, and evaluation.

of use to designers. They have also been influential in shaping perceptions of design. The operations research approach to design, developed in the 1940s and 1950s, was brought together in the 1962 Conference on Design Methods at the Imperial College, London (Jones and Thornley, 1963), and subsequently at the Birmingham Symposium of 1965 (Gregory, 1966) and the 1968 Design Methods Group conference in Cambridge (Moore, 1970). Since then we have seen the development of various "methodical" approaches to design, which come under various headings such as set theory (Alexander, 1964), decision making (Archer, 1969), and systems analysis and hierarchical decomposition (Mesarovic, Macko, and Takahara, 1970). Many techniques have been applied to the design process, including simulation (Forwood, 1971) and optimization (Radford and Gero, 1988). Many, but not all, employ mathematical procedures.

Numerous models of the design process have arisen from this work, notably those developed by Asimow (1962), Luckman (1967), and Markus et al. (1972). There has been widespread agreement, however, on the three-phase design model. The phases can be described as *analysis*, *synthesis*, and *evaluation* (Figure 1.4), as outlined by Asimow (1962). The first task is to diagnose, define, and prepare—that is, to understand the problem and produce an explicit statement of goals. The second task involves finding plausible solutions. The third task concerns judging the validity of solutions relative to the goals and selecting among alternatives. A cycle is implied in which the solution is revised and improved by reexamining the analysis. These three phases form the basis of a framework for planning, organizing, and evolving design projects.

Jones (1970) describes the three phases in a slightly different way as divergence, transformation, and convergence. *Divergence* is concerned with breaking the problem into pieces and defining the boundaries of a space in which a fruitful search for a solution can take place. *Transformation* involves putting the pieces together in a new way. *Convergence* is testing to discover the consequences of putting the new arrangement into practice.

This model has also been discussed in terms of goal decomposition, a concept largely attributable to the early work of Alexander (1964). According to this view, general goals can be arranged hierarchically. The top level goal may be vague—for example, to achieve personal satisfaction or the public good—and must be translated into a series of operational subgoals. The decomposition of the goal into subgoals is carried down to a level where the subgoals are in a form that enables us to evaluate their usefulness. That is, they have operational status. An example of goal decomposition appears in Figure 1.5.

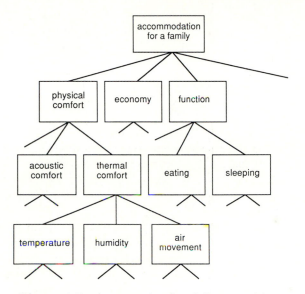

Figure 1.5   An example of goal decomposition.

Alexander suggested (although he later modified his view) that decomposition of the problem should be carried out to the "primitive" level—that is, to that level at which the solution is "self-evident." Further, he suggested a method of cluster analysis as an aid in synthesizing the various partial solutions (Figure 1.6). When looked at in isolation, however, many such techniques appear simplistic and of little value for certain design problems.

Certain properties of the three-stage process are important. The process is sequential, in that evaluation follows synthesis, which is preceded by analysis. It is cyclical in the sense that the sequence may be repeated many times. Evaluation results in a revised analysis of the problem, which leads to the synthesis of a new provisional solution. Asimow suggests further that each cycle is progressively less general and more detailed than the preceding one. To this extent the process may be described as recursive (Figure 1.7). Evidence is provided by current design practice. For example, the Royal Institute of British Architects publishes a plan of work for each project in which the stages of briefing, sketch design, detailed design, and documentation are clearly separated. Briefing is the analysis phase. A sketch is proposed, evaluated, revised, and worked into progressively greater detail. The formulation of a sketch design will also involve some sort of detailed analysis, the generation of various tentative proposals, and the evaluation of these proposals.

According to this model it is unnecessary for the design problem to be analyzed completely before synthesis can begin. Complete analysis is impossible in any but the simplest design task. Problem decomposition may work if we have sufficient foresight; that is, if we have all the necessary information at the start of the design task. In reality this hardly ever occurs, and we

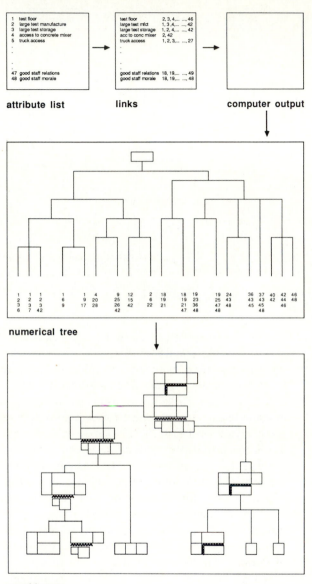

Figure 1.6 The use of cluster analysis to assist in synthesizing a design (after Alexander, 1964). All significant physical and non-physical desired attributes of the design are listed (the attribute list) and the other attributes to which they are linked are enumerated (the links). A computer program establishes a hierarchy in which groups of attributes with many links are grouped; these groups are linked to other groups (the numerical tree). In design, the conjunction of attributes at the extremities of the branches are considered and represented graphically before the attribute groups from other branches are combined (the graphic tree).

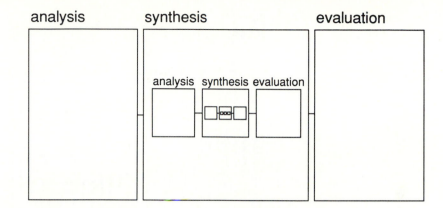

Figure 1.7   Analysis, synthesis, and evaluation as a recursive process.

proceed by carrying out sufficient analysis to allow us to start searching for a solution. This in turn brings to light various subproblems, and the same sort of processes are repeated. There are many subproblems, however, that cannot be considered until some attention is given to the overall solution.

The model has limitations. It says nothing of where designs come from—the synthesis part. Can synthesis itself simply be characterized as a process of analysis, synthesis, and evaluation? If the process is infinitely recursive in this way, we may find that there is, after all, no such thing as synthesis! Yet the model has proven useful and has formed the basis of many computer-aided design systems.

The task of cycling through the three phases of analysis, synthesis, and evaluation constitutes a kind of search procedure. Implicit in the idea of search are some important notions: the existence of goals so that the search has some direction; the existence of states or points along a search path that represent the position of the design is in its development; and the existence of mechanisms for moving from one state to another.

The challenging parts of design are the way these three phases might interact, and the process of exploring possible states. Let us now turn from this general model of design methods and focus on the problem-solving model of Popper and the importance of search.

1.2.2   Problem Solving and Design

Although we are wary of the view that science provides a model of design, we can learn from Popper's problem-solving model. Search is concerned with exploring a space of potential solutions and partial solutions; the generation and testing of hypotheses can be characterized as a process of search. Simon (1983) describes this process as analogous to exploring a maze (Figure 1.8). One has some destination—that is, a goal—in mind. The intersections between

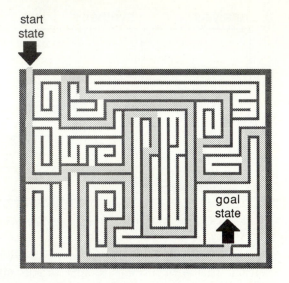

Figure 1.8   Problem solving can be compared to exploring a maze.

passages are solution states; it is necessary to negotiate the passages in some rational way until the goal is reached.

Design is seen as a problem-solving process of searching through a state space (Simon, 1969; Newell, 1973), where the states represent the design solutions. This state space may be both large and complex—large in that there will be many states, complex in that there exist many relationships between the states. The search will involve making decisions based on the goals and constraints that render certain solutions infeasible. For some large, complex problems a hierarchical approach may be useful (Mesarovic, Macko, and Takahara, 1970). Certain techniques in operations research are concerned with search within a numerical domain, but search techniques also apply to other forms of symbolic manipulation.

Much of the theoretical impetus for automated problem-solving systems has stemmed from the work of Newell and Simon (1972), which has influenced design research. They describe an automated problem-solving system called General Problem Solver (GPS). They state that the requirements of such a system are that it must be able to represent the current situation, the desired situation (goals), the differences between the two, and actions that change objects or situations to bring the system closer to the goals. The system must also be able to select actions likely to reduce or remove differences. GPS incorporates a table of connections between differences and actions that provides the means of evaluating progress and trying alternative paths. The means-ends mechanism is a way of making the search more efficient. Some such control mechanism is generally required in order to limit the amount of searching and direct the system's activities in the right direction. It is evident that design generally involves some sort of search process when considering the way in which designers tend to explore possibilities. The derivation of appropriate search strategies has occupied considerable attention in artificial intelligence research; Akin (1978, 1979) has observed search strategies at work in the behavior of designers that parallel the techniques employed in artificial intelligence, such as "hill climbing," heuristic search, and forward processing.

And what of the ill-structured nature of design problems, the fact that they are generally not well-formulated? The view of Simon (1973) is that while design is essentially ill-structured, a designer converts a task from an ill-structured to a well-structured problem by reducing it to suitably organized subtasks. A human designer calls up certain constraints, subgoals, and generators from his or her long-term memory. These are factors not specified in the problem statement. The formulation of the task is therefore dynamic. The problem representation can be revised continually to take account of the real situation so that the problem solver is faced at each moment with a well-structured problem, but one that changes from moment to moment. The formulation of the design problem can therefore be regarded as a problem-solving task in itself, one that is undergoing reformulation as the process continues. The operations by which states are changed are themselves in a state of change, as are goals, constraints, and evaluation criteria.

1.2.3   Simulation, Optimization, and Design

Problem solving implies search. What methods are available to control search? The two most prominent computer-aided design techniques employed in producing "good" designs are simulation and optimization. *Simulation* predicts the performance of a given potential design, while *optimization* identifies potential designs for a given performance goal.

In essence both approaches start with a design *goal* or *goals*, such as "a pleasure boat for a crew of four" or "a means of public transportation to the downtown business district." This design goal can be translated into descriptions such as "suitable size," "economy," and "comfort." In this model these are commonly referred to as *objectives*. Objectives must be satisfied for the goal or goals to be met. Certain variables such as areas, air temperatures, and cost, called *performance variables*, must acquire some values, certain ranges of which will satisfy the objectives. These ranges may be stated in specific terms as *constraints* or in general directional terms as *criteria*. A designer can assign values to *decision variables*, such as lengths and widths of spaces, types of components, and thicknesses of materials. These decision variables collectively describe the state of the design. By making decisions about the various values of the decision variables, the designer controls the values of the performance variables so as to satisfy the objectives and goals. The relationships between the performance variables and the decision variables are not always obvious. These causal relationships may constitute a kind of black box. The selection of the right kind of decision variables—the definition of the *state space*—is a

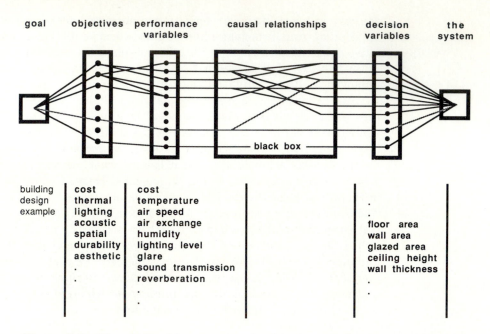

Figure 1.9   The relationships among goals, objectives, performance variables, and decision variables in optimization.

matter of the designer's expertise. Figure 1.9 shows the relationships among goals, objectives, performance variables, and decision variables. More generally, goals are related to required functions, and decision variables to the form or structure of the design.

Let us summarize the main features of simulation and optimization. Simulation is a problem-solving approach to design in which a design solution is postulated. That is, the decision variables are posited and their values are given. The values of the performance variables are obtained by calculation or other means such as observation; these values are examined or evaluated against given criteria to determine whether they are satisfactory. If the design is not found to be satisfactory, then some values of the decision variables are modified, and the process is repeated. In some cases the evaluation may provide clues as to what modifications will improve performance. In most cases the process ends when a seemingly satisfactory performance level is reached. One drawback of this approach is that there is no way to tell how good the first few postulates are in comparison with other possible designs. Another drawback is that the relationship between decisions and performances may not be obvious. Thus the necessary feedback from performances to decisions, required in simulation, may involve other processes.

Optimization is an approach that seeks not only a design, but the "best possible" design. Note the use of the term "possible" implying "best" under the available conditions of the given problem formulation—that is, the given criteria and constraints. In optimization a design is not postulated and then evaluated; rather, the performance criteria, the performance constraints, and the decision variables are stated. The decision variables are constrained to given ranges of values or discrete values. The performance criteria are used to guide the search through the state space so described in order to find those solutions that best meet the criteria. Where the solutions do not exist explicitly in the state space, they must be generated. The optimization approach to design can be considered as embodying both synthesis and evaluation, as opposed to the purely evaluative process of simulation.

It is worth considering the methods by which decisions may be made in optimization. A design solution (represented by a vector of decisions) may exist explicitly in the decision space. The task is to find the decision that matches some required performance. There may be a fairly direct mapping between the required performance and the solution. For example, the design task may be to join two metal sections. Here the performance requirements are bound tightly to a solution—to provide a welded joint. Alternatively, there may be an indirect mapping. The design task may be to refract light for some purpose. To accomplish this, the intermediate step of finding a medium must be taken. Glass is found to be a refractory medium.

Of course, the selection process may not be so simple. It is generally necessary to know of many mappings to determine whether some solution in fact links to a required performance. There may also be several alternative solutions, each vying for consideration. In order to determine whether a solution is suitable, we may have to interpret its performances and compare them with the required ones. Many design tasks appear to consist of this kind of selection—for example, the use of standard details or standard designs.

Optimization is a powerful tool. Once a problem—even a multicriteria problem—is mathematically formulated in terms of criteria, constraints, and decision variables it provides a powerful search strategy (Cohon, 1978; Radford and Gero, 1988). Optimization has failed to influence the field of design greatly, however, in part because it does not address the question of how to arrive at such well-packaged formulations. As a rule, designers do not think of their problems in optimization terms and do not easily relate the technique to their own design task. As shown in subsequent chapters, the application of knowledge to a problem stated in non-mathematical terms can provide a more comprehensible "front end" to the optimization approach.

1.2.4 Logic and Design

Since we are interested in formalizing design, it is useful to consider it in terms of logic, the abstract discipline to which all formal systems are subject. In calling upon logic to provide insight into the design process, we assume that design is essentially a reasoning process. One way to consider design in a logical scheme is to talk about it as a process of making logical deductions about

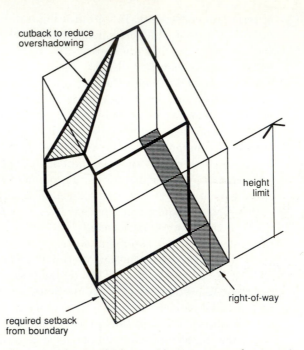

Figure 1.10   An example in which considering a set of constraints leads to a description of a design (after Broadbent, 1981).

a set of requirements—in much the same way as a mathematical or geometrical theorem is proven. We begin with axioms—the design requirements—and proceed along an inevitable course of reasoning to a theorem—the design. The procedure is incremental, and every state in our understanding improves on the last.

We can characterize this kind of process with reference to the methods discussed in the previous section. Let us consider the design of a building on a site for which planning regulations impose constraints. Typically, there may be a required setback from the street and a minimum distance from side boundaries. There may also be some complex incremental setback requirement—for example, that the developer may build higher if the walls of the building are a certain distance from the boundary. There may also be a right-of-way through the site that must be maintained. By imposing these constraints successively on the site, we arrive at an "envelope" into which the building must be placed. If we seek the largest building the site will allow, this envelope constitutes the shape of the building (Figure 1.10). The constraints of the problem have dictated the form of the building.

More formally, this process can represent a kind of constraint satisfaction model (Archer, 1970). The problem starts with the definition of a space of possible designs. All properties required to be present in the final artifact

to varying degrees in the initial problem state. The design task is concerned with describing an artifact in which certain desired properties are present to as satisfactory or high a degree as possible. This is accomplished by successively eliminating parts of the space of designs according to various performance criteria. The universe of all possible solutions is successively manipulated until a set of feasible solutions remains. Constraint satisfaction constitutes a set theoretic approach to design, and the process of manipulating sets is essentially deductive. Once set boundaries are determined, the computation of intersections and unions is essentially deductive. There are problems, however, with this approach. The task may be over- or underconstrained, or all the constraints may not be fully known. Constraints may also be in competition, and may interact in complex ways such that a complex family of designs rather than a single design is suggested.

This approach to "design by deduction" is not new. It was hinted at in the 18th century by Abbé Laugier, who coined the dictum "If the question is well posed, the solution will be indicated." It was espoused in the mid-20th century by the modernist architects who declared that "form follows function." There is little doubt that deductive processes, in which an artifact constitutes the theorem, are applicable to certain design tasks, but there is some question as to its appropriateness as a general model of design. A critique of the view that form logically and inevitably follows from analysis has been well aired, notably by Alexander (1964) and Hillier, Musgrove, and O'Sullivan (1972).

So far we have considered a design as something that can be derived from a set of design requirements, or a kind of complex theorem. Another way of looking at the analogy with logic is to say that a design is something that exists in the real world *about which* logical deductions can be made. When we are designing we do not have a clear picture of what the design is; if we did, our task would be needless. Even though we do not have the artifact, we would like certain things about it to be true. We say, "If we had the design here now, we would want to be able to deduce certain things about it, such as that it will function in a particular way or that it will invoke a particular response in the people who use it." In other words, a design is not the product of a deductive process but it is the kind of thing about which we would make deductions (Figure 1.11). Returning to the scientific analogy, a design is a thing about which one makes observations and predicts performance, given the appropriate theories. In design we are interested in reversing this process. We may have at our disposal all sorts of predictive theories, and we know what kind of behavior we want to be able to predict. All we lack is the design.

Can we run a deductive system backwards to arrive at a design form? Whatever the process by which we arrive at a design from a set of requirements, it is unlikely to be as simple as mere deduction. This view, that design can be seen as a reversal of the deductive process, has been advocated principally by March (1976), calling on the work of the philosopher C. S. Peirce (1839–1914). It is espoused in the model that is presented in the next chapter.

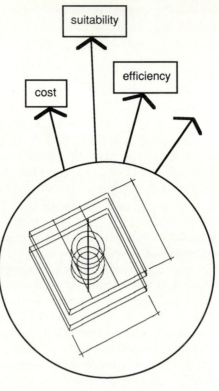

Figure 1.11   A design is the kind of thing about which we make deductions.

1.2.5   Language and Design

The task of proceeding from a set of required or desired performances to a concrete description of a design is a concern that design shares with natural language, which appears to have many features in common with design. It can be argued that design appears to bear a stronger resemblance to language than to science, problem solving, or logic. We are accustomed to talking about *vocabularies* of elements, *composition*, *style*, *context*, and even *meaning*. Just as there are rules about how words go together in language, there appear to be compositional rules applicable to the configuration of components in a design. This applies to the arrangement of gears and pulleys in a mechanical device as much as to the configuration of the classical orders in architecture (Figure 1.12).

The field of *Semiotics* takes this reference to language quite literally. Within this sub-discipline of linguistics the view is taken that artifacts serve as signs that communicate complexes of messages about function and usage. For example, the semioticist takes the extreme view that a house is not so

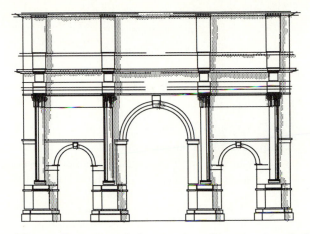

Figure 1.12 The configuration of elements in a Roman triumphal arch can be described in linguistic terms.

much a "machine for living in" (a well-known definition coined by the architect Le Corbusier) so much as a system for the production of "meaning" by those who use it (Preziosi, 1979). We can study the development of these artifact languages and the means by which they facilitate communication. As yet, however, semiotics does not appear to offer useful models in the context of operational design systems. In computer-aided design we are interested in manipulating symbolic representations of artifacts. We are therefore primarily interested in the meanings of abstract descriptions of designs contained in computer systems rather than the meanings of the artifacts themselves. Thus linguistics is used as an operational model relevant to the generation of designs rather than for theories of meaning within systems of signs.

There is another sense in which we can appeal to a linguistic model of design. At its most abstract, linguistics is concerned with the generation and interpretation of strings of symbols (Figures 1.13 and 1.14). Descriptions of designs in computer systems are just that—strings of symbols. Any theory that describes how such strings can be generated, and also what those strings mean, has a bearing on design. In the next chapter we develop a linguistic strand to the knowledge-based model that calls on aspects of formal language theory.

1.2.6 Typology and Design

We can extend the links between language and design developed in the previous section by considering design as a process that begins with prototypes and progresses to specific designs. Design involves *instantiation*. One of the concerns of language is in naming things. The original meaning of the verb "to designate" is to give a name to something, which has the effect of allocating

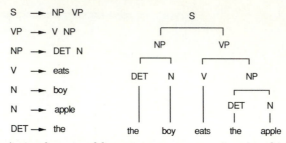

Figure 1.13 A simple natural-language grammar system in which a sentence (S) is transformed by a rewrite rule into noun phrase (NP) and verb phrase (VP), which are further transformed into a verb (V), nouns (N), and a determiner (DET).

it to a class. To designate an object as a cup is to place it within the class of objects that are cups. To designate someone as a lieutenant is to put that person within the lieutenant class. It can be argued that class divisions are human devices relating to our perceptions of the world and are not inherent in the objects themselves. In contrast, the holistic position would be that class distinctions are imaginary lines in a universe that is essentially homogeneous. Certainly, class divisions appear arbitrary at times and may change according to convenience. In the context of the Western-style kitchen a cup is a vessel with a handle, but we still recognize that there are cups without handles and that not all vessels in the kitchen with handles are cups (Figure 1.15). As with language in general, we make class distinctions—and complex ones at that—by social agreement.

In design we are accustomed to talking about types as well as classes. In Section 1.3 we distinguish between the nature of archetype and that of prototype, but for the moment let us regard a *type* as an abstraction of a class that accounts for the commonality of the members of the class. It is usually a conjunction of features rather than the embodiment of a set of alternatives. Useful types in design are often represented in terms of geometry, as diagrams (although diagrams are by no means the only form that a type can take). Hence we may consider the cruciform plan for a church as a type and depict it as two crossed axes of unequal length with perimeter walls of unspecified

Figure 1.14 A grammar for producing a row of alternating characters *a*s and *b*s in which an expression (E) is transformed by a rewrite rule into sub-expressions (X and A), which are further transformed into strings of characters *a*s and *b*s.

Figure 1.15 A set of kitchen utensils for which the definition of "cup" and "non-cup" pose difficulties.

dimension (Figure 1.16). Similarly, at a very simplistic level, we could depict the catamaran type of sailing boat as two elongated hulls connected by a rigid frame and a central mast (Figure 1.17).

It can be argued that designs can be described not only by their conformity to type, but also produced through them (Moneo, 1978). It is quite usual in design to start with a particular type and to bring elements together to produce an artifact by which the type is exemplified. In other words, design is a process of instantiation—proceeding from a class description to a description of an instance. Thus, according to this model, in design we select

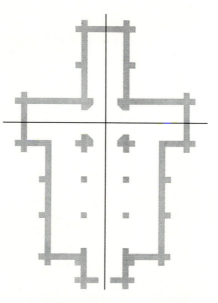

Figure 1.16 A diagram of a church type.

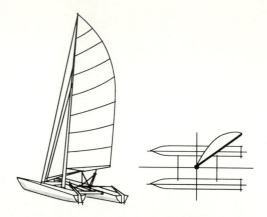

Figure 1.17   A diagram of a catamaran type.

a type and progressively refine it in response to a particular context. These ideas are expanded in our discussion of generic designs (Sections 1.3.4, 1.3.5, and 2.6.2).

1.3   Knowledge and Design

The idea of knowledge has been implicit in the discussion so far. In this section we will make the role of knowledge explicit and discuss its role in models of design.

Of the models described so far, optimization may offer the most systematic technique for automated design within the constraints of its method because it incorporates notions of description, goals, and evaluation. It is worth summarizing some of the key ideas in optimization before extending them to incorporate the idea of knowledge in general. We will then be in a better position to characterize design knowledge.

1.3.1   Knowledge and Optimization

In the terminology of optimization, the key to the design task is in mapping *decisions* to *performances*. Whether we choose to label something as a decision or as a performance depends on how we view the particular design task and what we have control over. For example, we may wish to design a rectangular room in a building. The decisions we choose to concern ourselves with may be the length, width, and materials of the room. The performances may be area and construction cost—the *performance variables* or *criteria*. Alternatively, at a different level of representation the decisions may be the area of the room

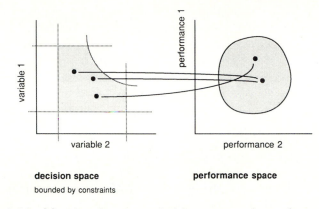

decision space   performance space
bounded by constraints

Figure 1.18 Mappings between a decision space and a performance space.

and the amount of money to be spent, and the performance may be stated in terms of the spaciousness or luxuriousness of the room.

We are not usually at liberty to make just any decision in relation to a decision variable. Generally we must make a decision which falls within a particular range—in the last example, that the width of the room must be greater than some minimum value. Our decisions are therefore *constrained*. We can represent the *space* in which decisions are to be made in the form of a graph where the axes are the decision variables and boundaries to the design space are depicted as constraints. We can also represent the performances on a graph where the performance variables are the axes. In realistic design tasks there will probably be many axes defining decision and performance spaces. Nevertheless the basic idea can be conveyed by considering only the two-dimensional spaces of Figure 1.18. A design is represented by a single point in the decision space. The other component required in this formulation is some function for mapping decisions to performances. In the case of floor area in our room example it may simply be the formula area = length × width. The mapping between the decisions and cost, however, will probably be more complicated.

In the design task, we utilize the properties of the performance space to arrive at appropriate decisions. For example, it may be desirable to maintain a large floor area while keeping the cost low. We can plot the surface of the space where both criteria are optimal and use the nature of its shape to arrive at a decision. If we can figure out some way of comparing cost and area (perhaps by means of a common utility measure), then the process of finding an optimal decision is relatively simple. In most cases, however, there is no measurable, single best solution, and the process involves subjective tradeoffs between different objectives.

Where is the knowledge in this view of design? In the previous sections we have intimated that the design process is a cognitive activity heavily reliant on the application of knowledge. We can generalize the problem of modeling design to that of devising appropriate structures for formalizing and capturing knowledge pertaining to design and rendering that knowledge operable. In knowledge-based systems, the knowledge must be able to be made explicit in such a way that it can be inspected and understood independently of the way in which it is controlled. Central to the idea of knowledge-based systems is the separation of knowledge pertaining to control, on the one hand, and knowledge pertaining to the particular task being undertaken, on the other.

1.3.2   Syntax, Semantics, and Design Spaces

The main problem with most of the models used in design methods, especially in optimization, is that they can deal only with quantifiable parameters stated in mathematical terms. In this book we wish to generalize these models of design. First, we would like to generalize the idea of decisions to include those that cannot be quantified. We would also like to recast the idea of decisions in terms of *descriptions*, allowing us to arrive at an explicit description of an artifact at some appropriate level of abstraction. The conventional product of the design process is a description in the form of a drawing or other suitable representation. In computer systems the description is generally contained in some kind of database. The important aspect of design descriptions, therefore, is that they are *explicit*. The performance, then, is simply that set of things we can declare about the design that is not explicitly stated, but that is *implicit* in the description. For example, if a design description contains information about the length and width of a rectangle, this information constitutes a description. The area is something that is not explicit in the description but must be computed. In this case, area belongs in the performance space. If we chose to describe our design in terms of properties such as areas, however, then area is part of the description space.

It may be helpful to consider these two categories of spaces, description and performance, in linguistic terms. Design descriptions are like sentences in natural language. They are explicit strings of symbols. Design descriptions therefore belong to a *syntactic* space. What is not explicit in a sentence in natural language can be regarded as the *meaning* of the sentence. The meaning of a sentence can be considered as all those objects and ideas to which the sentence refers, and must be derived by means of *interpretation*. We could therefore relabel the performance space the *semantic space* (Figure 1.19).

Design, then, is concerned with finding some design description in the syntactic space that maps onto some intended performance in the semantic space. Finding a design is essentially a matter of defining spaces. We argue this model more vigorously in the following chapter. Suffice it to say that, in natural language, in the most general sense mapping sentences to meanings is accomplished by knowledge. The syntactic space is therefore constrained by knowledge. The space of all legal sentences in a language can be defined by a form of knowledge known as a *grammar*. In the next chapter we argue that it is

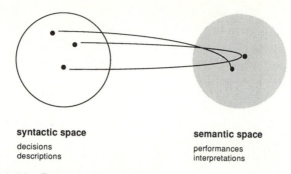

syntactic space
decisions
descriptions

semantic space
performances
interpretations

Figure 1.19   Restating the optimization model in terms of knowledge.

important to distinguish between the knowledge that defines syntactic spaces and that which provides mappings to the semantic space. The important point to make here is simply that knowledge relates syntactic and semantic spaces. Knowledge therefore incorporates the idea of constraints in design and how we map performances onto descriptions.

The task of knowledge engineering is to satisfactorily represent knowledge applicable to a particular domain and organize it so that it can be used efficiently in various ways according to the needs of the user of the knowledge-based system. Some of the important issues to be addressed in the rest of this book are appropriate methods of design-knowledge representation, how that knowledge is to be manipulated, and how it can be acquired.

1.3.3   The Characteristics of Design Knowledge

What is the character of knowledge about design? We may suppose that design knowledge covers a broad spectrum, including laws, rules, and formulas pertaining to the behavior of people, materials, objects, and spaces. Knowledge can be axiomatic in character, such that it can be stated in unequivocal terms: in a quadrilateral the sum of the interior angles is 360 degrees. It may be causal in character: ferrous metal rusts when exposed to air and water. It may also relate to conventions and experience with little basis in certainty or apparent logic. The distinction is also often made between knowledge that is certain and knowledge that is uncertain: if the foundation moves after a period of rain, then it is likely that the drain is cracked; if very low friction is required, then a pneumatic bearing is likely to be suitable.

The knowledge with which we are concerned in design may also apply to different abstractions of the design process and to different abstractions of design description. For example, we can talk of different control levels in design. There is knowledge about how components might fit together in a design. There is also knowledge about appropriate actions to perform in producing configurations, and knowledge about strategies. We must be able to represent and manipulate knowledge of this kind.

The aim of knowledge-based systems is not to mimic human cognitive processes but to provide structures that enable human beings to use them in ways that are intuitively appealing. We will attempt to demonstrate that a knowledge-based view of design is more appealing to human designers than was evident with the design methods and optimization view. It also appears that certain tasks that are difficult to model mathematically can be accomplished more feasibly using knowledge-based systems. There is also the promise (if not yet the realization) that designers will be better able to communicate with knowledge-based systems if they use a language with which they feel an immediate affinity.

1.3.4   Generic Designs

We have described design knowledge as that which defines spaces of designs. Knowledge is the most general term we can think of to represent the idea of defining spaces of designs. There are several ways, however, of looking at knowledge, and designers are accustomed to several means of formally articulating their knowledge—that is, several methods for defining spaces of designs—and passing it on from person to person. These methods include the use of existing designs, typologies, rules, and procedures; we survey these in detail in Chapter 3. The idea of typology is particularly useful, and worth raising at the outset.

A design space is essentially a delineation, or a definition, of a class of things—that is, designs conforming to particular meanings and a particular syntax. Knowledge is used to define classes of designs—that is, to define generic designs. It is often convenient to make the generic nature of knowledge explicit. For example, rather than use grammatical rules we might define a design space in terms of a class description, sometimes called a *generic model*. We could do this by listing all the properties of the class, including the ranges of properties that a design instance may take, and also the interrelationships between properties. If we wished to define a "small house" class, for example, we would probably list the kind of rooms it may have, and state that the class includes such things as a front door and a roof. We may also represent a generic model in terms of abstract diagrams depicting different aspects of a typical small house, such as an overall form (Figure 1.20). Such diagrams generally contain much implicit information about what is an essential property of the class and what is merely incidental, and their interpretation requires some prior understanding of the class.

In some cases a certain design instance is said to typify a class—that is, it embodies the features of its class in which we are particularly interested. Such a design is said to be an *archetype*. As an example, we may think of Beatrix Potter's house as the archetypical English country cottage, or of a red Ferrari as the archetypical sports car (Figure 1.21). Of course, archetypes are very

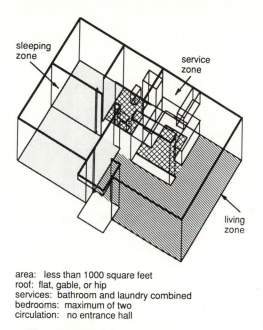

area: less than 1000 square feet
roof: flat, gable, or hip
services: bathroom and laundry combined
bedrooms: maximum of two
circulation: no entrance hall

Figure 1.20 A typical small house.

elusive, and rarely does a single object oblige us by exhibiting all the properties of the class to which it belongs. In considering objects as archetypes we generally bring knowledge to bear in discerning what is essential to the class, what is incidental, and what is not represented. Bound up with our appreciation of an archetype is the knowledge pertaining to the class of objects. The concept of an archetype is very useful because we seem to prefer thinking in terms of instances, especially representative instances, rather than in terms of the abstract world of classes of things.

In design we are also accustomed to thinking in terms of *prototypes*, which are generally taken as designs from which other designs originate. A prototype typifies, or exemplifies, a class of designs, and thus serves as a generic design. The idea of a prototype embodies three similar concepts. We may freely talk of a particular design being a prototype because it exemplifies a class of designs. A description of a class of designs may also be a prototype, as in the case of a diagram or description of a class. Or any other body of knowledge for defining a space of designs, such as a set of rules, may also constitute a prototype.

1.3.5 The Role of Prototypes

We have characterized a prototype as a representation of a class of designs. It is common to categorize the sophistication of design activity in terms of what we do with prototypes. These categories are refining, adapting, and creating

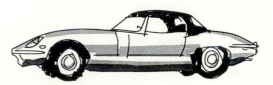

Figure 1.21   The archetypal country cottage and the archetypal sports car.

prototypes. It is important to see that in working with prototypes we do not merely take an individual design and modify it in some way. Rather we work within the constraints of our knowledge about the class of designs.

Design as *prototype refinement* involves working within the constraints of a particular class of designs—that is, the constraints of the knowledge that defines the space of designs. We can consider prototype refinement in the context of aircraft design. The conventions of aircraft design—that is, the conventional knowledge defining the space of aircraft designs—dictates the shape of the wings and where the engines may be located. Prototype refinement involves selecting from sets of candidate design decisions and adjusting parameters. Another way of characterizing this process is to take an exemplary aircraft design (of a 747, for example) and to refine it within the constraints of that class of designs (perhaps by adjusting the shape of the nose or the length of the wings) in order to produce a new design (Figure 1.22). In a different domain, the production of another Prairie-style house by the architect Frank Lloyd Wright would be seen as a process of prototype refinement. The prototype has been defined; a new design is produced from the space defined by the knowledge. The same process would apply if we took an exemplary

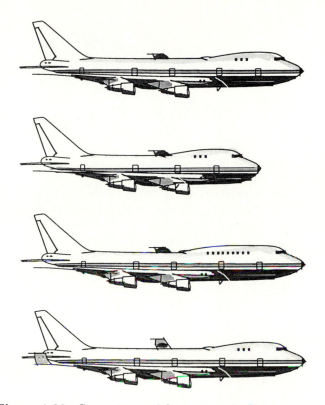

Figure 1.22 Successive modifications to the 747 prototype.

Prairie-style house and modified it according to the knowledge embedded in the style.

The second design category, *prototype adaptation*, involves extending the boundaries of a particular class of designs—that is, adjusting the concepts that define the space of designs. In aircraft design, for example, we elect to place the jet engines above the wings so that the exhaust gases pass over the wing surface and increase lift, an apparently successful adaptation. The designers responsible for this adaptation went beyond conventional knowledge about aircraft design. In the case of Frank Lloyd Wright's houses, we can regard the transition from the Prairie-style house to the "Usonian" style as one of adaptation. We can see the influence of the former in the latter style, and we may suppose that some transformation of knowledge has taken place to extend the space of designs.

The third category, *prototype creation*, or conceptual or original design, is where totally new prototypes emerge, as in the design of the first airplane (Figure 1.23) or the first photocopier. We may well question whether prototype creation is ever realized in practice. Prototype creation may simply be adaptation of sufficient magnitude to warrant the label creative. The automo-

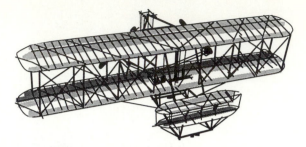

Figure 1.23 A product of apparent prototype creation, one of the first aircraft.

bile constituted a new concept different from the horse-drawn carriage, yet the first "horseless carriage" had the appearance of a horse-drawn carriage without horses. Over the years further modifications have taken place and today the generic automobile is indeed different in type from that of the horse-drawn carriage (Figure 1.24). Yet the process by which this new prototype came about has been essentially adaptive. We may indeed question whether original design is ever possible. Would human beings have ever thought of flying if there had been no birds?

The key to this aspect of a creativity continuum is the role of knowledge. The most interesting, innovative, and significant kind of designing occurs when we work beyond the confines of some particular subset of knowledge. According to this model, creative designers are those who can abrogate convention

Figure 1.24 The evolution of the car: the horse and carriage, the horseless carriage, and the modern automobile.

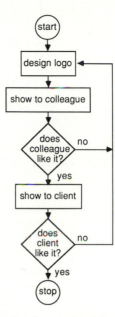

Figure 1.25   An example of procedural knowledge as a flow chart.

and call on other resources, either from within their own knowledge or external to it. Most of what we describe in the rest of this book will remain within the realm of prototype refinement, or at most prototype adaptation, but we will discuss the extension of ideas into the realms of innovation as we proceed.

1.3.6   Knowledge-Based Systems

We have discussed the key role of knowledge in design. Knowledge-based systems are computer systems in which an attempt is made to capture and render operable human knowledge about some domain (Buchanan, 1982). The goal is to represent knowledge in such a way that it is comprehensible to both human being and machine. Working systems that embody knowledge of a particular domain are often termed *expert systems* (Hayes-Roth, Waterman, and Lenat, 1983). Clearly, a knowledge-based system is different from a computer program as we normally understand it, in which knowledge about the domain under consideration is bound up in control statements such as variable declarations, procedure calls, conditionals, and loops (Figure 1.25). If we wish to inspect the knowledge in such a computer program, it is necessary to run it, to understand the language in which the program is written, and to disentangle control structures from domain knowledge. Furthermore, the operations of a program are predetermined. The knowledge does not exist as a discrete entity that can be processed in a variety of ways depending on what use we wish to make of it.

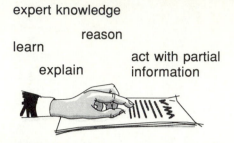

Figure 1.26   The behavior we expect of an expert.

In constructing an expert system the idea is to code knowledge into a computer so that it can be consulted in much the same way that one consults a human expert (Figure 1.26). The knowledge does not embody a unique expectation of how it will be used. Therefore, by consulting its *knowledge base*, an expert system should be able to perform a range of tasks—from answering simple queries and evaluating proposals to generating proposals and explaining how they were derived. There are many motivations for this.

One obvious motivation for using knowledge-based systems in design is the potential for increasing productivity by automating low-level design decision making. There is also much to be gained from making that knowledge available at the point where it is most valuable—that is, combined with the medium of design operation—in conjunction with a computer-aided design system. The technology has many important implications for the control structures of organizations. Knowledge-based systems may serve to centralize an organization's expertise and control its dissemination to individual designers. This may avoid the dilution of an organization's expertise. It may facilitate the application of that knowledge to an expanding workload without the delay of training staff. If such systems have the ability to explain their reasoning, then a human being can learn from using them.

The key to knowledge-based systems lies in distinguishing between knowledge and reasoning, or control (Figure 1.27). Ideally, we should be able to equip a knowledge-based system with a general-purpose reasoning facility, or controller, and then expose it to our knowledge base. Of course the knowledge may take any of the forms we have alluded to so far, such as rules or prototypes. As we describe in Chapter 3, it *is* possible to discuss reasoning independently of knowledge. For example, if our knowledge is represented as logical rules, then we can apply automated reasoning to process those rules and answer queries. It soon becomes apparent, however, that a key part of our knowledge, especially in design, is concerned with how we reason with what we know. Controlling knowledge is not just a matter of applying universal principles but involves a kind of *meta*-knowledge that is dependent on the domain. There appear to be hierarchies of knowledge and complex interdepen-

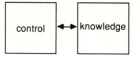

Figure 1.27 The key to knowledge-based systems lies in separating knowledge from control.

dencies between types of knowledge other than those suggested by the simple division of knowledge and control. Part of the art of formulating useful and effective knowledge-based design systems is in providing computational structures and systems of organization so that design knowledge of different kinds can be represented and made operable.

1.4   Conclusion

In this chapter we have discussed the nature of design. Design appears to encompass a wide range of behaviors and is essentially purposeful. Thus the first insight into design is that goals have an important role to play. A first approach to a design task is to work out what we want, then work out how to get it. But design sometimes appears as a process of exploration that leads to a redefinition of goals. Of course, goals are not the only means of determining the form of an artifact, and we must account for the idea of design conventions and style.

We considered the role of models in helping us to understand design. In modeling design we do not attempt to say what design is or how human designers do what they do, but rather provide models by which we can explain and perhaps even replicate certain aspects of design behavior.

There are various models of design founded upon understandings of problem solving, logic, and language. We considered designs as theories analogous to the results of scientific discovery. Although the processes by which scientific discoveries come about can be seen as similar to those by which designs are produced—particularly if we relate design to Popper's problem-solving model—the products of the process are very different. In science we are making observations and producing knowledge, essentially an inductive process. In design we are interested in the observation side of the equation. Designers wish to invent worlds about which observations can be made.

We then considered the contribution of the design-methods school to design understanding, particularly focusing on the three-phase design cycle of analysis, synthesis, and evaluation. A consideration of the complex interaction between these phases led us to focus on the process as one of exploring spaces of possible design states. This led us in turn to a discussion of the problem-solving model of design. We characterize design as complex problem solving, with the problem undergoing reformulation throughout the process.

Simulation and optimization offer specific problem-solving techniques involving evaluation and some means of producing the "best possible design." In order for these techniques to work we must be able to quantify the criteria by which we measure the performance of an artifact. Within the constraints of the method, optimization appears to offer the most systematic approach to automated design, though it does not necessarily provide the best vehicle for representing a designer's knowledge.

We then considered three basic design models that constitute the major foundation of the knowledge-based model of design that is developed in succeeding chapters. In discussing the first model, based on logic, we saw that it is more helpful to see a design as something about which one makes deductions rather than as the result of a deductive process. Then we considered a linguistic model of design, noting that design shares with language the concerns of generation and interpretation. Descriptions of designs in computer systems can be regarded as sentences for which there are a syntax and a semantics. This model maintains that generating designs is not dissimilar to the generation of sentences in natural language. We then considered the role of typologies, considering how we can also look at design as a process of instantiating from some understanding of the properties of a class of designs.

We then focused on the issue of knowledge. The ubiquity of knowledge can be demonstrated by recasting the optimization model in terms of knowledge that provides mappings between a space of performances and a space of designs. Knowledge serves to delineate design spaces. This led to the idea of generic designs or prototypes as vehicles for delineating design spaces. We introduced knowledge-based systems in which attempts are made to make knowledge explicit. The knowledge pertaining to a particular domain is separated from the reasoning mechanisms of the system. We concluded the discussion of knowledge-based systems by considering that knowledge about design includes knowledge about design reasoning. We need computational structures that facilitate the representation and implementation of different kinds of design knowledge. We considered design systems as falling along a "creativity spectrum" depending on how they deal with prototypes, with the refinement of prototypes at one end of the spectrum and the creation of prototypes at the other end.

Most of our concerns in this book can be regarded as routine design—that is, prototype refinement—though we tell our story within a framework that will account for some of the wider aspects of design. In Chapter 2 we focus on a design model that provides the basic structure of the rest of the book. The discussion so far provides the foundation for that model.

1.5   Organization of the Book

In the next chapter we argue that in formulating design systems there is considerable value in separating two types of knowledge, interpretive and syntactic. Various other issues flow from this distinction. We argue that these two types of knowledge can work together in a design system. If you disagree with our model and are interested only in tools and techniques, then the model still provides a useful way of organizing this material. The model we present is intended not only as a structure for organizing material, but to clarify the issues.

Certain themes reoccur throughout the book. Concepts like hierarchical decomposition and multiple abstractions have many manifestations and must be discussed in different contexts. We describe concepts in several ways.

Our approach is different from that in much of the design theory and methods literature. Rather than variables, constraints, and goals, the key words are objects, relations, knowledge, and reasoning. We utilize the terminology of logic and language. We hope that readers unfamiliar with terms such as abduction, interpretation, semantics, and syntax will soon come to appreciate their descriptive power.

We rely substantially on predicate calculus and logic as a unifying language for talking about knowledge-based systems and describing some of the complex concepts, particularly control. For those interested enough to experiment, concepts expressed in logic can readily be translated into the computer programming languages LISP and PROLOG (and a little less readily into PASCAL and C).

Exercises are provided to reinforce concepts and provide some direction for experimentation. Much can be learned from constructing a simple expert system shell and experimenting with a knowledge base. Several inexpensive expert system shells also are available that do not require sophisticated programming skills to use. Many of the ideas discussed here are beyond the scope of simple tools, however, and we hope that this book will prompt the reader toward further development and research in the field.

In Chapter 2 we describe our model, then look at how designs and design knowledge can be represented. Chapter 3 is about reasoning in general, Chapter 4 is about representing designs and design knowledge, and Chapter 5 is concerned with interpretation. The biggest challenge is using knowledge-based systems to produce designs, the topic of Chapter 6. Again the distinction between interpretive and syntactic knowledge proves extremely valuable in reconciling what appear to be radically different approaches to producing designs. This also carries through to a discussion of design processes, the topic of Chapter 7. In the final chapter, Chapter 8, we discuss what is needed for more effective design systems and provide some directions for further development.

1.6   Further Reading

At the end of each chapter we will indicate where the reader may find more information on the topics discussed. Where possible we point to books in which the ideas are elaborated in more detail. Otherwise, we point to papers in conference proceedings or journals.

If the reader wishes to delve still further, he or she should consult the research literature. Although widely dispersed, this literature falls into two categories: journals and conference proceedings. Very few, if any, cover knowledge-based design systems specifically, but those listed next contain material of relevance. Also, most journals in specific domains have occasional special issues related to knowledge-based design.

The following journals report research on knowledge-based design systems:

AI Magazine

Artificial Intelligence

Artificial Intelligence in Engineering

Artificial Intelligence for Engineering Design, Analysis and Manufacturing

Engineering Applications of Artificial Intelligence

Knowledge-Based Systems

Publication information appears in the journal list at the end of this book.

The primary applicable conference proceedings are the publications resulting from the series of International Conferences on the Applications of Artificial Intelligence in Engineering. Eight volumes have been produced so far, five edited by Sriram and Adey (1986a, 1986b, 1987a, 1987b, 1987c), and three edited by Gero (1988a, 1988b, 1988c).

The International Federation of Information Processing (IFIP) conducts a series of conferences on various aspects of computer-aided design. The following publications contain relevant material: Latombe (1978), Bø and Lillehagen (1983), Gero (1985a, 1985b, 1987a), and Yoshikawa and Warman (1987).

In addition two other conference series often contain material of interest: the biennial International Joint Conference on Artificial Intelligence (IJCAI) and the annual conference of the American Association for Artificial Intelligence (AAAI).

Simon (1969), an easy-to-read book that places the notions of externalizing design processes in a broader context, contains many seminal ideas. Also easy to read is Alexander (1964), an early book that describes design as a process and that has been highly influential in shaping views of design.

The view of design as partly modelable by optimization is presented in Radford and Gero (1988).

Good reviews of the issues in design, particularly from an architectural viewpoint, can be found in Broadbent (1981) and Rowe (1987). Similar reviews from an engineering viewpoint can be found in Pahl and Beitz (1984).

Exercises

1.1 What design concerns do you think are common to different fields such as electronics, mechanical and electrical engineering, structures, architecture, industrial design, and fashion design? What are the substantial differences? How relevant is the particular design domain to the study of design in general?

1.2 List as many different definitions of design as you can think of, and identify how, if at all, they can be operationalized in terms of computer-aided design.

1.3 Will the study of design ever enjoy the same status as a discipline with well-accepted theories and methods such as mathematics? Explain why or why not.

1.4 Discuss the analysis-synthesis-evaluation model of design. How is this model supported by our intuitions about designers at work?

1.5 What does it mean to describe design as "ill-structured problem solving"? Do you agree with this view? Why?

1.6 What are the strengths and limitations of optimization as applied to your own area of design interest? Why has optimization been successful, or unsuccessful, in your area?

1.7 Is design a logical process? What does it mean to say that design is or is not logical?

1.8 List the terms common to language and design. Why do you think such similarities have come about in certain design domains (long before the advent of formal language theory)?

1.9 What is a type? Describe some well-established prototypes in your domain and explain what it is about them that constitutes their designation as a type. How useful do you find these prototypes in design?

1.10 Define knowledge as applicable to knowledge-based design systems. Would this definition be commonly accepted in your discipline?

1.11 What should be our objectives in formulating a knowledge-based design system? How do these objectives differ from those that motivate conventional approaches to the design of computer programs and systems?

1.12 What is creativity? How would a design system have to perform to be labeled creative? Is this possible, or even desirable?

CHAPTER 2

A Knowledge-Based Model of Design

The purpose of this chapter is to present a conceptual framework that provides theoretical support for certain ideas about design. This framework takes the form of a knowledge-based model of the design process. It should be noted that models can explain only certain aspects of particular phenomena, those aspects that interest us because of some particular purpose. Our purpose here is to enable computers to assist in the design process.

The first chapter included a discussion of both the nature of design and what is expected from a model of computer-aided design. Initially, we consider three concepts on which to base a knowledge-based model: (1) *representation*, how we represent information in a computer system; (2) *reasoning*, what we understand by design reasoning; and (3) *syntax*, the role of syntactic knowledge in design and computer systems.

The argument is presented for a model of design in which two separable tasks are operative: *interpretation*, the mapping between design descriptions and their performance requirements, and *generation*, the composition of designs. From this simple idea we find that we can account for various processes in design and can specify what functions are required of a design system, which takes us part way to discerning how to construct such a design system.

This chapter therefore defines the basic structure of the rest of the book, setting the agenda for a discussion of concepts, tools, and techniques relevant to knowledge-based computer-aided design. When talking about general concepts it is helpful to tie ourselves to some conventions, which later on we can question. We begin by discussing representation.

2.1   Representation

Here we define facts, knowledge, and control, pointing to concepts that underlie various methods in the representation of information, rather than presenting a comprehensive explanation of representational methods. In presenting these definitions, we find ourselves describing a logic-based view of information processing that is abstract, rigorous, and has some links with the way we naturally talk about information. Our main interest here is in a formal view of knowledge as statements about the relationships between facts, which equips us to consider the role of knowledge in design reasoning.

2.1.1   Facts

How can information pertaining to design be organized in a computer system? We describe a system of organization that divides information into just two types: objects and relations.

Objects are simply units of information. It is important to distinguish between this definition and the way in which the term "object" is often used in design. Objects in design are usually spatially discrete entities, such as I-beams, piston valves, rheostats, and kitchens. Objects as units of information may conveniently map onto a designer's "objects" but this need not be the case. The objects in an information-processing system may also refer to other entities of interest to a designer, such as attributes, classes, and even processes. The nature of these entities is discussed in detail in subsequent chapters. For the time being it is necessary only to bear in mind the inclusive nature of the term objects as referring to units of information.

Relations, which are labels attached to groupings of information units, may also map onto what we sometimes regard as relationships between physical objects. Examples of such physical, or topological, relationships are "next to," "inside of," and "longer than." Again, in a designer's world the use of the term "relation" to describe groupings of objects is a liberal one. As we demonstrate here, "relations" can refer to object groupings beyond those suggested only by topology.

There are several ways of expressing this type of organization of information into objects and relations using symbols; this representation is considered in more detail in Chapter 3. One useful method is to enclose the names of objects that belong to a group in parentheses, and place the relation such that it precedes the parentheses:

Inside(PistonValve, CombustionChamber)
NextTo(LivingRoom, DiningRoom)
LongerThan(DriveShaft, RearAxle)

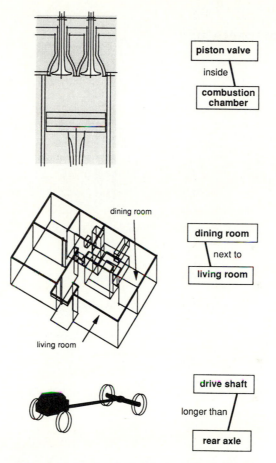

Figure 2.1   A set of objects and relations.

(We adopt the convention of representing atoms, or indivisible elements, as strings of alphanumeric characters, including the underscore character. The first character of each atom is a capital letter.) From now on we will regard these object and relation groupings as *facts*. They are translated into natural language as follows:

> The piston valve is inside the combustion chamber.
>
> The living room is next to the dining room.
>
> The drive shaft is longer than the rear axle.

Although these facts refer to physical objects and the physical relationships between them (Figure 2.1), facts can also be about other information units and their relationships, such as objects and attributes of various kinds (Figure

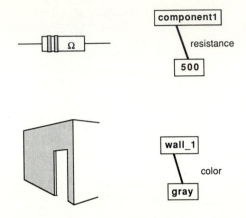

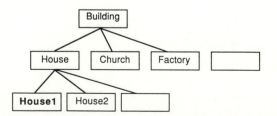

Figure 2.2 Physical and non-physical objects and their relationships.

2.2):

Resistance(Component1, 500)
Color(Wall_1, Gray)

These are translated as

The resistance of component 1 is 500.
The color of wall 1 is gray.

Note that numbers are allowed as objects in this notation. Also, using this notation, the same information may be recorded in several ways. For example, it may be convenient to represent the fact that a particular building, Building1, belongs to a class of objects called House (Figure 2.3). This can be represented in at least two ways:

Class(Building1, House)
House(Building1)

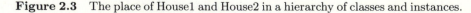

Figure 2.3 The place of House1 and House2 in a hierarchy of classes and instances.

Note that in the last example the class name is actually represented as a relation. The same method of representation can also be applied to attributes:

Attribute(Building1, SquarePlan)

SquarePlan(Building1)

We can also talk about a fact having a value: it is either true or false. A fact is true simply because we have declared it to be true, or it can be true because it is consistent with some knowledge that enables us to determine its truth status.

Dividing information into objects and relations is a simple idea, the richness of which can be enhanced if we apply this type of organization recursively. Objects represent units of information, which can also be facts. This enables us to create statements such as the following:

Joined(Element(1, Piston, 20), Element(2, Cam, 400))

Building(OfficeBlock(3), NoOfStoreys(6), Location(City(Cbd, South)))

which are, in effect, nested facts.

It is worth inspecting the last example in some detail. Notice that the relation Building is a class name, and that a fact can contain any number of arguments. This statement indicates something of the complexity and richness in representation afforded by the simple idea of distinguishing between objects and relationships.

One further point should be made about representation. When the names of objects are unknown, we can use variables to denote them:

Joined(Element(a, Piston, 20), x)

Building(OfficeBlock(n_1), NoOfStories(n_2), Location(loc))

The convention here is to represent variables as atoms beginning with a lowercase character. In order for the two "Joined" facts just shown above to be equivalent, the variables must be instantiated as follows:

$a = 1$

$x = $ Element(2, Cam, 400)

In order for the two "Building" facts to be equivalent, the following must be the case:

$n_1 = 3$

$n_2 = 6$

$loc = $ City(Cbd, South)

Notice that variables can be instantiated to facts as well as to objects.

What has been described here serves as a method of database organization, forming the basis of a *relational database*. It can also be argued that this method underlies all other methods of organization. Essentially we are talking about statements in a computer system that represent *facts* about the world. Whatever form of representation is chosen, we contend that facts can be described in terms of units of information and the relationships between these units.

2.1.2   Knowledge

We can regard objects and relations as the primitives of an information-processing system from which all other informational structures can be defined. Although objects and relations are very simple notions, they are powerful enough to provide a lever into understanding what we normally refer to as knowledge. *Knowledge* can be characterized as statements about mappings between facts. These mappings enable us to derive new facts from existing ones.

There are many ways of representing this mapping. We require a method that is both uniform, in that it fits within this organizational schema of objects and relations, and that also has some meaning in terms of the way designers are accustomed to talking about their knowledge. One of the most useful methods is to define a series of special relations that denote *logical* connections between facts. Examples of such logical relations are **and**, **or** and **if**. We will use a simple example to explore their utility as a way of representing knowledge.

An item of knowledge we may wish to see encoded in an information system is that by which we determine whether an object is supported by another object. We can, for example, say that an object is supported by another object if it is on the object. (There are other cases when an object is supported by another object but we will not consider them here.) An object is also above another object if it is on an object that is supported by another object. The logic of these statements will be apparent by considering a domain in which this type of knowledge might be useful, as shown in Figure 2.4. It is apparent that the car is supported by the bridge because it is on the bridge. We can see that the chimney is supported by the hill because it is on an object (the roof) that is supported by the hill. We know that the roof is supported by the hill because it is on an object that is supported by the hill—that is, the house. The house is above the hill because it is on the hill. How can we represent this knowledge in terms of objects and relations? One way is by means of the following nested facts involving the relations **If** and **And**:

$$\mathbf{If}(\text{SupportedBy}(a, b), \text{On}(a, b))$$

$$\mathbf{If}(\text{SupportedBy}(a, b), \mathbf{And}(\text{On}(a, c), \text{SupportedBy}(c, b)))$$

This representation, however, is cumbersome. Since **If** and **And** are special

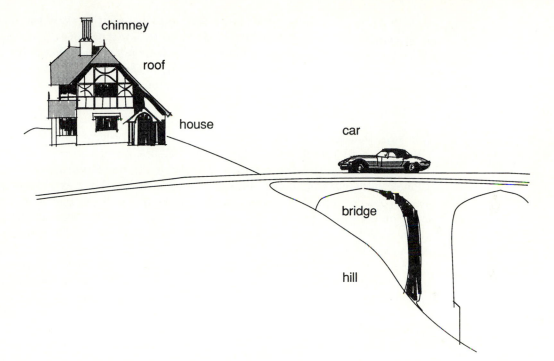

Figure 2.4   A simple world of objects and relations.

relations we generally drop the normal syntactic convention and place the relation between the facts to which they refer (that is, we adopt an *infix* notation):

SupportedBy(a, b) **If** On(a, b)

SupportedBy(a, b) **If** On(a, c) **And** SupportedBy(c, b)

These statements are now in a special form (a type of clause) following the pattern

head **If** *body*

where the *head* is a single fact and the *body* is a complex of facts arranged as conjunctions (**And**s) and disjunctions (**Or**s). Note that this mapping could also be expressed as

If *body* **Then** *head*

Both forms have certain advantages (for example, they are relatively simple to manipulate in a computer system), but there are other ways of depicting the same knowledge, which we consider in Chapter 3.

What we mean by knowledge as mappings between facts is the same as the idea of knowledge as generalization. A *generalization* is a statement about a class of objects that pertains also to members of that class. Clearly, generalizations represent an invaluable economy in the storage of information. An example of a generalization as a clausal statement is

Building(x) **If** House(x)

This says that an object x belongs to the class of objects called Building if it belongs to the class of objects called House. Even the rules defining the "supported by" relation can be described as generalizations: an object a belongs to the class of all objects that are supported by b if it belongs to the class of objects that are on b; or if a belongs to the class of objects that are on an object c that belongs to the class of objects that are supported by b. A generalization is simply a statement about the relationships between facts about class memberships.

We have discussed knowledge as logical clauses; other ways of representing knowledge are discussed in subsequent chapters. The advantage of the clausal form is that it makes explicit the idea of knowledge as statements about relationships between facts, and avoids the issue of how we process that knowledge. The processing of knowledge can be seen as an issue independent of the domain to which the knowledge belongs. It is possible, of course, to consider knowledge in terms of computational procedures—as programs containing control structures. The clausal form serves to highlight the essential nature of knowledge apart from the control component of computer programs.

Parenthetically, something should be said about the process by which logical statements such as clauses are manipulated. The normal reasoning process is that of *logical deduction*, manifested as *theorem proving*. Human beings appear to enjoy a special, intuitive relationship with the process of logical deduction, a process that can be implemented in various ways in computer systems. We therefore take the process of theorem proving as *a priori*. Descriptions of automated theorem-proving methods are provided by Nilsson (1982) and Kowalski (1979).

2.1.3   Control

We have described knowledge as concerned with mappings between facts. Here we characterize control in terms of mappings between items of knowledge.

It is possible to describe logical processes by means of logic itself. Logic thereby serves as a meta-language for describing its own operations, and also the operations of other logics. The advantage is that we can consider the way in which we control the use of knowledge in a computer system as an issue of meta-knowledge. *Meta-knowledge* is concerned simply with the relationship between knowledge statements—that is, the relationship between clauses (Figure 2.5).

	Facts	Knowledge Statements	Control Statements
Objects	physical objects classes attributes variables	facts	knowledge statements
	EXAMPLES Chimney, Car, Roof House, Hill, Bridge	*EXAMPLES* On(A, B), Above(A, B)	*EXAMPLES* Above(x, y) On(x, y) Above(x, y) On(x, z) Above(z, y)
Relations	object groupings topology	logical connections	meta-logical connections
	EXAMPLES On, Above	*EXAMPLES* **If, and, or**	

Figure 2.5 The roles of objects and relations in the three categories of information: facts, knowledge, and control.

Control knowledge can be described by defining special relations ("meta-logical" relations) that deal with relationships between clauses. A body of meta-knowledge to process logical clauses can also be represented by means of special clauses. The value of this type of representation is most evident if we deviate from strict logic and consider the computer implementation known as *logic programming*. As well as the normal declarative interpretation of logical clauses, logic programming is concerned with their procedural interpretation. Among other things this means simply that the ordering of the facts in the body of a clause is important. Meta-logic can be employed to direct how knowledge is manipulated procedurally. In subsequent chapters we demonstrate how these logic systems, such as expert system shells, frame engines, production systems, and other types of reasoning systems, can be described in these relational terms.

The idea can be extended to the consideration of meta-control—that is, to the knowledge pertaining to the relationships between control statements. This will be discussed in more detail in later chapters. The purpose here is to indicate the ubiquity of objects and relations at all levels of information organization and abstraction.

We have attempted to demonstrate a conceptual schema of how information can be represented in computer systems in terms of objects and relations. These two concepts pervade the way in which we represent information: as facts about designs; as statements about the relationships between facts, which we call knowledge; and as statements about knowledge, which constitute meta-knowledge or control knowledge. This schema addresses the issue of representation and neatly absolves us from a concern with the means by which knowledge is processed. If we wish to formulate a comprehensive model of knowledge-based design, however, we must go further. The next section addresses the issue of reasoning.

2.2   Reasoning

Here, building on Section 1.2.4, we discuss how design activity fits into how we understand reasoning in general.

2.2.1   Deduction

Logical deduction is the mode of reasoning we can usually discuss with assurance. This type of reasoning lends itself to verification, and we recognize it in good argument. It can also be readily formalized. Deduction provides the means to process the clausal statements in the preceding section. The operation of deduction was demonstrated by Aristotle (384–322 B.C.), who, as far as is known, was the first person to write about the syllogism. The syllogism provides a convenient vehicle for talking about the essential characteristics of reasoning as we understand them. An example of a simple syllogism is demonstrated here (Figure 2.6).

1. This is a house.
2. All houses are buildings.
3. Therefore this is a building.

The components of the syllogism have been variously labeled as (1) *case*, (2) *rule*, and (3) *result*, or (1) *premise*, (2) *axiom*, and (3) *theorem*. In terms of the information schema of the previous section it is convenient to label them as (1) *fact*, (2) *knowledge*, and (3) *inferred fact*.

The "this" of the preceding statements (1 and 3) refers to some existential entity. It is as if the person making the statements is pointing to some partic-

Figure 2.6   A syllogism represented graphically.

ular object, such as the house at 6 Cedar Street. The first statement typifies a very simple item of descriptive information in a computer database that maps onto an object in the real world. An example of this sort of information represented as a fact is

House(No6CedarStreet)

We normally regard the second statement as a *generalization*. It can also be seen as a statement about the relationship between facts concerned with classes of objects. It can be represented in clausal form as

Building(x) **If** House(x)

The statement "all houses are buildings" is the same as saying that an object, x, is a Building if x is a House, as defined in Section 2.1.2. This statement constitutes an item of knowledge. The final statement constitutes an inferred fact—that the particular object under discussion is a Building:

Building(No6CedarStreet)

Having established that there is a mapping between the items of the syllogism and those of the informational organization described in the previous section, we can focus on the issue of reasoning. In doing so we employ the syllogism as a simple explanatory device.

Deduction can operate in two ways. It can be concerned with the process by which it is decided that the third statement is consistent with the first two statements of the system. But it can also be the device by which new statements are derived from existing statements. Irrespective of the way in which deduction is operating, it can be said that the third statement, "this is a building," is deduced from the two preceding statements.

There is a further way we can characterize the process of deduction that we show later has a direct bearing on design. The third statement can be regarded as an interpretation of the first two statements. The interpretation of a set of logical statements is simply that which can be logically inferred from them. Putting it another way: in logic, the meaning of a set of statements is that which is not stated explicitly, but which is implicit in those statements. In terms of knowledge-based systems we can say that an inferred fact is a statement that is implicit in the statements constituting the facts and the knowledge of the system. The application of this idea of a knowledge-based system as an interpretation system is considered in more detail in Chapter 5.

2.2.2 Induction and Abduction

We have considered logical deduction. Are there other types of reasoning? Once we venture from the realm of deductive reasoning, the territory is less well-defined. Deduction is the basic building block of formal reasoning systems. It is generally recognized, however, that people have recourse to two other modes of reasoning: namely induction and abduction. Induction is the process by which the knowledge component of the syllogism is derived (such as the rule that "all houses are buildings," knowing only the other two statements).

Induction appears to be a vital human reasoning activity. In science it is regarded as the process by which theories are derived from observations of certain phenomena. There is a convenient mapping between the components of the syllogism and the way in which the components of scientific investigation are sometimes characterized (at least in the Baconian view of science). Within the realm of science the components of the syllogism can be seen as (1) *observation statement*, (2) *theory*, and (3) *prediction*. According to this view theories are generalizations, or bodies of knowledge, sufficient to describe certain classes of observable phenomena (Figure 2.7).

A third mode of reasoning has been proposed, principally by Peirce (1839–1914) (Feibleman, 1970). This type of reasoning, sometimes called *abduction*, is the derivation of statements about the world given logical rules and some logical consequences. In the syllogism illustrated previously, abduction would be operating if we knew the rule "all houses are buildings" and the conclusion "this is a building," but did not have at our disposal the original statement, "this is a house." The step by which we might decide that "this is a house" is abductive (Figure 2.8). That this reasoning step is by no means certain is evident if we consider that it is equally valid to conclude, from the information given, that this might be any other type of building, such as a railway station or a hospital. It is not generally correct to run deductive systems "backwards" in this fashion, and we generally regard an attempt to do so a serious logical error. Nonetheless, Peirce argued that abduction is a valid mode of human reasoning, Eco (1984) sees it operating in science, Charniak and McDermott (1985) use it to explain the process of medical diagnosis, and March (1976) views it as the key reasoning mode operative in the design process. How abduction can be modeled is discussed in Chapter 6. Here we will consider only why we should want to attempt such a task.

2.2.3   Interpreting and Producing Designs

In the previous section we discussed how a design can be described in a computer database in terms of facts. Such a description may consist of statements about the physical components of the artifact, such as their geometry, certain attributes, and the relationships between components. This constitutes a typical database as produced by a conventional computer-based drafting and modeling system. We can characterize deduction as concerned with the *interpretation* of such a database, the process by which we derive attributes of the design not explicit in the description. This is a common concern of computer-

Logic
1. case   minor premise   axiom 1   instance
2. rule   major premise   axiom 2   generalization
3. result   theorem   theorem   –

Logic Programming
1. fact
2. rule
3. goal

Knowledge-Based Systems
1. fact
2. knowledge
3. inferred fact

Science
1. observation statement
2. theory/hypothesis/conjecture
3. prediction

Natural Language
1. utterance
2. interpretative knowledge
3. meaning

Knowledge-Based Computer-Aided Design
1. design description
2. interpretative knowledge
3. performance/implicit attributes

Figure 2.7   Translation of the syllogism into different domains of knowledge.

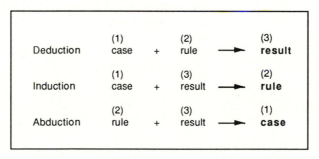

Deduction	(1) case	+	(2) rule	⟶	(3) **result**
Induction	(1) case	+	(3) result	⟶	(2) **rule**
Abduction	(2) rule	+	(3) result	⟶	(1) **case**

Figure 2.8   The three types of reasoning in relation to the syllogism.

aided design. For example, we may be interested in deriving the performance of some mechanical artifact described in a computer-based drafting or modeling system, in terms of its ability to withstand certain stresses or exhibit certain power-transmission characteristics. In structural design we may be interested in discovering the load that a given structure can accommodate before it exhibits a particular type of failure. These properties are what we have been calling interpretations, inferred facts, or implicit attributes. (In optimization and simulation we also call them performances.) We argue that interpretive knowledge can be made explicit in the form of logical rules or other representational structures, although this knowledge can also be concealed within procedural computer programs.

We therefore find it possible to represent the knowledge for interpreting artifact descriptions. Designers, however, are primarily interested in the *production* of descriptions. In design we do not usually start with an artifact description, but with some ideas about performance. We then endeavor to arrive at a description in terms of geometry and material attributes. This is entirely analogous to the abductive process described in Section 2.2.2. It also highlights something of the difficulty of design.

The knowledge that designers talk about with authority—which is the common currency of textbooks, and which is subject to verification—is generally of the interpretive kind. The question is, can this same knowledge be employed to produce designs? This would be tantamount to running a deductive system backwards in the manner demonstrated earlier with the syllogism. This issue is addressed in greater detail in Chapter 6. It is sufficient to say here that, provided the interpretive knowledge exhibits certain characteristics (specifically, that the knowledge represents a closed world), abduction serves to delimit a space of designs rather than individual designs. This space is generally defined in terms of bounds and constraints. There are also ways of arriving at individual designs by means of abductive reasoning, which are also discussed in Chapter 6.

Although interpretive knowledge is that which we are best able to articulate, it is not the only knowledge at a designer's disposal. Another substantial body of knowledge pertaining to design can be characterized as syntactic in character. Let us turn, therefore, to a discussion of the idea of syntax in design.

2.3   Syntax

We have discussed knowledge as concerned with mappings between facts, such that new facts can be inferred from existing ones. A rule may state that if fact A is true then fact B is also true. (We shall use rules as an example of a means of representing knowledge. Other means are described in Chapter 3.) In an interpretive system formulated on this basis, the number of facts increases as new inferences are made. There are, however, other types of mappings between facts. A rule can be formulated stating that if certain facts about a design are true, then they are replaced by certain other facts. Because this type of rule changes the state of a set of facts—that is, a design description—it can also be called an *action*. Another way of looking at actions is as *rewrite rules*. According to computational theory, rewrite rules form the basis of a grammar system that defines the syntax of a language. We need not pursue here the similarity between clauses and rewrite rules, except to say that grammars are a special kind of logic system in which facts can be said to be true in particular states.

This idea of a grammar appears to be very useful in design. Natural language is concerned with vocabularies of words and rules of sentence construction. Design is also sometimes characterized as concerned with vocabulary elements, such as gears, rods, transistors, struts, and windows (these are some of the designer's objects referred to in Section 2.1.1), and also concerned with rules that dictate how these elements are put together. A grammar provides a formal way of representing this knowledge about composition. The origin of generative grammars lies within computational theory and linguistics, particularly in the work of Chomsky (1957, 1963, 1971, 1975).

Formally, we say that a system defines a language if it consists of three components: (1) a vocabulary of elements, (2) a set of rules for transforming strings of vocabulary elements, and (3) a start state. One of the concerns of a natural language system, formulated along these lines, is the determination of whether a sentence is a "legal" sentence in the language. The rules of the system are the grammar rules, or rules of syntax. The rules determine the process of rewriting by which words in the sentence are replaced by certain non-terminal vocabulary elements (syntactic categories). If the sentence conforms to the syntax of the system, then the process of substitution eventually produces the start state. This process is called *parsing*.

There are other ways of representing this kind of knowledge, but rewrite rules provide a convenient formalism for making the nature of this knowledge explicit. We adopt them as a convenience, and in order to preserve uniformity, just as earlier we adopted clauses as a convenient way of representing interpretive knowledge.

There are parallels between this type of system (a syntactic system) and the syllogism discussed earlier (Figure 2.9). The components are (1) design description, (2) grammar, and (3) start state. The parsing of design descriptions is essentially deductive. As discussed later, the role of parsing in design is not obvious. How can induction and abduction be characterized in such a system? Induction is the process by which generative knowledge (a set of rewrite rules) is derived. The process of abduction results in the definition of a space of designs that conform to the language. In an interpretive system, abduction generally results in a space defined by bounds and constraints, but in a syntactic system the rules employed in parsing facilitate the *generation* of designs

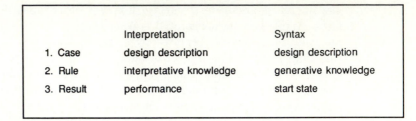

	Interpretation	Syntax
1. Case	design description	design description
2. Rule	interpretative knowledge	generative knowledge
3. Result	performance	start state

Figure 2.9 Mappings between the syllogism and components of interpretive and syntactic systems.

and partial designs. The transformation rules can be employed to bring about changes to the start state, and subsequent states, such that a space of states is generated. The transformations produce descriptions of designs conforming to the syntax of a language.

2.4   Design Spaces

Here we describe how the two types of knowledge just described, interpretive knowledge and generative (or syntactic) knowledge, can form the basis of a design system. It could be argued, of course, that interpretive knowledge and syntactic knowledge cannot always be separated. We may also suppose that design is concerned with other kinds of knowledge besides these two. Design knowledge can also be represented other than by rules. This is a useful starting point, however, and subsequently the model will be elaborated.

2.4.1   Interpretation and Generation

Consider a child's construction set (such as a set of Lego parts). Inside the box is a vocabulary of plastic parts which fit together according to a syntax of composition, determined by the plug and socket connections between parts. The designs that result can be interpreted as spaceships, moon buggies, cars, and houses. The interpretation works both ways. If the child wants to play with a "truck," he or she appears to translate this "meaning" into a particular set of design characteristics—four wheels, a platform, and a cab. Using knowledge of the vocabulary and syntax, he or she constructs a design with those characteristics. The analogy between design and reasoning leads us to see the interpretation of a design as corresponding to the logical process of deduction, while the translation of meaning into a design corresponds to the logical process of abduction.

Now let us consider two components of a design system, that concerned with interpretation and that concerned with generation. An interpretive system provides a mapping between some statements about a design (its description) and certain performances (or interpretations or meanings). Taken together, the statements about a design in a database constitute a sentence in some language. The second type of system is concerned with the rules that define the syntax of such a language. Mappings between sentences and meanings are a concern of semantics. On the other hand, the definition of what constitutes a legal sentence in the language is an issue of syntax.

These two types of systems can serve to define spaces of designs and spaces of interpretations. An interpretive system infers the meaning of a design. It maps an individual design onto a set of interpretations within a space of interpretations. It can also provide a mapping from particular meanings to a space of designs that can be interpreted as embodying those meanings.

A syntactic system also serves to define a space of designs—the designs in the language made possible by its grammar. We can therefore characterize a design system as concerned with the production of designs that belong to the space defined by the language, as well as to the space defined by the interpretive system—that is, the designs accord with some specified set of meanings.

The meanings of a design are simply the attributes of the artifact description that are not explicit in the description but must be inferred in some way—they are implicit attributes. Examples of such attributes include statements about the particular relationships between components (such as the adjacencies of rooms in a building), the cost of an artifact, or the performance of a machine in terms of energy efficiency. (Of course, these attributes are implicit only if they are not already stated explicitly in the design descriptions.) All possible designs that map onto a particular set of such meanings constitute a space of designs (Figure 2.10). In all likelihood such a space will be extremely large. Design spaces can also be defined in terms of syntactic systems. We would expect a design to constitute a member of a set formed by the intersection between the space defined by the interpretive system and that defined by the syntactic system (Figure 2.11). Another way to characterize design knowledge, therefore, is as that which defines design spaces.

This division between semantics and syntax (or interpretation and generation) is one of convenience. As demonstrated in later chapters, this distinction may be effectively concealed in the way the knowledge is represented.

The linguistic model will be developed in subsequent sections, but we are now in a position to discuss its application to design tasks and how it might contribute to the way knowledge can be organized within a computer-aided design system. We will consider two examples: the design of some mechanical artifact, such as a specialized machine to carry out some laboratory experiment, and the design of a house.

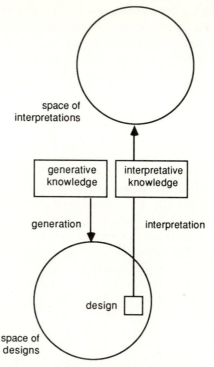

Figure 2.10   A space of designs is defined by generative knowledge. An individual design is interpreted to produce a set of interpretations.

2.4.2   A Design Syntax

We begin by discussing the syntax of the mechanical artifact. We consider how such an artifact can be described in a computer database. It will probably be made up of components, of types used in other machines. In other words, there is a *vocabulary* of parts that is known.

The database will contain representations of facts about the design. These facts will be such that they can be translated into a final artifact. It must be possible to manufacture the artifact from the database description, which generally means that the description can be translated into shop drawings or instructions for some numerically controlled manufacturing process. One normal way of ensuring this is to describe the artifact in terms of its components (vocabulary elements), their sizes, materials, and spatial relationships. Of course, not everything will be known about the candidate components. This "ill-definedness" is also encountered in natural language. The vocabulary of available words is not always completely known. Words can contain alternative suffixes and be of differing forms depending on their positions in relation to each other. Typically, in the design of a machine, the sizes of components may be unknown, though the ranges of sizes and the relationships

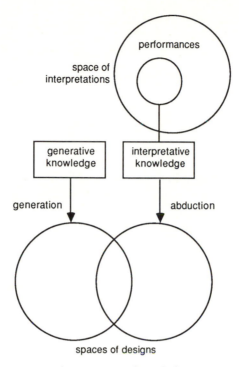

Figure 2.11   Generative and interpretive knowledge serve to define spaces of designs. Given a set of performances, we wish to produce designs that fall within the intersection of the two design spaces.

between sizes may be known. A method of representing partial information about components is discussed in Chapter 3.

There are also rules, or more generally knowledge, about how these components normally fit together, which we can describe as *generative knowledge*. We have discussed how a grammar may be expressed as actions. We can represent generative knowledge in terms of actions for putting components together, not necessarily in terms of how the design would be manufactured but how a designer might put the components together—perhaps on paper. Such actions might include various geometrical transformations, or more specifically, actions that connect components. The following actions are of a type that a designer might employ:

Connect power supply to transducer.

Double the radius of the drive shaft.

Place relief valve on top of pressure chamber.

These actions can be represented as statements about objects and relations, outlined in Section 2.1.1. In this case an action appears as a relation "operating" on objects:

> Connect(PowerSupply, Transducer)

The objects can also be represented as variables. (The representation of actions is outlined in more detail in Chapter 7.) Actions such as these facilitate the generation of configurations of components, therefore defining a space of designs. They constitute the grammar of a syntactic system.

Even though we are talking about a machine that has never existed before, we may be able to say that it will belong to a "language" of machine types—perhaps a language of laboratory equipment or desktop machines. For this task a design system must therefore be restricted to the syntax of this language. Assuming that such a language has been defined, its syntax acts as a constraint that removes from our concern certain irrelevant issues, such as weather protection, mobility, and the concerns of the languages of other machine types.

Our second example is of a house design. Components such as walls, windows, and the spaces bounded by walls constitute a loosely defined vocabulary. We could also consider the components to be primitives of lines, points, and labels appearing in space, or composites of these forming rectangles, triangles, parallelopipeds, and spheres. The syntactic knowledge about design may be characterized as that concerned with the configuration of these components. We could consider the full range of geometrical operations as defining the syntax of a design system—involving the union, intersection, and difference of components—but such a syntax is likely to produce a design space too large for practical consideration. Legal operations within the grammar may also define all possible ways in which rooms can be combined. Again, if we consider a syntax defining all possible house designs, the space of designs might be too large to be practical. In order to be useful, a designer's knowledge is restricted to a subset of those operations—that is, a more specific syntax, or "style."

We may therefore consider a grammar as that which defines the space of designs conforming to a particular style, such as the villas of Andrea Palladio (Figure 2.12) or Hepplewhite-style chair-back designs. We mention these examples because successful attempts have been made to capture grammars defining these styles precisely (March and Stiny, 1985). In these examples, knowledge has been captured in the form of rewrite rules that operate on vocabulary elements of lines, points, and labels. But we could envisage a space of designs being defined by a set of legal operations expressed as actions operating on vocabulary elements of rooms and fireplaces such as the following: locate the fireplace, position the living room east of the fireplace, and locate the kitchen adjacent to the dining room.

The syntax of a design system need not be limited to what we normally regard as style in a traditional, architectural sense, but to any system of spatial or even conceptual organization, any system for defining knowledge about how

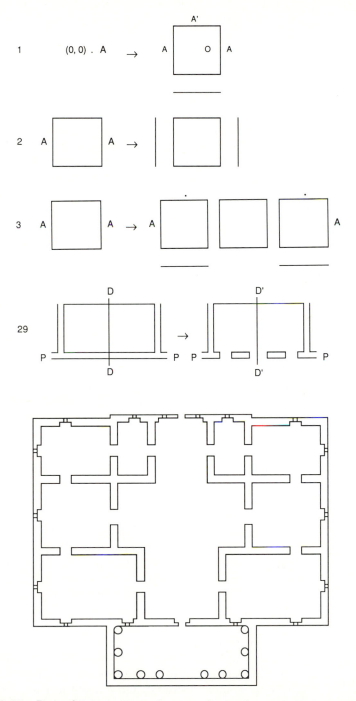

Figure 2.12   Part of a grammar and output for producing Palladian villas (after Stiny and Mitchell, 1978).

things go together—analogous to rules about how strings of characters may be organized to create a sentence.

2.4.3   A Design Interpreter

Let us now consider knowledge pertaining to interpretation. This is the knowledge by which the performance of an artifact, in this case a machine, can be inferred given some description in a computer database. Such knowledge can be represented in many ways, including the use of logic clauses as outlined in Section 2.1.2.

The performance characteristics that we require the artifact to exhibit may be known. In the example in which the artifact is an item of laboratory equipment, one performance characteristic may be to facilitate the mixing of substances in the form of test samples, combine them with certain chemicals, and produce spectral graphs on some display device. The machine must also be able to exhibit certain attributes, such as compliance with a cost threshold, dimensional and weight constraints, speed of operation, tolerances in calibration, accuracy of the display, and ease of use. The interpretive knowledge is that which enables these behaviors and attributes to be inferred from the design description.

We have characterized the knowledge concerned with the generation of designs and that concerned with their interpretation. We would expect a design system at least to contain this knowledge. The design task, then, is to produce a description of an artifact (for example, a machine) that conforms to the syntactic (generative) rules and that also facilitates the required interpretation. We could model the design process as one in which the syntactic rules serve to generate configurations of elements that are then interpreted to see whether they comply with the intended interpretations. Another way of describing this process is to say that the actions are employed to generate configurations of components that are then tested for conformity with a desired set of performances. This constitutes a naive *generate-and-test* approach. With design tasks of any complexity, we could not expect this approach to be of much assistance. Clearly, further knowledge is required (see Sections 2.4.5 and 2.6).

What of the knowledge pertaining to the interpretation of a design? This is the knowledge by which we infer the performances of a design from its description. In the case of a house design, performances may include the system of adjacencies between spaces or the character of spaces in terms of proportion, outlook, and use of natural light. Such performances might be expressed qualitatively in terms of "spaciousness," hardness of interface between inside and outside, and whether or not a room is sunny, or quantitatively in terms of ratios, percentages of perimeter walls, and solar insulation values. Performance may also be described in terms of how the occupants will use the dwelling and how the building functions in relation to their wants.

Design codes are a pervasive form of interpretive knowledge. We may be interested in a particular performance characteristic—that a design is safe and habitable. This characteristic can be summed up in the recognition that the dwelling complies with the building code—that it is a legal dwelling. Building codes, and codes generally, are therefore an interesting kind of interpretive knowledge. They do not usually tell us how elements may go together to produce a design (syntax), but rather embody the knowledge by which a design is to be interpreted as being safe and habitable.

With this kind of knowledge we can formulate a naive design system for producing safe and habitable dwellings. We require syntactic knowledge about how building elements may go together, as in the generation of Palladian-style villas, Frank Lloyd Wright Prairie-style houses, or any other system of actions in configuring building elements. With this knowledge we generate designs and interpret each of them to see whether they conform to the requirements of the code. The designs we are interested in are those contained within the intersection between the spaces defined by the syntactic knowledge (for example, all Palladian-style villas) and that defined by the interpretive knowledge (all safe and habitable dwellings). The generate-and-test method is impractical, of course, and further knowledge must be brought to bear in managing this kind of operation.

2.4.4 Representing Design Knowledge

We have defined design knowledge as that which defines spaces of designs and have focused on knowledge as explicit statements about mappings between facts. In the previous chapter we referred to descriptions of generic designs as a way of capturing design knowledge. Let us now relate this to our characterization of interpretive and generative knowledge.

Syntactic knowledge is essentially concerned with the configuration of elements. We have discussed grammatical rules, or actions, to delineate syntactic knowledge. Design knowledge, however, can also be depicted, in terms of generic descriptions, as *prototypes*. We may therefore expect that syntactic knowledge can be represented in terms of type descriptions; there should be some body of knowledge by which this generic description can be instantiated — translated into the description of a design instance. This knowledge could simply be embedded in the prototype description, in terms of acceptable ranges of possible values—for example, that a small house has somewhere between one and three bedrooms. Such generic design descriptions are used informally in design manuals and guides. For example, the cross-sectional view of the generic two-stroke engine is well-known. In building design a square plan with a centrally located square service core often serves as a prototype for a multistory office building. Of course, other types might be adopted, such as a service core at one end of the plan with a linear circulation pattern. These types are often cataloged in design guides (Duffy, Cave, and Worthington, 1976), so it is unnecessary to invent a new type with each design task

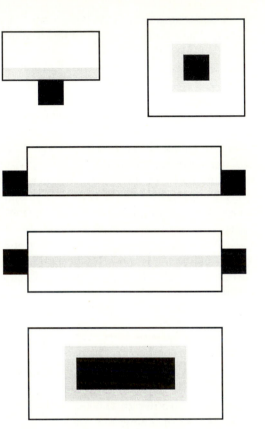

Figure 2.13 Some office building types (from Duffy, Cave, and Worthington, 1976). Service cores are black; corridors are gray.

(Figure 2.13). When design spaces are delineated as diagrams, a good deal of knowledge is required to bring about the instantiation process. Translating from such a description of a prototype to a design instance is not always straightforward. There are systems of representation that enable knowledge in the form of generic descriptions to be represented explicitly in computer systems (see Chapter 3).

We may also expect that interpretive knowledge can be represented in terms of generic design descriptions. We are used to type descriptions as diagrams, and diagrams imply configurations of elements. However, we are also used to the idea of interpretive knowledge in diagrammatic form—as acceptable tolerances in die casting and as constraints about distances from boundaries in building design, for example. The difficulty with translating interpretive knowledge into generic descriptions is that interpretive knowledge generally embodies many options. It is difficult to visualize the space of designs defined by the building code with a single diagram. We argue in Section 3.5 that there is some value in translating knowledge about what is safe and habitable into models of buildings, even though such models are difficult to visualize.

2.4.5 Design Processes

We have characterized design knowledge in terms of statements about mappings between facts and as generic descriptions, concentrating less on knowledge as existing designs and as procedures. What of knowledge about design processes? Presumably we can formulate rules about design processes; moreover, there are classes of design processes and procedures about design processes. We argue that operations can be treated in the design process as objects and that knowledge can be brought to bear in reasoning about these objects. These procedures apply to processes of producing design instances from generic descriptions, as well as to the process of abduction—reasoning from performances to design descriptions. Here, however, we shall explore the idea of reasoning about generative knowledge.

In Section 2.1.3 we introduced the idea that knowledge about the control of a logic system can be represented in terms of relationships between rules. Rules are treated as objects that are manipulated by means of meta-rules. We have also introduced the idea that generative rules can be seen as actions that change design descriptions in some way. Here we discuss the idea that facts about actions may constitute "vocabulary" elements in a design control system. What we have said about a design system in terms of the interpretation and generation of compositions of vocabulary elements also applies to the interpretation and generation of action sequences.

Generally, we regard the vocabulary elements of a design system as physical elements. Descriptions of configurations of elements can be interpreted by means of interpretive knowledge. Descriptions that conform to the syntax of some language can also be generated as outlined. There are other types of entities on which a system could operate, such as those that do not map directly onto physical objects.

It is possible to devise a system that controls a design system as one in which the objects are actions, and in which vocabulary elements are manifested as statements about the relationships between actions. Therefore knowledge exists that allows us to interpret sequences of actions, and syntactic knowledge exists that defines a language of such actions. The sequence of actions generated by such a system constitutes a *plan of actions*; the system effectively serves as a controller for a generative system.

Designers appear to possess a ready vocabulary of permissible actions— for example, to put objects in place, to move objects, erase them, and enlarge or reduce them. The full range of geometrical transformations is available. But actions can also be described in relational terms—for example, to place an object next to another object, or to locate an object inside another object. There are, of course, further transformations by which objects change

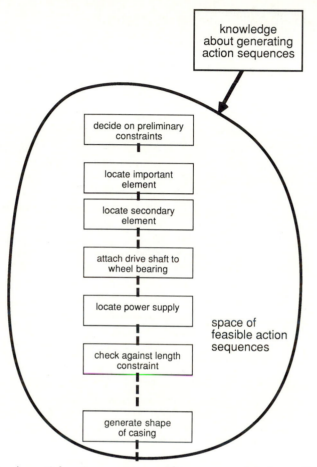

Figure 2.14   A partial action sequence within a space of design action sequences defined by generative knowledge.

their meanings such that abstract geometrical entities become design components. Taxonomies of actions can be devised that indicate the relationships between different actions in much the same way that taxonomies of relationships between objects can be drawn up. The implications of this approach are discussed in Chapter 7.

The control of actions becomes important when we consider that design can also be seen as the selection and sequencing of appropriate actions to produce descriptions of artifacts with particular attributes. The ordering of actions inevitably involves the resolution of conflicts among competing actions. Designers seem to adopt strategies that let them actively avoid situations where conflicts occur, or detect conflicts at an early stage, when they are relatively easy to rectify. The syntactic rules of this language may therefore be concerned with the generation of appropriate action sequences (Figure 2.14). As is demonstrated in Chapter 7, these rules provide an opportunity for knowledge about design processes to be made explicit.

2.5   Redefining Spaces of Designs

It has been argued so far that design knowledge facilitates interpretation, generation, and control at various levels of abstraction. Here it is argued that there is another body of important knowledge relevant to design systems— that is, knowledge that facilitates the acquisition of other knowledge.

In Section 2.2 we introduced three types of reasoning. We can model the derivation of performances by means of deduction. Abduction can serve to define spaces of designs. Induction is seen as the process by which knowledge is acquired.

2.5.1   Knowledge Acquisition

We have suggested that design knowledge can be made explicit as rules or clauses, which bear some intuitive relationship to the way designers often think about their knowledge. This form of knowledge can also be rendered operable in a computer system. However, design knowledge can be recorded and made operable in other ways. One common method is as procedures, which constitute a "medium" to represent knowledge that combines knowledge and control. It can be said that the knowledge component of procedures is implicit, as opposed to logical rules, in which the knowledge is explicit.

Design knowledge is also implicit in various other media; one such device is *existing designs* or *design episodes*. This implicitness is apparent when we think of the importance of precedence in design. Designs are rarely produced merely from a statement of intent and a body of knowledge, but rather takes place within a "design environment," against a background of examples of similar artifacts. Existing designs can be used in several ways: (1) by their simple imitation, in whole or in part, (2) by their treatment as prototypes in the manner outlined earlier, (3) as a source of knowledge—of rules or principles that have contributed to the production of particular designs, and that can be extended to the production of new designs, and (4) as examples from which analogies can be drawn.

The last two methods are of the most interest here. In considering the use of existing designs as a source of knowledge, we see the process of learning as one in which existing designs are subjected to scrutiny, in such a way that the knowledge that brought them about, or could bring them about, can be extracted. This knowledge then forms part of a design system. Such generalization from examples is the way that knowledge acquisition is often understood (Winston, 1970; Michalski, 1983). If we extract knowledge, we can never be certain that it is accurate or complete, a problem commonly recognized in the acquisition of scientific knowledge. Knowledge as theories can enjoy only the status of a conjecture or a hypothesis, and can never be "proven" to be true, only refuted (Popper, 1972).

Storing knowledge in the form of generalizations is obviously more efficient than remembering every single design encountered. In making generalizations, however, a lot of information is lost. We cannot anticipate every use to which the knowledge will be put in making generalizations, nor can we be sure of the right abstractions to consider. So we make assumptions, which may be inaccurate. Thus it is interesting to consider the fourth use of existing designs, as contributing to analogical reasoning. Such reasoning lets us see something similar in the design task currently underway that reminds us of previous design episodes; we learn as we go, extracting the most useful lessons from existing designs appropriate to the current situation.

What is the role of learning in a computer-aided design system? One approach is to store existing designs as a resource in a computer system from which knowledge can be extracted as required. Such designs may also be the products of the system's own operations, including the "histories" of certain designs. A design system may therefore learn from its own activities. The knowledge that is acquired includes both interpretive and generative knowledge, as well as knowledge about design processes. Once acquired, this knowledge constitutes part of the system's knowledge base.

This ties in with the idea of creating prototypes, discussed in the previous chapter. Carrying out this design activity, we acquire the knowledge that extends the space of design possibilities.

2.5.2   Innovative Designing

Learning serves as a way to improve the performance of a system from one task to the next, but learning also continues throughout the design process. Spaces of designs delimited by knowledge undergo redefinition during the design process as knowledge appropriate for the task at hand is learned and refined.

The existence of this key process in design suggests that design knowledge is by nature ill-defined. Searching a space of designs is especially difficult because the boundaries of the space are determined as much by the process itself as by a body of knowledge given *a priori*. Searching a design space can result in modifications to the character and boundaries of the space. The definition of design spaces is therefore dynamic, and the design process involves not only search within such spaces but the acquisition and modification of the knowledge that defines those spaces.

Having extolled the primacy of learning in design, we may conclude that a design system must contain knowledge about how knowledge is acquired. In Chapter 8 we briefly consider how certain aspects of learning applicable to design can be modeled in a computer system. Learning is a process that is imperfectly understood at present; mastering it is expected to constitute one of the most significant advances in knowledge-based computer-aided design.

2.6   A Formal Model of a Design System

We are now in a position to describe a formal model of design systems that provides a coherent framework for the remaining chapters. Being an abstraction, this formal model cannot account for all the nuances of the design process, but rather to structure the way in which we consider the process. It is therefore a unifying model for the remaining chapters. Our formal model is based on an analogy with language and with the three reasoning processes of deduction, abduction, and induction. We suggest that a linguistic model is particularly useful in the context of computer-aided design and that there is a direct analogical morphism between these three forms of reasoning and the three classes of design processes: interpretation, producing designs and learning.

We can make the following analogies between language and design:

	Language	*Design*
Vocabulary	words	parts
Syntax	grammar	actions for configuration
Utterances	sentences	designs
Semantics	meaning	interpretation of designs as performances

2.6.1   Design Processes

We can formally define a set of terms appropriate to design as follows.

Design descriptions (D): the results of the design process, produced as a result of design decisions.

Vocabulary (V): the elements manipulated to construct a design.

Knowledge (K): the design-specific knowledge open to and utilized by the designer or design system; relates the vocabulary and interpretations.

Interpretations (I): either goals or requirements (intended interpretations) when used to drive a design process, or the result of analyzing or evaluating an existing design (*actual* interpretations).

What is a design and what is a vocabulary element depend, of course, on our viewpoint. There are hierarchies of vocabulary elements. The vocabulary element "spoke" may be a part of the design "bicycle wheel," which is in turn a vocabulary element of the design "bicycle."

We can map from the design domain to the logic domain as follows:

theorem = interpretation (I)

premise = design description (D)

Of necessity, designers begin designing before all the relevant information is available. Indeed, since design is often characterized as an exploration process, what is relevant manifests itself only as the design proceeds, and changes direction even from the same starting point. Thus designers need to begin their work with only a minimal set of information, which is used to guide the introduction of more detailed information.

Human designers may form their individual design experiences into generalized concepts or groups of concepts at many different levels of abstraction—that is, they schematize their knowledge. These concepts consist of knowledge generalized from a set of like experiences and form a class from which individuals may be inferred. Thus there exists generalized design knowledge that applies to the commencement of a design for a particular domain and context. Similarly, there exists generalized design knowledge that applies to the continuation of a design in a particular domain.

Here we present a conceptual schema for this knowledge called a *prototype*. Our notion of prototypes is related to the prototypes of Bobrow and Winograd (1977b) and Aikins (1979), as well as to the prototype theory of Osherson and Smith (1981). Prototype theory construes membership in a concept to be determined by its similarity to the concept's best exemplar. Our notion builds on this theory by abstracting the prototype to allow it to become the generalization of the concept (Gero, 1987b). Prototypes are related to *scripts* (Schank and Abelson, 1975) but have additional functional and structural features.

Thus a prototype denotes a description of a class of generalised designs that embeds a notion of the design description to be produced (D_p) perhaps in a parameterized form; vocabulary; interpretive knowledge; syntactic knowledge; and the interpretations or requirements:

$$P = (D_p, V, K_i, K_s, I) \tag{2.8}$$

Although the contents of a prototype are developed by individual designers, like-minded designers will tend to agree on the specific boundaries. Thus in the design of a house, the initial prototype is likely to include such notions as style, location on the site, cost, and the existence of a variety of spaces based on their functional activities (Figure 2.15). A prototype used later may be concerned with the design of the kitchen (Figure 2.16) and is likely to include such vocabulary elements as an oven, a stove, a sink, a refrigerator, countertops, plumbing, and lights. As a design proceeds from an initial prototype, other prototypes are brought into play. In this way new requirements are added to the system, requirements that were not manifest earlier. The process of selecting additional prototypes that produce increasing detail and additional requirements provides a basis for exploration in design.

Using prototypes, we can now formalize the three types of design activity outlined in the previous chapter: prototype refinement, prototype adaptation, and prototype creation.

HOUSE PROTOTYPE

INTERPRETATION   (I)

function:	to provide a living environment for a family
style:	(modern, Spanish, colonial, ...)
.	
.	
cost:	(low_cost, medium, high)

DESCRIPTION   (Dp)

a_kind_of:	building
number_of_storeys:	1 to 3
length	maximum length < site length

VOCABULARY   (V)

parts:	(kitchen, living_room, bedrooms, bathrooms, ...)

INTERPRETIVE KNOWLEDGE   (Ki)

if	house has arches and
	external finish is cement render
then	style of house is Spanish

SYNTACTIC KNOWLEDGE   (Ks)

if	kitchen located
then	locate dining room next_to kitchen

Figure 2.15   House prototype.

Prototype Refinement

Here a prototype is selected, perhaps on the basis that its performance adequately matches the desired design performance (goals) in some general sense. A design is produced within the values of the variables allowed for in the pa-

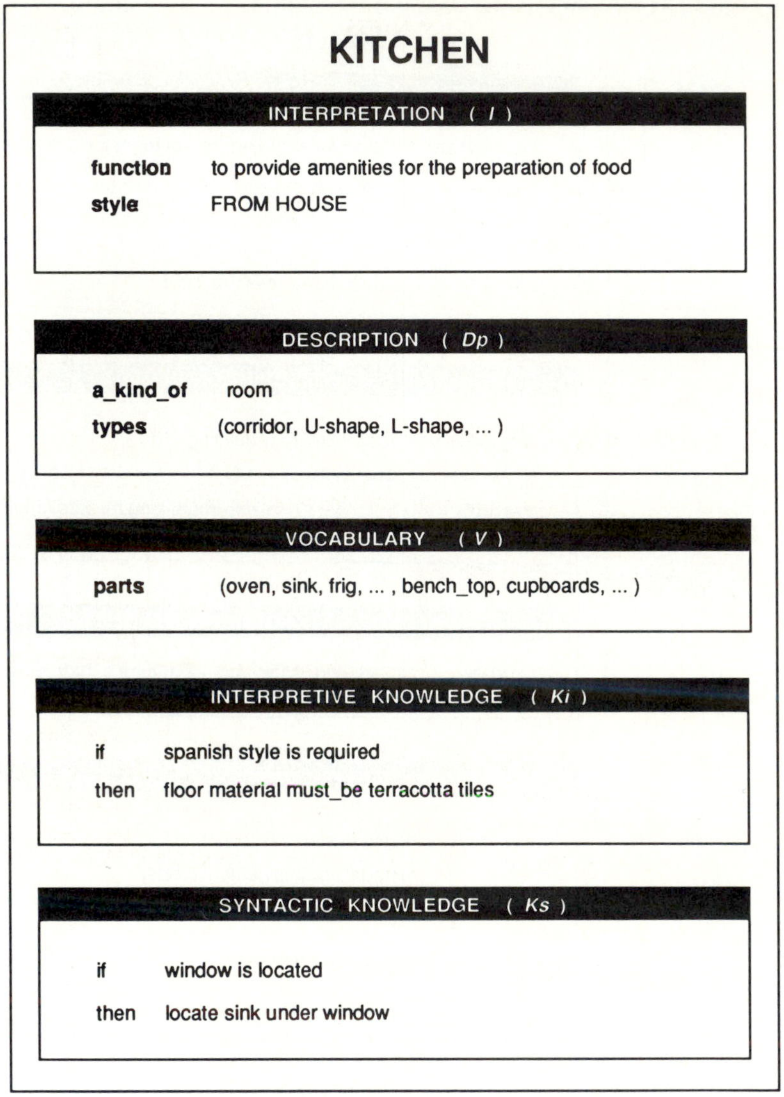

KITCHEN

INTERPRETATION (I)

function   to provide amenities for the preparation of food

style   FROM HOUSE

DESCRIPTION (Dp)

a_kind_of   room

types   (corridor, U-shape, L-shape, ...)

VOCABULARY (V)

parts   (oven, sink, frig, ... , bench_top, cupboards, ...)

INTERPRETIVE KNOWLEDGE (Ki)

if   spanish style is required

then   floor material must_be terracotta tiles

SYNTACTIC KNOWLEDGE (Ks)

if   window is located

then   locate sink under window

Figure 2.16   Kitchen prototype.

rameterized design description, or as allowed by some design-description generator. These values are such as to produce the required interpretations. In
other words, prototype refinement involves instantiating the variables in the
prototype and determinating their values (Figure 2.17). Design by prototype
refinement can be modeled by modifying Equation 2.4.

$$D = \tau_8(P, I) \tag{2.9}$$

If these values are numeric only, and if the required interpretations can be

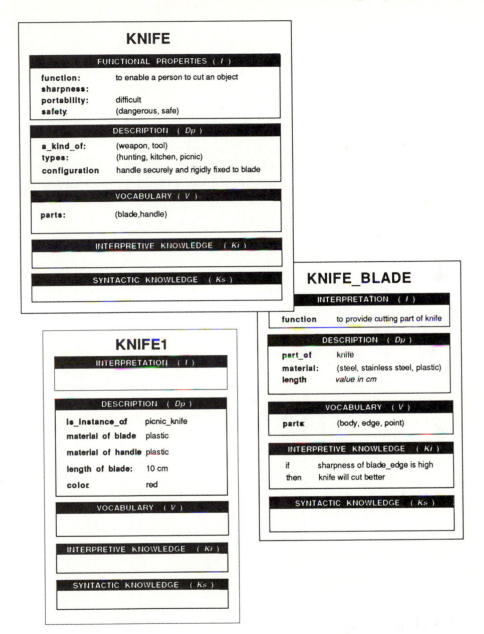

Figure 2.17   An example of prototype refinement.

expressed numerically, then optimization methods can be used. If the values are symbolic, other approaches are required. *Routine design* falls into the category of prototype refinement.

Prototype Adaptation

If a prototype is found inadequate in some sense, the designer or design system may be able to adapt it to make it more useful. Modifying the vocabulary, V, syntactic knowledge, K_s, interpretation, I, or interpretive knowledge, K_i, results in new variables for the prototype (Figure 2.18). Thus designing by prototype adaptation becomes

$$V' = \tau(V)$$

and/or

$$K'_s = \tau(K_s)$$

and/or

$$K'_i = \tau(K_i)$$

and/or

$$I' = \tau(I)$$

and

$$D'_p = \tau(D_p)$$

hence

$$P' = \tau(P)$$

and

$$D = \tau_9(P', I) \tag{2.10}$$

Once a prototype is adapted and found useful, the remainder of the design process becomes one of prototype refinement.

Prototype Generation

Generating new prototypes is the highest of the design endeavors. In a formal sense, very little is known as to how this occurs.

A prototype defines a state space within which a design description can

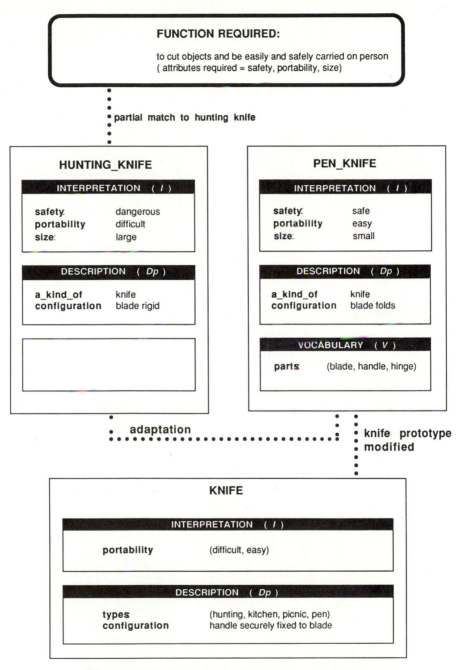

Figure 2.18   An example of prototype adaptation.

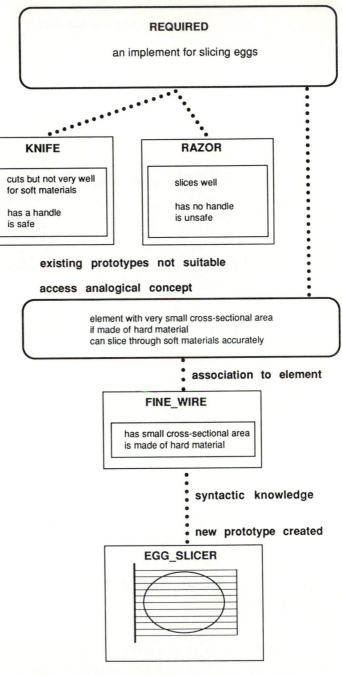

Figure 2.19   An example of prototype creation.

be produced. Prototype refinement commences by accepting this (implicitly or explicitly) defined state space and searching for descriptions. Prototype adaptation changes the state space prior to refinement. Prototype generation, which is concerned with producing a new state space, therefore constitutes a kind of abduction, where the artifact of the process is not a design but a prototype—the creation of an entirely new vocabulary and syntax combination (Figure 2.19).

We can summarize these three processes with a simple example from automobile design. The knowledge by which certain interpretations of cars are made (such as mass appeal, simplicity, low cost, and ease of production) can be brought to bear in selecting a suitable prototype. So if we are interested in designing a car with "mass appeal" we may use this broad goal to select the "Volkswagen" prototype for consideration. Other, secondary goals (such as "suitable as a city car for a young executive") may assist in the instantiation of the prototype. This prototype refinement involves selecting from among a set of available options, such as two-tone paint, stripes, turbocharger, or cloth seats.

The prototype may match the general requirement of catering to a mass market, but the satisfaction of certain other goals may lie beyond the values allowed for its parameters. For example, the requirement that the car should also be suitable for a small family may not be allowed because it is difficult to fit a child's car seat into the back seat. In this case knowledge must be brought to bear in transforming the vocabulary or syntax implicit in the prototype to make an adapted prototype, perhaps modifying the body to a wagon, which produces a different interpretation ("suitable for a young executive with a small family"). The new prototype so adapted is then subjected to prototype refinement.

Accounting for the origins of the Volkswagen prototype—prototype generation—presents difficulties. Clearly prototype generation is the most difficult of the three basic design tasks. Superficially it appears to occur with changes in styles of furniture and buildings, or the invention of alternative approaches to satisfying some performance objective, such as the replacement of the pistone engine by the jet engine. On closer inspection, however, such developments are frequently the result of a rapid sequence of prototype adaptations.

2.7 Summary

In this chapter we have presented a conceptual framework that provides theoretical support for certain ideas about design. We began by discussing facts as a means of describing designs. Facts are statements about the relationships between objects, where objects are any unit of information. Knowledge can be characterized as statements about the relationships between facts, specifically how new facts can be derived from known ones. The most explicit way to represent this mapping is as rules of the form "A is true if B is true."

Knowledge serves to define spaces of designs. There are other ways to define design spaces than by using rules. The methods we are most familiar with include existing designs, descriptions of generic designs, and procedures. We have suggested that these methods of knowledge representation are isomorphic; implicit in each is this notion of mappings between facts.

In design systems we must consider how knowledge is controlled. Many new facts can generally be derived from existing facts, and it is necessary to navigate through this knowledge in order to make the right inferences. We have argued that control can also be represented as knowledge—as metaknowledge or statements about knowledge statements. How this can be done (and why we should try) is discussed in Chapter 4. It should be possible to construct computer systems in which knowledge about different operations is made explicit.

We then turned our attention to reasoning as it is traditionally understood. We use our knowledge to reason from premises to conclusion by a process known as deduction. But two other reasoning processes, induction and abduction, are less well understood. Induction involves the acquisition of knowledge, generally given several examples of premises that produce similar conclusions, while abduction involves reasoning to premises given the knowledge and the conclusions. We argued that the latter process most accurately characterizes design reasoning. We know what conclusions we want—the design must exhibit certain performance characteristics—but we do not have the description of an artifact that meets the requirements.

We employed this characterization to define two categories of design knowledge, that concerned with interpretation and that concerned with design syntax. Although interpretating a design is by no means simple, we can readily model one with deduction. Design shares with language a concern with interpretation and also with composition, how things go together—in the case of language, compositions of words; in design, compositions of design elements. Another way of looking at this linguistic view is to observe that design descriptions in computer systems are strings of symbols. We are interested in both the interpretation and the generation of those strings of symbols. Syntactic knowledge can be expressed as grammatical rules or actions—mappings between facts—but also by other means.

We therefore characterized design as a search within a space defined by these two categories of knowledge. We are interested in designs that map onto a particular performance and that map onto to a particular system of organizing elements. As we shall see in Chapter 6, we can model certain kinds of design activity by focusing on abduction alone. But the simplest model of design is generate and test, where we employ our knowledge about syntax to produce designs and knowledge about performance to interpret designs. We keep generating and testing designs until we achieve one that matches our goals. Of course, this model does not get us very far.

We then discussed how we can consider knowledge about design processes in a similar manner. If design syntax can be represented in terms of actions, then there may be knowledge pertaining to the "language" of actions that can be brought to bear on the design task. We considered actions as objects and how there may be knowledge about interpreting and configuring actions in order to produce designs.

Even when knowledge about design is available at various "control levels" we must account for other aspects of design. We considered knowledge acquisition and how interesting and innovative design occurs where we extend the boundaries of our design spaces. Design appears to be concerned with the definition of the spaces being searched as much as with the search process itself. This would suggest that we must capture the knowledge by which design spaces are defined during the design process. Another issue is the acquisition of goals, or intended performances, as the design process is underway. We will not devote much attention to these last two, difficult issues until Chapter 8. Their resolution may hold the key to truly effective knowledge-based design systems.

This chapter concluded with a formal symbolic model of a design system and a formal description of prototypes. These formal models provide a coherent framework for a comprehension of knowledge-based design.

2.7.1 Further Reading

Four of the International Federation for Information Processing's specialty conferences contain material of interest here: Gero (1985b), Gero (1987a), Latombe (1978), and Yoshikawa and Warman (1987). March (1976) is an important paper that places design in the context of reasoning with logic and introduces the notions of abductive logic in design. More information on prototypes in design, including their application, can be found in Gero, Maher, and Zhao (1988), and in Lansdown (1987). The fundamental notion of scripts are elaborated in Schank and Abelson (1977).

Exercises

2.1 Model a simple scene in terms of facts about objects and relations. Devise some knowledge statements in natural language for inferring new facts. How easy is it to formalize these statements as logic clauses? Can most design knowledge be represented as logic clauses?

2.2 How does the division of all things into objects and relations pervade a logic view of knowledge? Discuss the idea that control is just another kind of knowledge.

2.3 Translate the facts and knowledge of Sections 2.1.1 and 2.1.2 into a logic program (in PROLOG, LISP, or some similar language). Experiment by adding new facts and clauses.

2.4 Devise some examples of syllogisms applicable to different domains, such as diagnosis and scientific observation. Compare these with a syllogism pertaining to the interpretation of a design. Discuss the value of the syllogism as a simple means of characterizing reasoning.

2.5 Discuss the link between generalization and knowledge.

2.6 Investigate formal logic and formal language theory (or computer theory). Standard texts such as those by Lemmon (1965) and Cohen (1986) may help. What is the link between logic rules and rewrite rules? What do logic and formal language theory have to say about the representation of knowledge?

2.7 Define generate and test as a model of design. What are the strengths and weaknesses of generate and test as the basis of a design system?

2.8 Choose a design task and describe it informally in terms of interpretive and generative systems.

2.9 Describe how hierarchies of control may operate in a design system.

2.10 What is the role of learning in design?

2.11 How does the model presented in this chapter account for the nature of design? What does it fail to account for?

2.12 What is the value of the prototype idea? Explain why a schema such as a prototype is needed.

Representation and Reasoning

Part I discussed design in general terms, dealing with its broad characteristics and processes. Part II explores in greater detail the information—facts and knowledge—required for design, as well as the methods and tools available for representing and reasoning with this information. It looks at the attributes of design that must be described and the ways in which knowledge-based methods can describe them.

Knowledge-based systems deal with the processing of knowledge. In order to process knowledge, there must first exist ways to represent this knowledge and methods to control its processing. Before knowledge can be represented in design, the type of knowledge involved must be identified and classified. Chapter 3 therefore reviews the kinds of knowledge with which design is concerned. The existing formalisms with which design knowledge may be represented are described and illustrated by examples. Chapter 4 then proceeds by describing the various ways in which reasoning can be carried out in design. Control of reasoning processes is fundamental in knowledge-based systems. Design systems require a variety of reasoning mechanisms, which are presented and described here before being used in the elaborations of Part III.

CHAPTER 3

Representing Designs and Design Knowledge

In this chapter we begin to amplify the model of the previous chapter. In focusing on representation we are not concerned with the distinction between interpreting and producing designs, which is treated in subsequent chapters.

We have characterized design as concerned with the production of a description of the form of an artifact in response to a statement about desired performance. That is, design is concerned with the production of descriptions that can be translated into physical reality. Thus we need a mechanism for describing designs—that is, a description language. This language will consist of a vocabulary and some system of organizing the "words" in the vocabulary. In a designer's world this vocabulary is usually seen as a collection of design elements that must be brought together in some way.

A design language should be rich, allowing the full range of design expression pertinent to the particular domain. Such a language should enable us to depict reality within an appropriate system of abstraction, one that facilitates communication between different stages of the design process and between the design system, the designer, consultants, the client, and the manufacturer. The language should also enable us to embody design intentions as design form.

We can speak of the issue of representation more specifically in terms of information. Designers appear to make extensive use of information about elements, relationships between elements, properties, and classes. This information is generally about both the design and the design intentions, or goals. The other information with which designers are concerned is knowledge. Designers appear to use knowledge about how to manipulate elements, how to proceed in a given situation, how to produce and interpret designs, and how to control the process.

Traditionally, designers have externalized designs, goals, and knowledge in terms of graphical and textual symbols. In knowledge-based systems we need appropriate methods of representing this information. The way information is represented and organized will bear on the correctness and ease with which design intentions can be realized, and with how human beings can communicate with the design system.

In our symbolic model of a design system we have characterized design as concerned with design descriptions (D), a vocabulary of elements (V), interpretations (I) and design knowledge (K). We must be able to represent these components: the description and its elements $(D$ and $V)$, the intended interpretations (goals) or the actual interpretations (I), and the knowledge $(K_i$ and $K_s)$ by which the action of translating this interpretation and the vocabulary into a design can take place.

We emphasize three representation issues in this chapter. The first issue is how designs and their vocabularies can be represented. We identify objects, properties, relationships, and the use of generic descriptions as major representational issues. One powerful reasoning device that we discuss is the idea of inheritance. Classes of designs pass on properties to subclasses and instances of designs.

The second issue is the representation of design goals, the intended interpretations or performances. We argue that the representation of goals is essentially the same as the representation of designs. The difference is that in the case of goals our description is partial, incomplete, and at a level of abstraction that cannot readily be translated into a physical artifact. In design we take such partial descriptions and transform them into complete descriptions that constitute designs.

The third issue is knowledge about design and how it can be represented in design systems. Design knowledge is needed to interpret and produce design descriptions, matters considered in depth in later chapters. The representation of designs and the representation of design knowledge are often bound together, particularly when we talk about class and instance hierarchies. Thus both are considered here.

3.1   Representing Designs

In this chapter we consider different ways of representing a particular design example that is used to explore issues in representing designs and design knowledge. All readers live in some kind of dwelling and know at least a little about buildings, so our example consists of some very basic information about a house. Some of the information we wish to represent is about "my house", a particular house that exists in a physical location. Some of the knowledge is about houses in general. Here is a set of loosely structured sentences about the domain:

My house is a particular house.

A house is a kind of building.

A house is used for living in.

A house can be described in terms of certain properties:
    it has a front door,
    a certain number of rooms,
    and a color.

My house has a front door called door1.

The number of rooms in my house is 7.

The color of my house is white.

The color of door1 is blue.

The height of door1 is 2,050 mm.

The kitchen is adjacent to the dining room.

If there is no information to the contrary,
    then a house is assumed to contain 10 rooms.

Yellow is a bright color.

Gray is a dull color.

How can we formalize this information? We shall see that the description of a house is essentially that of a class description (which matches part of our definition of a prototype) and that "my house" is an instance of this class.

We want to represent designs in such a way that we can make useful interpretations from our representations. How we describe designs is essentially a linguistic concern. Much has been said on the role of language in describing the world and thereby coloring our perception of it (Wittgenstein, 1922; Quine and Ullian, 1970). As yet, there is no agreed theory of representation. For our purposes it is not necessary to complete a critique on epistemology before we can start to use symbols in computer systems to describe things, but it is useful to be aware of some of the assumptions and issues that underlie our methods. We will attempt to bring these to light as we proceed.

In this section we look in detail at the issues in design representation. We assume the need for some system of decomposition, some way of describing a design in terms of component parts. We then need some means of representing configurations of parts in a computer.

3.1.1 Objects, Relations, Properties, Classes, and Instances

In natural language the basic vocabulary unit that conveys meaning is generally taken to be the *word*. In design the selection of the basic vocabulary

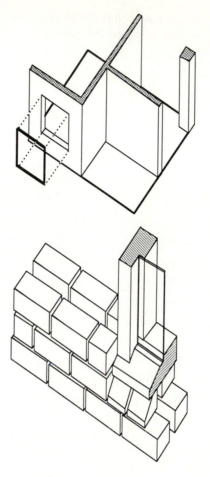

Figure 3.1 Designing at different levels of detail. At one level the elements are walls, columns, and windows, while at another level we are dealing with bricks, frame members, and panes of glass.

unit—which we call an *object* or *element*—is more problematic. Designs appear to be describable in terms of component and subcomponent hierarchies. The level on this hierarchy from which we select our vocabulary of objects depends on the appropriateness of that view to the particular purpose. For example, it makes little sense to begin to design a whole building by thinking of the individual bricks of which it is composed; if we attempt to locate every brick, then the task of producing and representing designs will be extremely cumbersome. In designing a window opening, we may choose to design at a level of detail at which the brick is a basic unit, along with the window frame and glass (Figure 3.1). Thus the selection of what we choose to call objects must be given careful consideration.

As discussed in Section 2.1.1, we are also interested in the *relationships* between objects. There are many ways that elements can be related, but one important type of relationship is that which manifests itself spatially in some way. Groups of elements may therefore exhibit topological relationships, an example being two adjacent rooms in a house.

Further, we are interested in the *properties* of these objects. A convention in computer-aided design is to talk about the *name* of an element (such as brick) and a list of statements attached to it that are its properties, such as weight and size.

The configuration of elements in space is an important consideration in design. Hence it is convenient to represent properties that are geometrical in nature, such as size and location according to some kind of spatial system. We can also identify some properties as functional in nature; they describe what an element does or how it performs. We can devise computer-aided design systems that contain libraries of elements. The design then constitutes a configuration of elements describable in terms of their names and properties, including their spatial locations.

We may also divide a designer's world into classes and instances. We can speak generally about collections of things, such as houses, and their properties. This means that we can be economical in describing instances. Knowing that something is a house, we can transfer all the properties of the house class onto the instance.

We therefore have three convenient systems for dividing a designer's world (Figure 3.2). As we shall see, each can be used to represent similar concepts, they can be combined in certain ways, and each affords certain advantages in the representation of designs. We present some examples of how we can make statements about designs based on the idea of objects and relations, objects and properties, and instances and classes.

Objects and Relations

These are statements about the physical and non-physical objects that go to make up the design, and the names we attach to ordered groupings of objects, which we call relations. For my house, some of the objects are as follows:

> door1
>
> white
>
> 7
>
> my house

Note that we are using the term "objects" in the conventional information-processing sense as any unit of information, and not just in the conventional design sense, which restricts its meaning to that of physical elements. (We use the term "element" when we wish to make it clear that we are referring

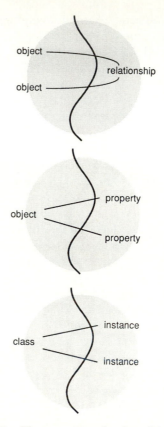

Figure 3.2 Three systems for describing designs.

to specific physical objects.) So the number "7" and the color "white" are objects in this representational system. Some statements about relations are as follows:

The dining room *is adjacent to* the kitchen.

My house *has* 7 rooms.

Yellow *is the color of* door1.

In this example the parts of these statements in italic type are the relations.

Objects and Properties

Here we choose to describe the world in terms of objects and properties. We can easily make statements about design objects (physical elements) and their properties. For my house, such a description might be the following:

Object	*Property*
my house	color white

where color white is a property of the object called my house. We can be more precise and represent a property in terms of attributes and values:

Object	Attribute	Value
my house	color	white
room1	color	blue
	height	2,400mm
room2	color	green
	height	2,400mm
room3	color	red
	height	2,400mm
room4	color	red
	height	2,400mm
room5	color	blue
	height	3,200mm
room6	color	yellow
	height	2,400mm
room7	color	white
	height	3,600mm
door1	color	gray
	height	2,050mm

This is a very simple but extremely useful system of describing designs and design components. Now let us extend the idea so that we can treat properties (attribute and value pairs) as objects. We can then depict the "properties" of properties:

Object	Attribute	Value
yellow	appearance	bright
gray	appearance	dull

Classes and Instances

We can make statements about a design in terms of a hierarchy of classes, subclasses, and instances:

My house is a kind of house.

Room1 is a kind of room.

Yellow is a kind of color.

The entities "my house," "room1" and "yellow" are instances of classes. Classes can also belong to "super classes":

A house is a kind of building.

These various systems of describing designs overlap in interesting ways. We can talk about relations between properties (attributes and values)—for example, length 3,600 mm *is greater than* length 2,050 mm, and the color yellow *is brighter than* the color gray. We can also talk about classes of properties—for example, color red and color green belong to the class primary colors, colors belong to the class of non-geometric properties, and the attribute weight is a member of the class of attributes that are additive. There are also classes of relations—for example, *bigger than* and *smaller than* belong to the class of relations that are transitive, and *next to* is associative.

Not all of the possible combinations of these systems of description we could consider are meaningful to a designer. For example, is it ever necessary to talk of an instance of a property? And there are similarities between certain categories. For example, at times the idea of a relationship can be seen as a link between two objects, one of which is a particular type of object called a property. In many ways these systems of description are isomorphic. The method we choose depends on how useful it is and how it appeals to our intuitions about the design task.

3.1.2   Representation Issues

In this section we review some of the important issues in the representation of designs and design knowledge, including the idea of hierarchies of representation, redundancies in representation, and emergent form.

Given the constraints of the computer medium, how can we make statements about designs in these terms? Conventionally, we would probably describe a design by means of graphic symbols. The representation is such that someone with a trained eye and the necessary background knowledge can inspect the design and identify elements, trace through the workings of the design, analyze, evaluate, and eventually manufacture, fabricate, or construct the artifact depicted. There are also things about the design that we would find difficult to represent graphically, such as an element's weight, or its cost, or its code number in a catalog. We make use of textual labels to minimize ambiguity and to describe features of the design that are otherwise hidden.

This kind of graphical representation is fine if the reasoning about the design is being carried out by a human designer. Behind the graphical images of elements is some idea about what they represent in the real world. A human being can interpret the graphic symbols. In a knowledge-based design system, we must make explicit the information that is implicit in a designer's drawing, or provide the knowledge for doing this.

There is generally an implicit hierarchy in the way we describe designs, and we can think of these different descriptions as each being an abstraction of the design. Some descriptions can be derived from other, more primitive descriptions. For example, if we know about the geometry of a design, in terms of the dimensions and spatial locations of components, then we can generally infer the topology of the design—that is, the network of adjacency

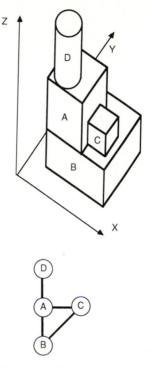

Figure 3.3   A network of adjacency relationships can be inferred from the dimensions and spatial locations of design components.

relationships exhibited by the design (Figure 3.3). This hierarchy is directed in certain ways. If our description is purely topological, then the design's detailed geometry, in terms of dimensions and locations in space, does not follow inevitably—many geometries of components achieve the same set of topological relationships (Figure 3.4).

This idea of hierarchies of abstraction is important. It is common to refer to the level of description of a design from which all other descriptions can be derived as a *canonical description*. As pointed out by Bobrow and Winograd (1977a), this turns out to be an elusive concept because we can never be sure that our level of description is not too detailed or coarse for our purposes. Designers are used to the idea of describing the same design in many ways. Generally it is useful to think in terms of multiple representations (or multiple views) of the same description rather than a canonical description from which all other descriptions can be derived, even at the cost of introducing redundancies. This is appropriate if the description will be put to different uses and employed by different players in the design process, who have different perceptions of the design. There are problems with redundancy, however, in that some consideration has to be given to how to maintain consistency in the representation of a design that is undergoing change.

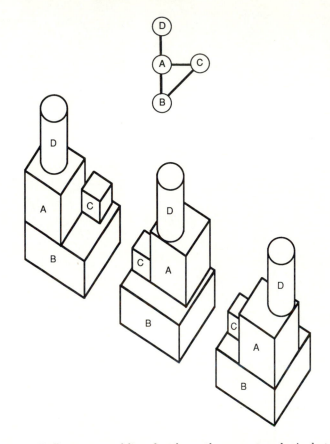

Figure 3.4 Different assemblies that have the same topological structure.

Design cannot always be reduced to the problem of selecting, configuring, and refining elements. If we take elements and combine them in certain ways, we often produce something that exhibits properties quite different from the mere summation of the properties of the constituent elements. Components interact with one another. This certainly applies to the products arising from the mixing of volatile chemicals; it applies equally to the combination of shapes and volumes in design. A simple example occurs when two shapes are overlaid; a third shape emerges that is the intersection of the first two (Figure 3.5). What results is emergent form (Stiny, 1982). Although we can model such spatial operations in computer systems, this process is cumbersome and expensive. It is difficult to model the fluidity with which designers are able to represent and reason about spaces.

In the rest of this section we consider descriptions of designs in terms of objects and relations, objects and their properties, and instances and classes. In each case we discuss commonly used methods of representation, including facts, semantic networks, rules, and frames. In the following section, we discuss the representation of formal and spatial relationships in design.

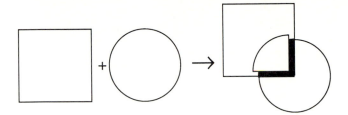

Figure 3.5 Overlaying two forms produces a third form.

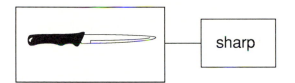

Figure 3.6 An object and a property as two related objects.

3.1.3 Objects and Relations

We look first at the physical and non-physical objects that characterize a design and its components, and the relations among them. Some relations refer to individual elements, as in ascribing the description "sharp" to the object "knife" (Figure 3.6). There are also other relationships of a comparative nature—for example, we say an object is larger or smaller or cheaper than another object (Figure 3.7), and others that could usefully be described as structural in character—knowing that an element is a sub-component that goes to make up a larger element (Figure 3.8). For example, a spoke is part of a wheel, and a wheel is part of a bicycle. Or we may want to express a taxonomic relation, that an element belongs to a family of elements—for example, that a

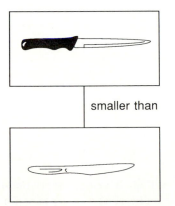

Figure 3.7 A comparative relationship between two objects.

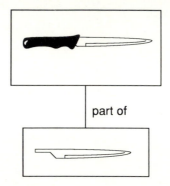

Figure 3.8   A structural relationship between two objects.

transistor belongs to the family of elements known as electronic components (Figure 3.9).

Relationships are essentially labels we attach to *ordered groupings* of objects. However, a relationship is not simply an attribute attached to such a group; the ordering can be significant. Relationships themselves can exhibit properties, such as *commutativity* and *transitivity*. For example, knowing that A is the same size as B is the same as knowing that B is the same size as A. The relationship "same size as" is commutative (Figure 3.10), but the relationship "bigger than" is not. Stating that A is bigger than B is certainly different to stating that B is bigger than A (Figure 3.11). Some relationships are also transitive: if we know that A is bigger than B and that B is bigger than C, this implies that A is bigger than C (Figure 3.12).

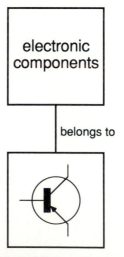

Figure 3.9   A taxonomic relationship between two objects. A transistor belongs to the family of electronic components.

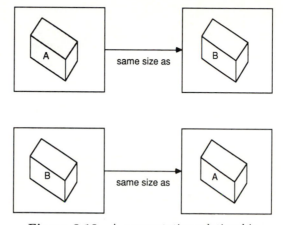

Figure 3.10   A commutative relationship.

Objects and Relations as Facts

As discussed in Section 2.1.1, one of the simplest ways of representing a relation between two objects is as a fact. Indeed, we can define facts simply as statements about objects and relations. A convenient and consistent method of representing facts is to state the relation, followed by the objects that are related, as in the following predicate calculus notation:

$$Relation(Object_1, Object_2, ..., Object_n)$$

Some examples of facts are as follows:

NextTo(LivingRoom, DiningRoom)

Connected(FuelLine, CombustionChamber)

We may also be interested in the relationship between objects other than physical elements, as in

Color(Door1, Blue)

where Color expresses a relation between Door1 and Blue. (Of course, this idea

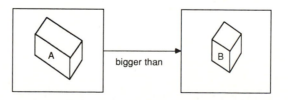

Figure 3.11   A non-commutative relationship.

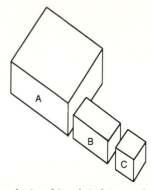

Figure 3.12   A transitive relationship. A is bigger than B and B is bigger than C implies that A is bigger than C.

maps onto that of representing objects and properties.) The idea of relations is sufficiently rich to enable us to represent many facts about designs. We will return to this subsequently. We may need many facts to adequately express the network of relationships that can exist between a group of objects.

Objects and Relations as Semantic Networks

If we draw a diagram representing objects and relations, we may represent the objects by labeled circles and the relations between the objects by labeled arcs between them (Figure 3.13). We can use such a diagram to express the relationship of adjacency between the rooms of my house, producing a kind of "bubble diagram" of the plan. This diagrammatic analogy is known as a *semantic network* because the semantics or meanings are attached to the arcs. (Note that the term "semantic" alone is used in a different way elsewhere in this book.)

Design objects, then, can be represented by nodes, and the relationships between objects can be depicted as labeled arcs or links (Quillian, 1968). The nodes can be any kind of object, including concepts, instances, classes of physical objects, and property descriptions. This representation provides an understandable and modular system of organizing information about designs,

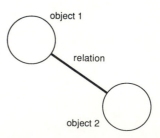

Figure 3.13   The basic components of a semantic network.

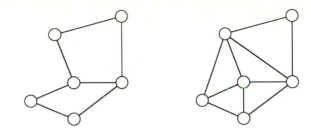

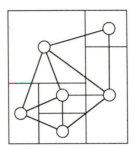

Figure 3.14 An illustration of how manipulating a semantic network can lead to the solution of a spatial problem. A graph representing desired connections between elements is triangulated. A floor plan can be generated as the "dual" of the adjacency graph. The example makes use of graph theory as elaborated by Steadman (1983).

and new objects and relations can be added at any time without altering the existing structure. Where nodes are objects and arcs depict the various spatial relations between components, then the spatial organization of a semantic network can lead to insights into the spatial organization of the design. The network can help solve spatial design problems (Figure 3.14).

The disadvantage of this representation formalism is that we need some system of translating semantic networks into computer-manipulable form, so we must rely on some other method of knowledge representation. Most often, semantic networks are expressed using the idea of frames, discussed in Section 3.1.4, or as facts. Another disadvantage of semantic networks is that network diagrams can become cumbersome if there are many connections. Certain concepts are also difficult to express in this medium, such as the idea of alternatives—for example, that a door is either blue or yellow.

3.1.4 Objects and Properties

Here we consider in detail the description of designs in terms of objects and their properties. There is little difficulty in accepting the convention that in-design we are concerned with describing objects, and that we can use labels to identify objects. We can also describe properties of objects such as topo-

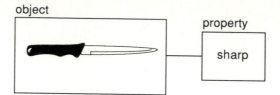

Figure 3.15   An object and a property.

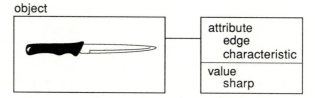

Figure 3.16   The object, attribute, and value triple.

logical, geometrical, physical, temporal, or functional (Figure 3.15). Each of these properties may be described in terms of attributes and values (Figure 3.16). We therefore describe designs in terms of labels that refer to objects, their attributes, and the values of those attributes. A building element may carry the name brick, with the attributes of color, texture, and weight, the values of which are gray, coarse, and 4.1 kilograms, respectively (Figure 3.17).

Although certain attributes may be common to most objects, not all of them are of interest. For example, it may be useful to be explicit about the texture of a brick because bricks come in many different textures and texture is a significant determinant in environment quality. It would be pointless, however, to be specific about the texture of an electrical cable even though it can be measured. On the other hand, the electrical resistance of a brick is usually of minor concern, whereas the resistance of a cable is of great importance. In describing objects for design we are therefore careful to select attributes over which we have some control or that are major determinants in the performance of the final artifact. Unfortunately we can never be entirely sure what attributes are going to be important. In the same way that objects "interact" to produce new objects, attributes can "interact" to produce designs with properties that are different from the simple summation of the attributes of all their component objects. As an obvious example, the configuration of two

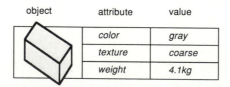

Figure 3.17   An example of an object described in terms of attributes and values.

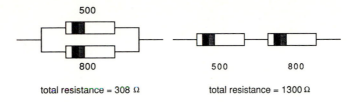

total resistance = 308 Ω   total resistance = 1300 Ω

Figure 3.18   The property of two resistors in parallel is different from the sum of the two properties.

electrical components in parallel does not result in a circuit whose resistance is the sum of the resistance of the two components (Figure 3.18). Another example is the surface area of two boxes joined together (Figure 3.19). In these examples the results are predictable and within the nature of the geometry, topology, and functional performance characteristics of the objects. A less obvious example is when two different metals inadvertently form a galvanic couple.

Attributes can themselves be categorized. We identify attributes that are common to all solid physical objects, such as dimension, mass, and volume (Figure 3.20). There are also hierarchies of attributes such that what is known about certain attributes leads to the derivation of other attributes—volume from dimensions, for example (Figure 3.21). It is also sometimes useful to make a distinction between what an object is and what it does. Thus we distinguish between *physical* and *functional* attributes. The properties of dimension, mass, volume, texture, and color may be regarded as physical. Functional attributes, often described in terms of behavior, say something about what the object is used for or what it does. Thus we can describe an object such as a knife in terms of the sharpness of the edge, the hardness of its material, and its weight (physical attributes), but we can also assign to it the attribute that it cuts—and the value of this attribute is that it cuts precisely, or coarsely, or to a depth of 1.0 mm. Cutting is a functional property (Figure 3.22).

Functions are unusual attributes in that they generally cannot be measured by physical inspection alone but must be considered in some kind of temporal context. We have to see how the objects behave or how they are

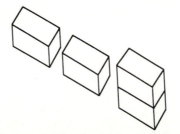

Figure 3.19   Combining two boxes.

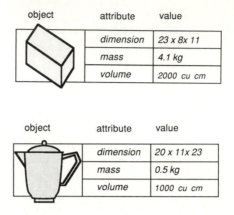

object	attribute	value
	dimension	23 x 8x 11
	mass	4.1 kg
	volume	2000 cu cm

object	attribute	value
	dimension	20 x 11x 23
	mass	0.5 kg
	volume	1000 cu cm

Figure 3.20   Very different objects with the same attributes.

used. Sometimes objects function in ways other than those intended, as a result of complicated interactions with their environment that cannot be readily predicted. Sometimes functional attributes would appear to require special consideration in design systems. In some cases it is convenient to assign function as a property of an object that stands apart from its attributes. Thus an object has attributes and functions (Figure 3.23).

It should be clear now that there is no single way of describing objects that accounts for the phenomenon of property. The convention we adopt will depend on how we wish to operate on the design description in our design system.

Objects and Properties as Facts

Perhaps the simplest way to represent this triple of object, attribute and value is as a set of discrete facts about the components of the design. We can state

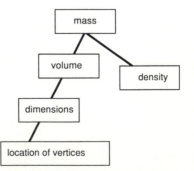

Figure 3.21   Some properties arranged in a derivational hierarchy.

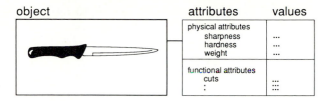

Figure 3.22   Examples of physical and functional attributes.

these in the following form:

Color(Door1, Blue)

where Door1 is an object, Color is an attribute, and Blue is a value. In this case the attribute is placed first, treated as the name of a relation between Door1 and Blue. Of course there is nothing to prevent us from defining the fact as

Door1(Color, Blue)

or any of five other possible configurations. The advantage of one representation over another is that in the formalism known as first-order predicate calculus it is permissible for objects to be represented by variables, but it is not directly permissible for relations to be represented as variables. We can exploit this in the way facts are formulated. We can therefore construct a computer system that facilitates the following exchange. We begin with a list of facts expressed in this way, such as

Color(Door1, Blue)
Color(Door2, Green)
Color(Roof, Blue)

We can then present a statement to the system in the following form:

$\exists x[\text{Color(Door1, } x)]$

(The $\exists$ symbol means "there exists.") Here we are saying there exists an x such that the color of Door1 is x. We then ask, is this statement consistent with the list of facts, and if so, what is the value that will be instantiated to

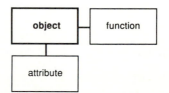

Figure 3.23   The object, attribute, and function system of representation.

x? We would expect such a system to return

$$x = \text{Blue}$$

We could also consider the statement

$$\exists x[\text{Color}(x, \text{Blue})]$$

This is equivalent to asking, what is blue? The answer would be

$$x = \text{Door1}$$
$$x = \text{Roof}$$

Thus there are two possible instantiations of x. The point we wish to make here is simply that the predicate calculus enables this kind of exchange. It is therefore necessary to consider what may be represented as variables in the way we formulate these facts.

Objects and Properties as Semantic Networks

We can represent the idea of objects and properties in a semantic network where objects and values are nodes, and attributes are the labels attached to arcs joining nodes. The preceding examples are depicted as semantic networks in Figure 3.24, which shows information about simple objects and their attributes.

Objects and Properties as Frames

It is often difficult to impose structure on information organized as facts and semantic networks. We often want to say similar things about different objects, such as the fact that all doors in my house have a particular color, even though we do not yet know the value of that color. We need some kind of framework by which information about an entity can be grouped into a uniform structure. In a *frame* (Minsky, 1975; Fikes and Kehler, 1985) the different properties of an object are each represented by a *slot* and a *value*, sometimes called a *filler*, in the slot:

> *frame*
> *(slot_1)*   *(value)*
> *(slot_2)*   *(value)*
>   ...

An example of a frame for our description of a house is

> my house
>   number of rooms   7
>   color   white

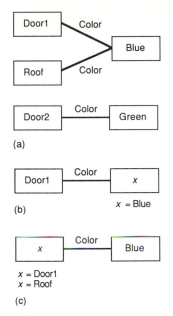

Figure 3.24   Information about simple objects and their attributes as semantic networks constituting a rudimentary design description (a). A query to the system involves mapping an incomplete network onto the design description and finding appropriate instantiations for any variables (b and c).

Frames may be compared to semantic networks, with the frame name ("my house") analogous to a node, and the slots to arcs. Frames facilitate the ordered and succinct representation of large amounts of information.

The frame formalism exploits several important and useful ideas that are pertinent to design: inheritance, possible values, defaults, bounds, and attached procedures. We shall discuss the idea of inheritance in a later section. *Possible values* are the values of properties that it is possible for an object to have. For example, in the design of a structural beam only certain beam profiles may be available. These would be listed in this slot. *Defaults* are an expedient for representing values that are not yet known. A designer may begin by assuming that a design component has a particular property, then subsequently revise this assumption. For example, in the design of a structural beam it is necessary to assume that it has a particular cross-sectional area in order to take its own weight into account. The cross section of the beam could be taken as a default that is subsequently revised as the calculation proceeds. By *bounds* we mean simply some indication of the range of values that may appear in a particular slot, such as a range of possible beam depths—say 100 to 1200 millimeters. *Attached procedures* are instructions on how to calculate values not forthcoming from other sources. Thus the formula for calculating the bending moment of a beam might occupy a slot, or a call (pointer) to a procedure for performing that calculation.

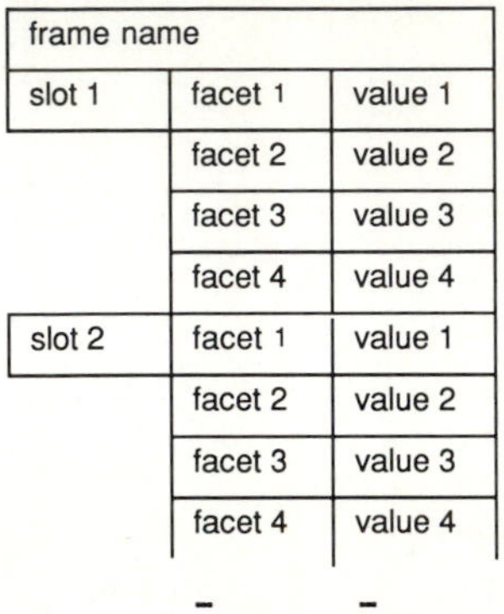

frame name		
slot 1	facet 1	value 1
	facet 2	value 2
	facet 3	value 3
	facet 4	value 4
slot 2	facet 1	value 1
	facet 2	value 2
	facet 3	value 3
	facet 4	value 4

Figure 3.25   The components of a frame system.

According to the frame formalism, slots can be divided into *facets* to accommodate the information about values (Figure 3.25). For the house example, we can consider the "number of rooms" slot and give it four facets. The slot can have a value, but if no value is known, then at least we can expect the value to fall within a particular range or bound. This facet can be called a value class—an integer between 2 and 12. If there is no information to the contrary, then we may assume the default facet—that the number of rooms is 10. If we really want to know how many rooms there are and a floor plan is available, the procedure is to "count from plan." A pointer to this procedure falls within a facet named "if needed." The "number of rooms" slot has four facets:

```
my house
      number of rooms    value          7
                         value class    integer 2 to 12
                         default        10
                         if needed      count from plan
```

Thus frames allow us to represent a range of information about objects (such as my house) and attributes (such as the number of rooms), and to say several specific things about the values of those attributes.

3.1.5   Classes and Instances

Designs can also be described in terms of instances and classes. A *class* is simply a collection of one or more entities, where an entity is itself an instance or a class. It is possible to construct hierarchies of classes such that classes form subclasses and superclasses of each other (Figure 3.26). Classes may belong to more than one distinct superclass (Figure 3.27). An *instance* is an entity

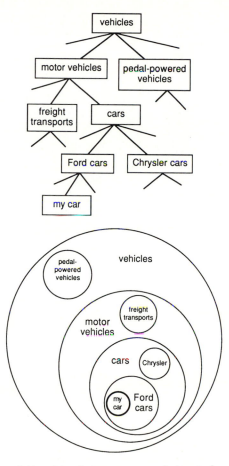

Figure 3.26   The relationships between some abstract classes and instances represented graphically in two ways.

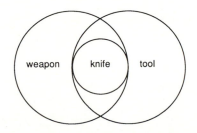

Figure 3.27   A class may belong to more than one superclass.

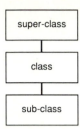

Figure 3.28   The relationship between a class, superclass and subclass.

that belongs to what we choose to regard as the real world whereas the class is an abstraction. It is worth remembering that we are here concerned with representations in design and that our instances can belong to hypothetical worlds. We can talk about my house as an instance of a design even though such an entity may not really exist.

Figure 3.28 shows the relative nature of superclasses, classes, and subclasses. It is also important to note that we can say things about classes we cannot say about instances. We can say of a class of beams that it has the property of length and that the length is a range—for example, 2.0 to 7.0 meters. The length of an instance of a beam, however, cannot be a range. It may lie within a range or be unknown, but only a single figure is the actual length of the beam instance. For some instances we may not know all the values and these may appear uninstantiated, but if the object (or concept) actually exists, the values do exist even if they are unknown to us (Figure 3.29).

We can utilize the idea of classes and instances in describing designs in

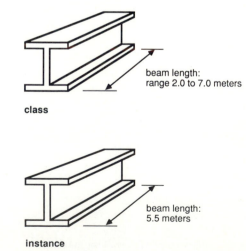

Figure 3.29   Classes can represent a range but instances have unique values, though they may be unknown.

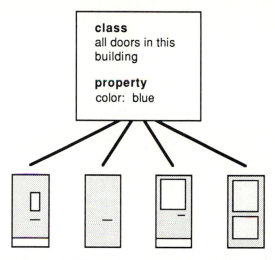

Figure 3.30   A class and instances of doors.

three ways. The first, to identify classes and instances of design components, is often implicit in the way we talk about designs—such as when we say that all the doors in this building are blue. The collection of all doors in this building is a class, and the members of that class all have the common property that they are blue (Figure 3.30). The use of the idea of class here is as a shorthand convention that saves us from enumerating all the instances of door in the design and labeling them with the property blue.

In design we can construct many class/instance hierarchies, such as part-whole hierarchies, in which we are interested in the relationships of groups of components to other groups. This can be useful in organizing information about a design. For example, door handles belong to the class door furniture, which belongs to the class hardware (Figure 3.31). We can also talk about the idea of objects and properties in this context. Classes serve to group objects with common properties. In this sense we can talk about a class having a property. Second, we can also consider a design to be a description of a class of artifacts the instances of which are instances of that design. A design is not uniquely tied to one artifact; rather, it is in the nature of designs that they can result in the production of many artifacts. In this sense a design describes a class of artifacts (Figure 3.32).

Third, we can talk about a design as a particular instance of a class of designs (a prototype or generic design), and of the components of a design as instances of component classes. This is the use of classes that interests us most here. In a hypothetical design world we could therefore choose to consider design decisions, and design descriptors, as instances—even though, strictly speaking, the design may be realized many times as instances of artifacts (Figure 3.33).

The idea of classes is particularly valuable. If we know the properties that

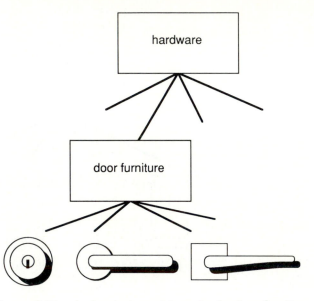

Figure 3.31   A class-instance hierarchy for door furniture.

formed the basis of a class's organization and if we know that an object or instance belongs to that class, then we know that the object has the properties of that class. The transfer of properties between classes and instances is generally described as *inheritance.* An instance or subclass inherits the properties of the class of which it is a member. This process enables us to produce new facts about the world. Let us demonstrate inheritance with a very simple example. We need some method for depicting classes and for depicting class, subclass and instance connections. We can do this by means of tables of simple pairs:

Superclass	*Class*
building	house
house	single-story dwelling

Class	*Instance*
single-story dwelling	my house

These mappings can also be depicted graphically, as in Figures 3.34 and 3.35. The first two pairs are about relationships between classes. The third pair is a relationship between an instance and a class. By representing class relations in this way, we can easily state something about my house that was not explicitly stated—that is, that my house is a building.

We can reason with classes in this way. Much of what we want to say about a design could be represented in terms of classes and instances, but

Figure 3.32 A design as a description of a class of artifacts.

this can be cumbersome. For example, we can make the statement that fire engines belong to the class of all objects that are red (Figure 3.36). But are we at all interested in the red objects superclass? Does an instance of a fire engine inherit any properties from this superclass other than color? It is generally more convenient to consider those properties whose ancestry (or class relationships) we are not interested in simply as properties of the class we are representing. So color red is simply a property of the fire engine class (Figure 3.37).

We can adapt the preceding tabular form to include properties (see also

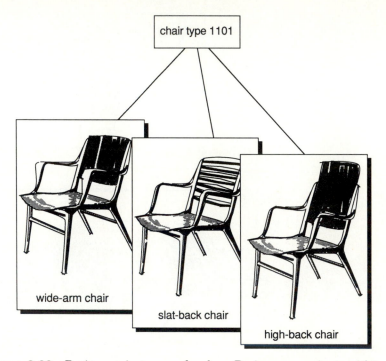

Figure 3.33   Designs as instances of a class. Designs are variants within a family of designs conforming to some generic description.

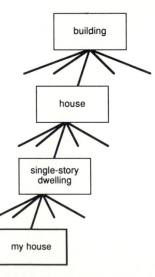

Figure 3.34   A class-instance hierarchy for houses as a tree.

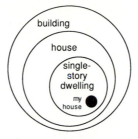

Figure 3.35   A class-instance hierarchy for houses.

Figure 3.38):

Superclass	Class	Properties
building	house	has at least one door
		used for living in
house	single-story dwelling	has one floor

Class	Instance	Properties
single-story dwelling	my house	has brick walls
		has red roof

We find that my house inherits the properties of the superclasses of which it is a member. Thus we can reason that my house has at least one door, is used for living in, and has one floor, even though these features of the instance are not stated explicitly. A separate instance of a single-story dwelling, such as your house, will have the same properties, though it will not necessarily have the properties peculiar to my house, such as brick walls and a red roof (Figure 3.39). Inheritance is useful to provide properties to the instance when specific facts about properties are not available. We will subsequently demonstrate in detail how we can make use of this method of describing instances, classes, and superclasses both to describe designs and to represent knowledge.

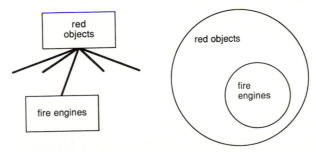

Figure 3.36   The class of red objects.

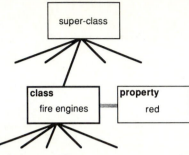

Figure 3.37   Red as a property of a class.

Classes and Instances as Facts

We can represent this information about instances and classes as facts:

AKindOf(House, Building)

AKindOf(SingleStoryDwelling, House)

IsA(MyHouse, SingleStoryDwelling)

where "AKindOf" denotes that a house belongs to the class of objects that are buildings—a house is a kind of building—and "IsA" denotes that MyHouse is an instance of the SingleStoryDwelling class. We can construct similar statements about properties:

Property(House, HasAtLeastOneDoor)

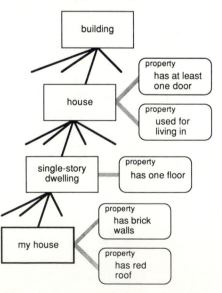

Figure 3.38   A class-instance hierarchy with properties.

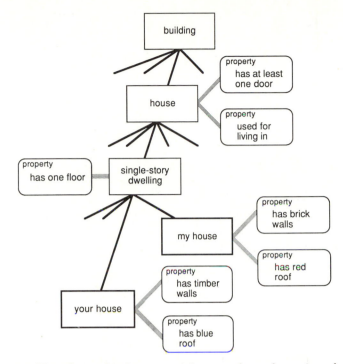

Figure 3.39 The relationship between my house and your house in a class-instance hierarchy.

Property(House, UsedForLivingIn)

Property(SingleStoryDwelling, HasOneFloor)

Property(MyHouse, HasBrickWalls)

Property(MyHouse, HasRedRoof)

Instances and Classes as Semantic Networks

Lists of facts such as the preceding ones can become a little confusing. A semantic network can make it clear how the properties of classes are inherited by subclasses and instances. A semantic network for describing a house is shown in Figure 3.40. The "ako" label, which stands for "a kind of," indicates a link between subclasses and classes, while the "isa" link indicates that an individual is an instance of the class to which it is connected.

Instances and Classes as Frames

Frames provide an extremely useful tool for expressing information about instances and classes. We need two types of frames: class frames and instance frames. Initially, the information available to a designer may be only of the

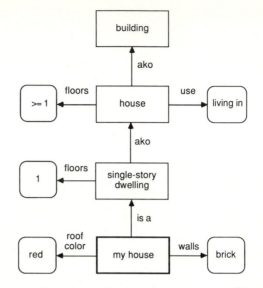

Figure 3.40   A semantic network for describing classes and instances of a house.

class frame variety—that is, information about classes and class properties. Any one design is represented in terms of *instance frames*. Each instance frame is an instantiation of the class frame to which it is connected. This can be made clear with a simple example. Figure 3.41 shows a class frame containing a description of the class of objects that are houses, including slots, facets, and values. The "ako" slot provides a link with a superclass, which is buildings— all houses are buildings. If we decide that we want to describe a particular house, my house, then we need only know that my house is a house in order to construct a frame for it. The instance frame inherits all the slots of the class frame to which it is connected. This copying of frames provides a concise and well-structured system of organizing information about design objects.

We can also consider another example from structural design. Consider the representation of knowledge about structural beams. Beam types can be arranged as a hierarchy of classes, as shown graphically in Figure 3.42. Figure 3.43 shows classes and instances of beams represented as a set of frames.

frame name	house	
ako	value	building
number of floors	value class	integer 1 to 3
	default	1
use	value	living in

Figure 3.41   A class frame for a house.

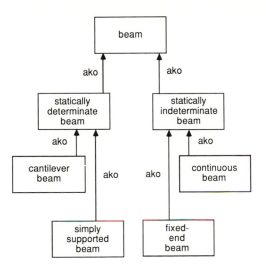

Figure 3.42 A hierarchy of structural beam types.

frame name	beam		
	sub-classes	value	statically determinate beam statically indeterminate beam
	load type	value	non-axial
	orientation	value	non-vertical
	number of redundants	if needed	calculate redundant
	material	possible values	timber, steel, prestressed concrete, reinforced concrete

frame name	statically determinate beam		
	ako	value	beam
	sub-classes	value	cantilever beam, simply supported beam
	number of redundants	value	0
	moment	if needed	calculate moment
	shear force	if needed	calculate shear
	deflection	if needed	calculate deflection

frame name	cantilever beam		
	ako	value	statically determinate beam
	number of supports	value	1
	support type	value	rigid

frame name	beam 1		
	is a	value	cantilever beam
	load	value	100
	span	value	3000

Figure 3.43 Class frames for structural beams.

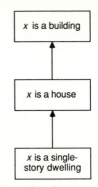

Figure 3.44   The relationship between two rules about houses.

Instances and Classes as Rules

Information about instances and classes can also be represented as rules (Figure 3.44):

> **If** x is a house
> **then** x is a building.

> **If** x is a single-story dwelling
> **then** x is a house.

We must use rules in conjunction with simple statements, facts:

> My house is a single-story dwelling.

This last statement relates an instance to a class. (We could have used the predicate calculus notation to depict this fact.) Again, the information about classes and properties can be represented as rules (Figure 3.45):

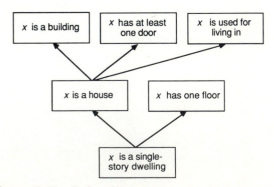

Figure 3.45   Rules about classes and their properties.

If x is a house
then x is a building **and**
x has at least one door **and**
x is used for living in.

If x is a single-story dwelling
then x is a house **and**
x has one floor.

We also need facts about classes and properties:

My house is a single-story dwelling.

My house has brick walls.

My house has a red roof.

The form of a rule for linking instances and subclasses to classes is as follows:

If x belongs to class y
then x also belongs to class z.

where x is an instance or a class. Examples of such rules follow:

If x is a house
then x is a building.

If x is a cam shaft
then x is an engine component.

The entities house, building, cam shaft, and engine component are generic terms, or class labels. We can also use rules to link instances and subclasses to properties. The form is generally as follows:

If x belongs to class y
then x has the properties z.

where z is a list of properties exhibited by the class y. For example,

If x is a house
then x has bedrooms.

If x is a cam shaft
then x is connected to the pistons.

In the first rule, if we substitute for x some instance of a house, such as my house, then we can deduce that my house has bedrooms because that is one of the properties of the house class. The second rule tells us that a component that is labeled on a set of working drawings as a cam shaft must be connected to the pistons.

We can also define classes in terms of properties by means of rules:

> **If** x has property y
> **then** x belongs to class z.

For example,

> **If** x has bedrooms
> **then** x is a house.

> **If** x takes in fuel and expels exhaust in two strokes
> **then** x is a two-stroke engine.

We could therefore employ rules to capture the kind of knowledge we represent in frame systems. The tabular form of frames, however, provides a more succinct way of representing information about instances, classes, and properties.

This example demonstrates something of the blurring of distinction between design descriptions and knowledge about designs. Design knowledge can be partially explained in terms of descriptions of classes of designs, or generalizations about designs. There is a mapping between generic descriptions and rules. This ties in with the idea of design knowledge as parts of prototypes—generic descriptions of designs.

Note that in this discussion we have made certain assumptions about designs. The description of a design in terms of physical elements and their properties is a convention. At times it seems more appropriate to assume that the universe is not divisible into such discrete objects. We have already mentioned how elements can be arranged hierarchically in a compositional sense; it is difficult to identify those that are to form the basic units of our design. But we have alluded to another problem in that the boundaries between things appear so fuzzy that the idea of a universe of discrete entities is not helpful. An alternative entity to discrete elements has been proposed—that is, "individuals," where an individual is a definite entity without definite parts (Goodman and Quine, 1947; Stiny, 1982). We will not pursue this idea here because the manipulation of individuals is extremely difficult. Stiny (1982) discusses the issue of a "calculus of individuals" in the context of design.

We also make certain assumptions in talking about classes. Whereas manipulating classes is relatively straightforward, identifying classes often poses difficulties. Sometimes we wish to ascribe properties to a class, such as house, only to find that not all of those properties are necessarily inherited by subclasses and instances. We can often find instances of a class that do not exhibit all the properties of the class. For example, we have said that a house has at least one door. Some houses, such as one might find in the tropics, may have no door, but a continuous open perimeter, and some houses, such as doll's houses, tree houses, and opera houses, are not used for living in. These issues in no way make it harder to divide the world into classes, particularly if we keep in mind the limitations of this approach.

In this section we have demonstrated how information about designs in terms of objects and relations, objects and properties, and instances and classes can be represented as facts, semantic networks, rules, and frames. Let us turn now to a specific concern in design—the representation of form and spatial relationships.

3.2   Representing Form and Spatial Relationships

So far in this chapter we have concentrated on the representation of non-geometrical characteristics of designs. In design we must often specify the size and shape of design elements, how they are related, and where they are positioned in space. The issue of representing and reasoning about geometry and topology in knowledge-based design systems could fill an entire book. Here we only present some of the most direct applications of knowledge-based systems to representing form and spatial relationships.

We can describe design form in terms of objects and relations, what is known as the *topology* of a design. The objects are the design elements; the relations, the connections between elements. We may wish to specify many relationships between elements, such as "touching," "connected to," "on top of," and "underneath" (Figure 3.46). These relationships may facilitate complex communications between elements that are significant in determining the behavior of the design—for example, the traffic flow between spaces in a building, the flow of current in an electrical circuit, and the distribution of forces in a truss (Figure 3.47). In many cases some medium provides the link between elements, as in the case of conductors connecting electrical components, or pipes connecting water tanks (Figure 3.48). In such cases the spatial constraints on the location of elements being connected are different from building structures with very little spatial flexibility in how loads are distributed from one element to another.

A very simple example of a topological description is the building plan in Figure 3.49. We can represent the linkages between spaces, through wall openings, with the following set of facts:

Link(A, B)	Link(D, G)
Link(B, C)	Link(G, H)
Link(B, E)	Link(H, F)
Link(C, D)	Link(H, I)
Link(C, F)	Link(F, E)

This can also be depicted as a semantic network (Figure 3.50).

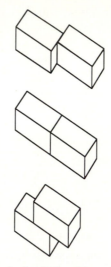

Figure 3.46   A set of topological relationships: "touching," "connected to" and "on top of."

Spatial information about design elements can also be represented in terms of objects and properties. A very simple example is the representation of the spatial location and size of a rectangular building element in terms of x and y values in some coordinate system; the name and coordinates are treated as properties of the object. We can formulate a fact that corresponds to the room in Figure 3.51:

Room(Office(1), 3.0, 1.0, 4.0, 2.0)

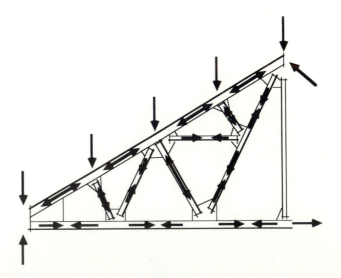

Figure 3.47   Relationships that facilitate communication between elements, in this case the distribution of forces in a truss.

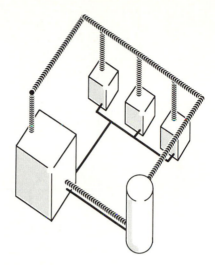

Figure 3.48   Relationships for which there is a communicating medium.

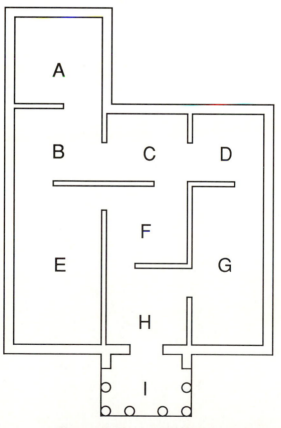

Figure 3.49   A building plan.

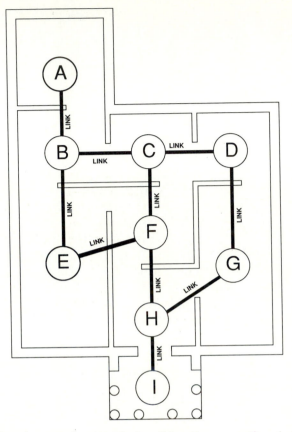

Figure 3.50 A semantic network describing the connections in Figure 3.49.

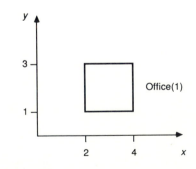

Figure 3.51 The geometrical representation of a room.

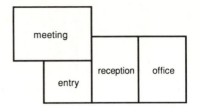

Figure 3.52   A configuration of rooms.

The first argument of the Room predicate contains the name of the room (as a nested predicate), while the four other arguments give the the common values of the vertices on the north, south, east, and west faces, respectively. An entire room layout, such as that represented in Figure 3.52, can be described in this way.

Objects can also be described hierarchically, as in the simple plan in Figure 3.53. The office is a composite object made up of components such as columns and walls. Each of these components is made up of line segments; these segments are defined by end points, which have coordinate values. We can make use of a list construct (using infix notation) to depict a list of elements of any length. The list has the following appearance:

a_1 **and** ... **and** a_n **and** *nil*

The atom *nil* indicates the end of the list. In prefix notation a list with three elements would have the following form:

and $(a_1,$ **and**$(a_2,$ **and**$(a_3, nil)))$

Each of the following predicates has two arguments. The first argument is the name of the object: a room, component name, or graphic primitive in the form of a line or point. The second argument is a list in each case:

Composite(Office(1), Column(1) **and**
     Column(2) **and** ... **and** Wall(1), Wall(2) **and** ... *nil*)
Object(Column(1), L(1) **and** L(2) **and** L(3) **and** L(4) **and** *nil*)

...

Line(L(1), P(1) **and** P(2) **and** *nil*)

...

Point(P(1), 0.00 **and** 0.00 **and** *nil*)
Point(P(2), 0.00 **and** 0.50 **and** *nil*)

The advantage of this representation is that it enables us to reason about objects at higher levels of description in terms of objects and compositions, with the intricacies of low-level descriptions (lines and points) effectively concealed from view.

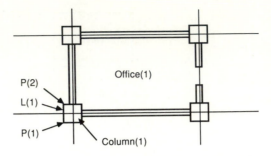

P(2)

L(1)

P(1)

Office(1)

Column(1)

Figure 3.53   A room that can be described hierarchically. *P* indicates points or vertices and *L* indicates connecting lines.

We may also want to represent information about spatial form in terms of instances and classes. We can construct class hierarchies of forms. For example, there is a class of objects called polyhedra. One subclass of this group is the set of objects whose eight vertices exhibit the topology of Figure 3.54. Within this class are parallelopipeds, truncated square-based pyramids, and a host of other forms. Within the parallelopiped class are cubes, slabs, and "shoe boxes." We can list properties attached to each of the classes in the hierarchy. Any parallelopiped can be completely described in terms of nine numbers, height, length, breadth, the x, y, and z coordinates of its origin,

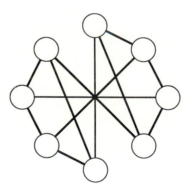

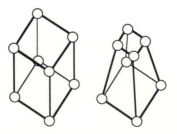

Figure 3.54   Part of a class hierarchy of geometrical forms: two of the general forms and their underlying topology.

and three rotation factors. The volume of the form can readily be calculated from these numbers. A cube, as a subclass of parallelopipeds, "inherits" these general properties with the added property that height, length, and breadth are equal. We can exploit this information about classes to structure our knowledge about formal elements. For example, we need not record how to calculate the volume of a cube once we know that a cube is a parallelopiped.

Using these systems of design representation, it is possible to create design descriptions at various levels of abstraction. This redundancy of representation bears some relation to the way designers think about artifacts. For one purpose, relationships may be important, but for another, geometry may be important. An ideal medium for representing knowledge about design is one that enables any thought to be represented but that does not anticipate the properties of the things described (Swinson, 1983). The medium should be unconstraining and yet powerful enough to facilitate operations that are useful to designers.

3.3   Representing Design Intentions

In this section we consider the representation of design intentions—the intended or desired interpretations or performances we need to describe to a design system in some way. Design intentions are the goals of the system. We argue that goals can be considered simply as partial, incomplete, and unresolved descriptions of a design. As such, their representation need be little different from the way we might describe designs in a knowledge-based system. In the same way that we can describe designs hierarchically, we can have hierarchies of goals. However, there are differences. We argue that, because of interactions between goals, goal hierarchies are not necessarily or readily decomposable. Another distinction is that we often talk about objectives as particular kinds of goals. Objectives might be expressed in terms of wanting to maximize or minimize some property of a design. Finally, we consider the fact that goals are not always known at the outset of a design task, neither do they necessarily remain static during the design process.

3.3.1   The Nature of Design Descriptions and Goals

We have discussed the description of a design as residing explicitly in some storage medium, such as the database of a CAD system, or the facts base of a knowledge-based system. We have discussed how this description belongs to the syntactic realm— explicit statements about the design—in a similar way that we talk about sentences in natural language. In a design description these statements can be linked in various ways, not the least of which is

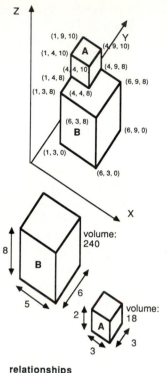

Figure 3.55 The role of a canonical design description: dimensions, volumes, and relationships derived from a design description in terms of coordinates of vertices.

hierarchically, according to some kind of derivational status. Thus certain facts, such as facts about the area of a surface, can be derived from facts about the dimensions of the surface. Descriptions in terms of dimensions and spatial location tend to constitute a canonical description: many other geometrical properties of the design can be derived from them (Figure 3.55). If a facts base contains both kinds of information, there is some redundancy in the representation. That is, the facts base contains a canonical description of the design in terms of dimensions, from which other properties of interest can be derived, but it also contains some of those descriptions explicitly. The main problem with redundant descriptions is maintaining consistency (Figure 3.56).

Does a non-redundant description, one containing only derivable information but not the facts from which it can be derived, constitute a design description? One of the few stipulations we place on a design description is that it must contain sufficient information for purposes of interpretation and manufacture, given a certain level of background knowledge. For example,

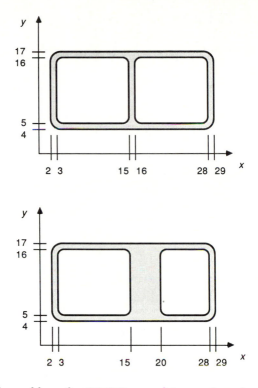

Figure 3.56   The problem of maintaining consistency where there is redundancy in the description of a design. Changing one parameter also changes the cross-sectional area of the object.

a design description may consist of code numbers of standard components from catalogs and information about their spatial locations. If a canonical description in terms of the precise geometry of all the components is needed to manufacture the artifact, and such a description is missing, then the design description is incomplete.

We can also term this derivable or interpreted information the performance of the design. We have characterized design as a process in which we know the performance but lack the description of the artifact from which the performance can be derived. This initial derivable information is generally referred to as the goals of the design system. By *goals* we mean those properties that we want a final design to exhibit—simply some statements describing what we would like the design to be like.

As with design descriptions, goals can be expressed in terms of objects and relations, objects and properties, and classes and instances. One of the goals may be that the elements within the design exhibit certain topological relationships, as in desired room adjacencies in a building plan. Or perhaps other relationships are to be evident in the design—the front seat of a car is to be bigger than the back seat. We may also represent goals in terms of objects and properties. Individual elements are to have particular properties—the dining room is to be 12 square meters—but we will probably also require that the entire design configuration exhibit certain properties—that it cost no more than $50,000, or that it will "delight the eye." Goals can also be expressed in terms of classes and instances—for example, the goal may be to produce a design that fits within a class of objects called "house."

The way we express goals need be little different to the way we represent designs. Just as there are hierarchies of abstraction in the way designs are described, there can be hierarchies of goals. In the next section we discuss some of the major differences between representing goals and representing designs.

3.3.2   Goal Hierarchies

Design goals can generally be reduced to subgoals. We may consider the goal that a designed artifact is to be portable. Portability implies several other properties such as light, compact, and has a handle. These properties can constitute subgoals; if all are achieved, then the design should be portable. Of course, there will be other subgoals. Compactness may imply that the artifact has a "simple profile" and does not have any sharp projections. In formulating such a hierarchy, we soon discover that there will be disjunctions of subgoals (Figure 3.57). There are many different ways in which something can be said to have a simple profile: it can be a cube, a pyramid, a sphere, or any of an infinite range of shapes. This is a problem because in satisfying one subgoal we may hinder the satisfaction of another. Were it not that goals often have subgoals that frequently interact, we could characterize design as a process of incrementally reducing goals to subgoals until we find design elements that satisfy them. The overall design would constitute a synthesis of each of these elements (Figure 3.58). These issues will be pursued in greater detail in Chapter 6.

Goals can be seen as interpretations of designs that do not yet exist. As argued in Chapter 6, we can express interpretive knowledge with greater conviction than we can express knowledge about goal decomposition. In the preceding example we dealt with knowledge about inferring whether a design is to have a handle. An artifact is portable if it is light, compact, and has a handle. There are other ways in which an artifact may be portable that this statement does not preclude. There will be little argument on this, but in goal decomposition we turn the statement around and say that for a design to be portable, it must be light, compact, and have a handle. In creating goal hierarchies we must be aware that we are assuming that our knowledge says everything about our design world that there is to know (the closed-world assumption).

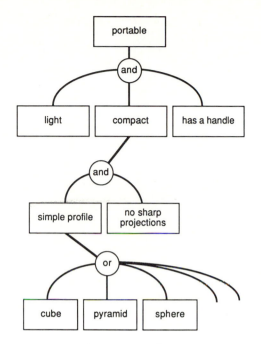

Figure 3.57   A hierarchy of goals and subgoals.

3.3.3   Goals and Objectives

We can make comparisons between interpretations of different artifacts, which amounts to interpreting performance characteristics or specifications. We bring knowledge to bear in deciding that one design is better than another by comparing their performance characteristics. For example, we may prefer one design because it costs less to manufacture, or there may be some more complex knowledge for evaluation involving interactions between performance characteristics. We may prefer one design because it is less expensive than most alternatives but will be easier to maintain than the cheapest alternative. Interpretations of performance characteristics are sometimes termed *evaluations*.

It is not always possible to produce a design that exhibits intended performance characteristics. Sometimes no design satisfies all the design goals, in which case we may be content with the design whose performance comes closest to the goals. Again, this may require some knowledge for interpreting the performance specifications of alternative designs and the ideal performance characteristic embodied in the goals. For example, we may have the goal that the manufacturing cost of a design be as close to $50,000 as possible—that is, produce a design closer to $50,000 than any other design that could have been produced were cost not taken into account (Figure 3.59). Goals of this type are

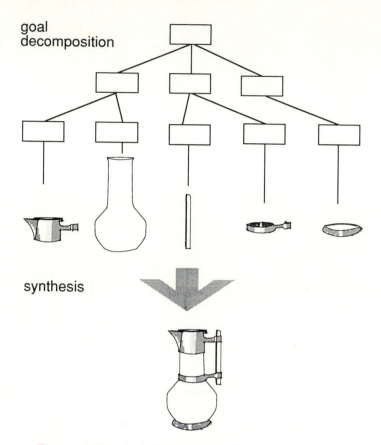

Figure 3.58   Goal decomposition followed by synthesis.

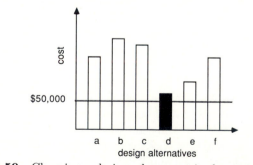

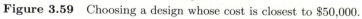

Figure 3.59   Choosing a design whose cost is closest to $50,000.

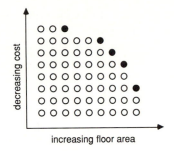

Figure 3.60 A graph relating cost to area. The "best" designs are those whose cost and area properties lie on the outer edge of a set of feasible solutions.

often termed *objectives*. Usually they are expressed in terms of optima—for example, to minimize cost and maximize floor area (Figure 3.60).

The knowledge by which designs are produced to meet objectives may be difficult to formalize, especially when the performance specifications being evaluated are not quantifiable and are not measured in the same terms, or when there are many interacting objectives. The techniques for meeting multiple objectives expressed numerically are addressed by multicriteria optimization (Radford and Gero, 1988; Cohon, 1978). Optimization will not be addressed in detail in this book except when knowledge-based techniques can be used to formulate optimization problems and interpret the results.

3.3.4 The Revision of Goals

Having established that we can use a language for stating goals that is similar to the way we make statements about designs, let us turn to other characteristics of design goals. We have defined goals in design systems rather rigorously, saying that they are statements that must be consistent with a design description. Everyday use of the term goal, however, includes the implication that goals may be something we strive for even if we do not achieve them. Thus we may posit the goal that our design can be manufactured for less than $600. If the artifact costs $700, then clearly the design description is not consistent with the posited goal. We could search for a different design description, of course, but if some other properties of the design are attractive we may accept it. We would say in this case, however, that the goal may be considered a constraint that has been relaxed. This highlights one of the key problems in design—the changing nature of goals. Design goals are not always static but are often reappraised according to what is possible. They appear to be as much a product of the design process as the design descriptions that follow from them.

Subsequent chapters deal with how goals can be satisfied. Here we have attempted to show that the representation of goals is essentially the same as the representation of designs. The interacting nature of goals demands, however, that goals be treated differently, particularly in the way goal hierarchies are formulated and manipulated. Goals constitute descriptions of a design that are incomplete and unresolved.

3.4   Representing Design Knowledge

Knowledge provides the means by which we progress from known facts to new facts. Knowledge may also embody general descriptions of designs—descriptions of prototypes rather than of design instances.

In Section 3.1 we introduced the representation of knowledge as rules and as generic design descriptions. In this section we review traditional forms of knowledge representation, which bear on how we represent knowledge in knowledge-based design systems. Then we consider various ways in which design knowledge can be categorized.

3.4.1   Traditional Means of Knowledge Representation

Design knowledge is embodied in whatever serves to direct and constrain the search for designs—for example, physical laws, codes, standards, principles of good practice, and style conventions—as well as in experience from previous design tasks. Knowledge also provides a facility for passing on expertise from one designer to another. Thus it is worth considering briefly the mechanisms traditionally at the disposal of designers for producing designs, learning about design, and passing on design knowledge. A great deal of the design process is embodied in the use of existing designs—analogical design as well as the use of typologies, design rules, and procedures.

Existing Designs

An individual design task generally takes its place within the context of other, similar designs. Existing designs that are the result of attempts to address similar problems serve to define a space of design possibilities. This does necessarily suggest that existing designs are there to be copied. Rather, they may provide analogies from which partial design solutions can be drawn. A new design may represent some kind of interpolation within a continuum defined by existing designs. Existing designs may also serve as a source of general rules or principles that can constrain the production of new designs, as instances from which prototypes are generalized.

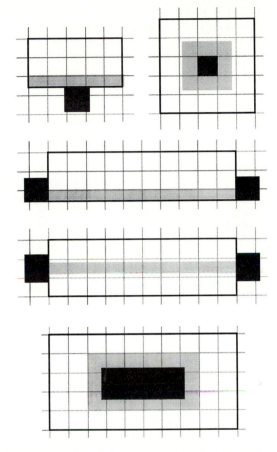

Figure 3.61   Some examples of design types of office floor plans, where there are different relationships between office space (white), circulation (gray), and elevator/stair core (black).

Types

Types are systems of abstractions that serve to define classes of design descriptions. It is quite usual in design to decide rapidly on a type that meets the design requirements. For example, if the required design performance is to provide accommodation for a family of four, then the type house will probably be selected; if the house is to be built on a narrow site, then type row house may be selected. These types carry with them information about overall form, possible numbers of rooms, and possible room configurations. Types become parts of prototypes (see Section 2.6.2) in knowledge-based design systems. Reference works on design are often full of type descriptions in diagrammatic or textual form (Figure 3.61).

Design Rules

The idea of rules generally suggests codes and regulations, but a rule is just a particular form that permits generalizations to be expressed—for example, for a very long span use, or consider using, a suspension structure, or when lifting heavy objects consider using a hydraulic system. Rules may appear as imperatives, but generally there is an implied (if not a stated) precondition to a rule. So the imperative always provide waterproof membranes beneath concrete floor slabs positioned on the ground has preconditions such as when building a waterproof structure and when using a concrete floor slab.

We may utilize rules to externalize knowledge pertaining to the interpretation of designs, as well as the knowledge by which designs are produced.

Procedures

The same methods applied to the representation of knowledge about form can be applied to the representation of process in design. Just as there are existing designs that define spaces of possibilities, there are also existing procedures (Figure 3.62). Having completed one design, we may employ some memory of the process to produce a new design. This may take place at various levels of abstraction. Design reference books often contain outlines of procedures to be followed in producing a design. One such example, at a very coarse level, is the Royal Institute of British Architects' plan of work for use by architects, a generic model of a design process. We may observe classes of design procedures that are implicit in the way designers work, although these are rarely externalized. Of greater interest perhaps is the use of rules relating to procedures—for example, do not do y until x has been completed.

In addition to the five traditional means of knowledge representation discussed, design knowledge can be discussed usefully in other ways, particularly in relation to various mental models that have been proposed, such as cognitive maps (Figure 3.63) (Lynch, 1960; Kaplan and Kaplan, 1982).

We are concerned here with ways to externalize design knowledge; how designers represent their knowledge internally is an interesting but not necessarily connected question. It has been argued that non-experts may reason using rules, but as they become more expert the rules are superseded by stored experiences. Reasoning takes place by matching the patterns of a new situation with those of past experiences (Dreyfus and Dreyfus, 1984; Schank, 1982).

3.4.2   Categories of Design Knowledge

To give some order to the scope of our concern in design knowledge, it is helpful to consider some useful classifications of knowledge. A variety of classification schemes can be adopted:

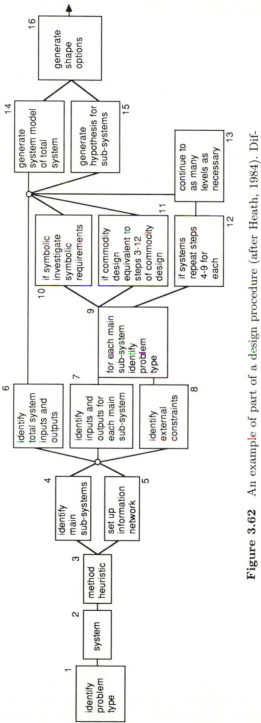

Figure 3.62 An example of part of a design procedure (after Heath, 1984). Different procedures are proposed for different building types: commodity buildings, symbolic buildings and systems buildings.

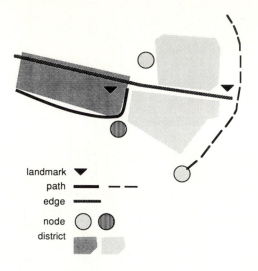

Figure 3.63   A spatial cognitive map (after Lynch, 1960).

1. Axiomatic and heuristic

2. Domain specific and domain independent

3. Conventional and idiosyncratic

4. Deterministic and nondeterministic

5. Certain and uncertain

6. Precise and imprecise

Each of the categories is labeled according to the extremes of a continuum, and any single "item" of knowledge will sit somewhere along each continuum. Thus according to this organization, an item of knowledge may effectively fall anywhere within a space defined by six axes (Figure 3.64). In the following subsections we discuss each of these categories, using examples of knowledge in the form of rules in order to make comparisons.

Axiomatic and Heuristic

Axiomatic knowledge consists of definitions, axioms, and laws. These may be *a priori* or immutable, or at the very least the result of scientific investigation and scholarship. The word "heuristic" sometimes refers to the process of search utilizing "rules of thumb." Because design is so heavily bound up with search, we assume that a heuristic is any knowledge whose origins are not easily justified in that they are never *a priori* nor can be proven from *a priori* knowledge. That is, heuristic knowledge is not deducible from any axiom. Heuristic knowledge is generally gained from experience. An example of a very popular rule of thumb used by structural designers for calculating timber joist

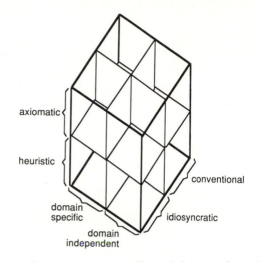

Figure 3.64   Three of the six categories of knowledge as a three-dimensional space.

sizes for a roof is "span over two plus one" (assuming that joist depth is in inches and span is in feet). This rule standardizes established practice based on experience (Figure 3.65).

To illustrate the representation of heuristic knowledge, let us present two rules for determining appropriate floor systems in multistory buildings. These examples are taken from the HI-RISE structural design system for making preliminary structural engineering design proposals (Maher and Fenves, 1985).

> **If** occupancy is commercial **and**
>   material is steel
> **then** assume 4 inches concrete on a steel deck **and**
>   beams in both directions using span/depth ratio of 24.

> **If** occupancy is residential **and**
>   material is reinforced concrete
> **then** assume concrete slab with span/depth ratio 28 **and**
>   beams in long directions only using span/depth ratio of 18.

We can also describe this knowledge as "shallow": it cannot furnish us with sophisticated explanations of why certain design decisions should be made. Axiomatic knowledge is "deep" knowledge, in that it is deduced from fundamental principles or axioms.

Domain Specific and Domain Independent

The distinction between domain-specific and domain-independent knowledge is relative, depending on what we regard as the boundary of the domain of the task at hand. In a design system in which the domain is hydraulic

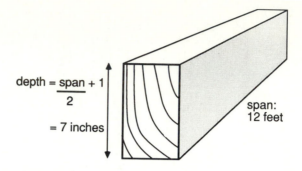

$$depth = \frac{span + 1}{2}$$

$$= 7 \ inches$$

span:
12 feet

Figure 3.65   A heuristic for calculating the size of a timber joist.

pumps, the domain-specific knowledge is knowledge relevant to the design of that particular artifact. Domain-independent knowledge, on the other hand, relates to the physical laws and general design strategies common to all design problems. Of course, we could also say that our domain is design and that the domain-independent knowledge relates to reasoning in general (Figure 3.66).

It is generally desirable in knowledge-based systems to separate out knowledge according to its relevance to what we choose to regard as the domain of expertise. So we can focus on externalizing the knowledge of the expert in hydraulic design without being concerned with how that knowledge will be processed by the knowledge-based system.

Conventional and Idiosyncratic

Conventional knowledge is knowledge that is generally accepted. Again we are dealing with a relative term. Normally we think of conventional knowledge as that accepted by a group of people, such as a body of design experts (electrical or civil engineers, for example); we think of idiosyncratic knowledge as that accepted by some individual but not necessarily shared by colleagues

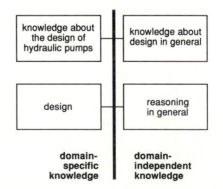

Figure 3.66   The relative nature of domain-specific and domain-independent knowledge.

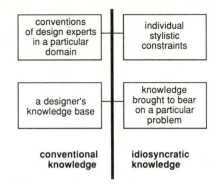

Figure 3.67   The relative nature of conventional and idiosyncratic knowledge.

(certain stylistic constraints, for example). Of course, we may also refer to the knowledge at the disposal of an individual designer as conventional, and the knowledge the designer brings to bear on a particular problem as idiosyncratic (Figure 3.67).

Of the three categories of knowledge just discussed, none appears to require any peculiarities in the way knowledge is treated. The sources of the knowledge bear little on the way the knowledge is handled, at least from a computational point of view—although those sources may bear on which knowledge is assigned greater credibility in the event of contradictions. The following categories have more significant implications for the way the knowledge is controlled.

Deterministic and Non-Deterministic

In a technical sense, it is common to refer to non-determinism in the context of decision-making systems in which certain operations result in different decisions but in which there is no mechanism for selecting between decisions. For example, a system may contain the rule that fact a produces decision b. The same system may also contain the rule that fact a produces decision c. When there is no knowledge applicable to the control of reasoning in such a system, it can be said to be non-deterministic—the outcome may be decision b or decision c, with no way of predicting which (Figure 3.68). We can apply a similar distinction to certain categories of knowledge, particularly in the realm of diagnosis. We might consider the following example of knowledge non-deterministic (Figure 3.69):

If   there is a water stain on the wall
then   it is produced by rising damp **or** by a leaking roof.

Because it is difficult to deal formally with the idea of non-determinism, we usually expect a computer system to adopt some strategy based on heuristic

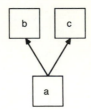

Figure 3.68 An abstract non-deterministic rule.

knowledge for selecting between alternatives, such as always taking the first alternative or generating every possible outcome. The problem may also be treated as one of manipulating uncertainties, as explained in the next subsection.

Certain and Uncertain

We have referred to knowledge as able to be expressed in terms of mappings between facts. Uncertain knowledge is that in which some measure of probability is associated with the truth of the statement. Thus an uncertain rule is one that says if a is true, then b is true with a particular probability. This form of knowledge is particularly applicable to diagnosis, for example (Figure 3.70):

> **If**   there is a water stain on the wall
> **then**   it is produced by rising damp (0.8).

> **If**   there is a water stain on the wall
> **then**   it is produced by a leaking roof (0.4).

The values of 0.8 and 0.4 are certainty factors on a scale of 0.0 to 1.0, where 1.0 is absolute certainty. But this probabilistic approach could also apply to how we represent knowledge about making design decisions (Figure 3.71):

> **If**   wind speed exceeds 100 kph
> **then**   building should have core-type lateral restraint (0.8).

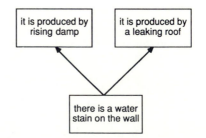

Figure 3.69 A non-deterministic rule about water stains.

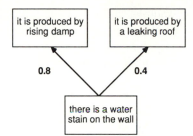

Figure 3.70 A probabilistic rule about water stains.

The numbers may be probabilities based on statistical information, or more likely some intuitive perception of the uncertainty. This approach is applicable to knowledge about design decisions as well as to diagnostic knowledge. Techniques have been developed for reasoning with this kind of knowledge and for propagating values through networks of inferences (see Chapter 6).

Precise and Imprecise

Imprecise knowledge is knowledge that deals not with the truth or falsity of a statement, but rather with its degree of membership. For example,

    If   loss of property is highly critical
then   the design must be very safe.

There is no question that the loss is critical or that the design must be safe, but the imprecision is in the degree of criticality or safety. Imprecise knowledge involves mappings between facts that involve imprecise terms, such as very, a lot, and severe. We present an example of a rule applicable to structural damage assessment:

    If   structure needs *a lot* of repair **or**
             structure is *slightly* reusable **or**
             structure must be rebuilt
then   damage is severe.

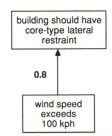

Figure 3.71 A probabilistic rule about lateral restraint.

In this example, the terms in italics are difficult to define precisely. Terms such as highly, very, and severe can be treated as linguistic variables. We can reason with these linguistic variables, and we may choose to quantify them using the formalism known as fuzzy set theory (see Chapter 6).

3.5   Translating between Representations

We have discussed different ways in which knowledge can be represented. Here we discuss why we might want to translate our knowledge from one form to another, and how it can be done, in the process reinforcing some of the concepts of this chapter. We concentrate on knowledge in the form of interpretive rules and how this can be translated into a frame-based representation.

We have considered the way in which interpretive knowledge enables us to discover how a design performs. Interpretive knowledge also serves to define design possibilities—spaces of designs. We focus here on the knowledge itself and how it can be translated into other forms to serve useful purposes even when there is no performance to be met or no design description to be interpreted. The useful purpose is that this knowledge can be used to construct a model. The model serves as a framework onto which we can affix information about a particular design. This in turn facilitates communication with other computer systems.

We are interested in a model of the domain to which our knowledge applies. Here we consider a model as a description of a class of designs. The idea of a model is similar to that of a prototype described previously in that it delimits a space of designs. However, we expect a model to contain much less information than a prototype description. If we instantiate a model, then we produce only a partial description of a design, one that may contain very little information about the form of the artifact. A model is something we might readily derive from interpretive knowledge. Intuitively, we can visualize the idea of a model of an artifact such as a house that conforms to the building code. It has at least one external door, several habitable rooms, the floor area is generally larger than 150 square meters, and the rooms have windows that are a certain percentage of the floor area. We are not considering any particular house, but a class of houses.

It is quite usual to talk about models in the context of expert knowledge. A model provides common ground among experts and is ubiquitous in any area of expertise. Medical experts generally share common models of the human body and common models of diagnostic practice, and have little basis for communication with those who support less conventional models. The same applies to experts dealing in automobile maintenance; a substantial part of their knowledge can be expressed in terms of generic descriptions of automobiles and generic descriptions of common faults and solutions. Not only do

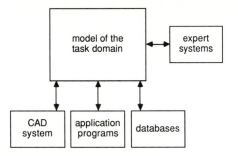

Figure 3.72 The relationship between databases, a CAD system, expert systems, and a model.

such models enable a mechanic to work on a vehicle not seen before, they also enable him or her to talk with another mechanic about that vehicle. A model assumes a common vocabulary for labeling the vehicle's parts and processes, as well as a common structure for describing its systems and subsystems.

A knowledge-based system that incorporates a model of its domain includes an explicit description of the objects and concepts in its knowledge base. This description provides a uniform representation to which other sources of expertise can be linked, facilitating communication between different computer systems such as CAD systems, application programs, and external databases (Figure 3.72).

This model could be provided explicitly *a priori*, in which case it could be used as a means to check the consistency of any knowledge stored. The model must be sufficiently comprehensive, however, to account for any objects, attributes, relationships, and classes in the knowledge base. As new rules are added to the knowledge base, it is necessary to update the model. Another approach is to extract the model from the rules.

The idea of translating from one representation to another was introduced earlier in this chapter. Different methods of representation serve different purposes and define spaces of designs in different ways. Different systems of knowledge representation are also isomorphic. We expect to be able to translate knowledge from one system of representation to another. Here we focus on the idea of translating between rule-based and frame-based representations of knowledge because these methods are the most powerful currently available for describing designs and representing design knowledge (Manago and Gero, 1986).

We have stated that one method of knowledge representation can be translated into another. We take the view that a generic, frame-based description of a design provides a good medium for depicting a model. Furthermore, any method of knowledge representation is implicit in any other representation. Even if we are not representing the knowledge as frames, a frame-based representation is implicit in our knowledge. A generic, frame-based description is implicit in a rule-based representation. We therefore say that there is a model implicit in the rule-based representation (Figure 3.73).

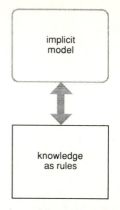

Figure 3.73   A model implicit in a knowledge base.

3.5.1   Design Models Represented as Frames

The frame-based method of knowledge representation is an appropriate method of describing a design model. One advantage of such a system is that it can contain information about what could be in a design description as well as what is in the description. For example, if a system for building design cannot find the description of a laundry in its design description, we would expect it to conclude that the building does not contain a laundry—a reasonable response to a lack of information on the matter. If the system cannot find in the description the type of roof covering, however, the system should know to find out about the roofing material. This is more sensible than concluding that there is no roof covering, or that the roof covering is not of any type. These responses can be achieved by making explicit in the model what can and cannot be included in the design description. One method involves the construction of a generic frame with *may-have* slots and incorporating commonsense reasoning about the uniform degree of detail present (Bobrow and Collins, 1975).

A model-based system can also provide a facility using the same knowledge to derive the performances of several components of the design, as in the case of a building design in which each room must be checked for compliance with similar sets of rules. The model provides a structure for information as it is derived. Facts such as the floor area of the kitchen and the fact that the kitchen complies with the code can be stored in the kitchen frame of the model.

3.5.2   Extracting Implicit Models from Rules

We now discuss some examples of how the implicit knowledge contained in a rule-based representation can be extracted and expressed explicitly as frames. The following rule serves as an example:

If x is a house
then x is a building **and**
          x has at least one door **and**
          x is used for living in.

This can be readily translated into a generic description:

Superclass	Class	Properties
building	house	has at least one door
		used for living in

In turn, this can be translated into a representation using the conventions of frame-based systems:

Class name	house
A kind of	building
Number of doors	must have at least 1
Function	living

Here the knowledge is specifically about classes, superclasses, and their properties. Of course, the knowledge with which we are dealing may involve more than classes and properties.

By what knowledge can we extract a model from a knowledge base? The process is essentially one of interpreting a knowledge base; we can formalize this interpretive knowledge as rules. To make explicit the representation of the objects referred to in a knowledge base, it is necessary to analyze some known patterns in the preconditions and consequents of the rules. Some rules for achieving this are given here:

If   a rule contains a pattern in the form A **is a** B
then   translate this into a frame description:
          A **is an** object,
          B **is an** object,
          B **is a kind of** A.

If   a rule contains a pattern in the form A **is a** B **and**
          A or B do not already exist
then   translate this into a frame description:
          frame name A,
          frame name B.

If   a rule contains a pattern in the form A **of** B **is** C
then   add a slot A to frame B **and**
          record C as a possible value for slot A.

If   a rule contains a pattern in the form A {**contains, has, have**} B
then   translate this into a frame description:

A **is an** object,
B **is an** object,
B **is a_possible_part_of** A.

This kind of analysis provides a frame representation containing information in terms of objects, attributes, and values. It is likely, however, that the rules in the knowledge base also carry implicit meaning that cannot be extracted syntactically in the manner just shown. While it is possible to partly construct such a model from the rules of the knowledge base, this representation may have to be augmented and refined manually. Certain definitions may also have to be added.

In Chapter 5 we present an example involving the construction of a model from a rule base that is subsequently integrated into a proprietary CAD system.

3.6   Summary

We began this chapter by reviewing three systems of organizing information about designs: (1) as objects and relations, (2) as objects and properties, and (3) as instances and classes. We discussed some of the issues surrounding these systems. We also showed how each system could be facilitated by knowledge-representation methods commonly used in knowledge-based systems—facts, semantic networks, rules, and frames. Some of these ideas were reinforced with examples of representing design geometry.

We then discussed some of the major issues in representing design intentions—that is, required interpretations or design goals. Goals differ from descriptions of designs most substantially in that generally they cannot be decomposed in the same way as design descriptions. We distinguished between goals and objectives and considered how goals are not always well-stated at the outset of a design task and cannot be expected to remain the same throughout the design process.

In discussing the representation of classes of designs and their properties, we found that we were already considering knowledge. We highlighted some of the important issues in knowledge representation, including traditional ways of representing knowledge and categories of design knowledge. Finally, we provided an example of a system for translating between a rule-based representation of knowledge and a frame-based system.

3.7   Further Reading

Bobrow and Collins (1975) discuss many of the issues in the representation of knowledge. Brachman and Levesque (1985) is a collection of many of the important papers in knowledge representation. A good summary of the predominant knowledge representation techniques can be found in Barr and Feigenbaum (1981). Representation pervades many introductory texts on artificial intelligence, such as Charniak and McDermott (1985).

Exercises

3.1 Discuss the idea that most of what we want to say about a design can be expressed in terms of objects and relations, objects and properties, and instances and classes. Do you agree with this notion? What is the relationship between these three systems of representation?

3.2 In formulating knowledge-based design systems, what are the implications of the fact that the properties of objects sometimes interact?

3.3 Choose a particular design domain and discuss the appropriateness of facts, rules, semantic networks, and frames as representation tools.

3.4 Devise a hierarchical description of a simple design along the lines presented in Section 3.2. Discuss the strengths and limitations of this approach. Translate this description into a program (in PROLOG, LISP, or some other language) and experiment with different ways of extracting information.

3.5 Goals are partial, incomplete, and unresolved design descriptions. Do you agree with this statement? Describe the similarities and differences between designs and design intentions. How do these similarities and differences affect the representation of goals in design systems?

3.6 How are methods of knowledge representation used in knowledge-based systems similar to or different from traditional non-computer methods?

3.7 Provide examples of design knowledge appropriate to your domain for each of the categories defined in Section 3.4.2. Indicate any item of knowledge that falls within more than one category. Are there any combinations of categories that would never occur?

3.8 Explore the issue of maintaining consistency in design descriptions. When does consistency present a major problem in a design system? Investigate various strategies for maintaining consistency in a facts base.

CHAPTER 4

Reasoning in Design

In Chapter 3 we explored the representation of different types of knowledge relevant to design. Because in most knowledge-based systems it is possible to consider the representation of knowledge independently of its control, we can focus on control as a separate issue. Reasoning can be defined as the processing, or control, of knowledge. In this chapter, we mainly discuss reasoning connected with interpretation of design descriptions. The ideas will be extended to the production of design descriptions in Chapters 6 and 7.

We argue here that there is a uniformity in the way in which we can talk about knowledge and its control. Knowledge is concerned with the relationship between facts, and control is concerned with the relationship between knowledge statements. Control is therefore discussed first as an issue of meta-knowledge. We then investigate types of control knowledge appropriate to interpretation, including backward chaining, forward chaining, goal-directed forward chaining, and knowledge as actions that change the state of a facts base. This brings us to a consideration of the control of production systems, including scheduling and focus of attention.

We then consider the formulation of expert systems. As well as reasoning, these systems are concerned with the "dialogue" between computer systems and the outside world. We discuss how an expert system can be formulated to reason specifically about objects, attributes, values, and classes. Finally, we consider some of the limitations of formal reasoning and the role of non-monotonic reasoning.

In this chapter we utilize formal logic to explain certain control concepts. To assist readers unfamiliar with the logic formalism, concepts are also explained informally and with diagrams when possible.

4.1   Control Issues in Design

We may well ask why we should regard control as an issue of the design process at all. Should we not be able to hand over design knowledge to automated reasoning systems that then take care of the control? The answer is that control appears to be an important aspect of design knowledge that can help make design tasks tractable. The built-in control of automated reasoning systems provides control for a language with which we can define the meta-knowledge appropriate to the task at hand. Different design tasks may require different control strategies. For example, there may be several candidate interpretations of a design, which may conflict. The knowledge statements that support the various interpretations generally constitute a web of interconnecting constraints; we must provide mechanisms for effectively propagating truth values through this web. The type of control mechanism adopted also depends on what we wish to determine from the design system. The nature of the queries we wish to put to the system, or its relationship to other systems, will also determine the character of the control mechanism.

The psychology of designers at work is an issue of great interest, one bound to furnish rewards in formulating better computer models (Akin, 1986; Lawson, 1980; Dietterich and Ullman, 1987). This issue is not addressed in this book. In knowledge-based systems, however, models must be devised to accommodate our understanding of design knowledge. It is necessary to provide theories and computational devices that facilitate the representation of human knowledge in a form that bears some similarity to the way human experts feel comfortable about articulating their knowledge. This process will inevitably involve compromises with what is technically feasible and with what can be mapped onto models of information processing. Control is such an issue. We must consider both how we understand the issue of control in design and what computational devices are available for exercising control in automated reasoning systems.

We have considered the representation of knowledge in the form of logical rules, which serve as a reasonable starting point for modeling design knowledge. Can we find similar models that account for the application of that knowledge to particular design tasks? In order to answer this question, it is worth considering what is meant by "control knowledge" from the point of view of the designer.

We can identify several important control tasks that are desirable to

model: search strategies, goal satisfaction, failure handling, constraint manipulation, non-monotonic reasoning, multiple knowledge sources, multiple
abstractions, and multiple control levels. Each of these strategies is discussed
in the following subsections; we also refer to them throughout the remaining
chapters.

4.1.1   Search Strategies

In Chapter 2 we considered knowledge as defining spaces of designs and spaces
of interpretations. Very simply, we can characterize the design process as one
that attempts to encounter a description in the design space that maps onto
some intended performance in the interpretation space. These spaces can be
defined in various ways—that is, there are various ways of representing design
knowledge—but the process of producing a design from that space seems to
involve some kind of progression from an incomplete, unresolved, and possibly
contradictory description of a design to a complete description. In the most
general sense this process can be regarded as *search* (Figure 4.1).

Search applies to both the interpretation and the generation of designs.
In interpretation there must be a way to traverse the space of possible interpretations and partial interpretations so that we arrive at only those of most
interest. We can model the generation of designs as a process in which a design
appears to pass through various states (Figure 4.2). Any one state may lead
to one of several new states. There are therefore decisions to be made, and it
becomes necessary to select among competing courses of action. Clearly it is
necessary to model the knowledge that enables designers to make such control decisions about interpretation and generation. We label such knowledge
control knowledge or *meta-knowledge*.

4.1.2   Goal Satisfaction

The term search suggests an endeavor to find something. In our model of
generation this can be characterized as a quest for an artifact description, the

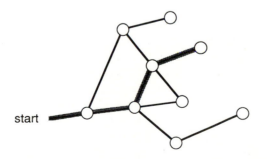

Figure 4.1   Producing design descriptions involves a search through a design space.

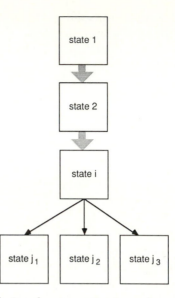

Figure 4.2   Producing designs by generation is analogous to searching a maze by moving from state to state.

interpretation of which matches some understanding of desired performance. To this extent the generative component of design can be said to be goal-directed.

Goal-directed search in the realm of interpretation can be characterized as a "top-down" strategy. We start with some kind of statement and attempt to prove whether the statement is true of the design—whether the design possesses a certain set of properties (Figure 4.3). But design interpretation need not always be driven by the quest for goals. For example, the activity of finding out all you can about a design suggests a kind of "bottom-up" or data-driven approach to search (Figure 4.4) rather than a "top-down" or goal-directed strategy.

In producing designs we may also proceed in a top-down manner: starting with goals and decomposing them into subgoals (Figure 4.5). These subgoals are further decomposed until we find a set of subgoals that we know can be satisfied by certain design decisions. In the process we must resolve conflicts between goals and must consider how decisions made in response to individual goals can be combined, or synthesized, to produce a whole design. Essentially we start with some understanding of what we want and then work out how to get it. We can also adopt a bottom-up, data-driven strategy (Figure 4.6), which may amount to experimenting with the elements in the vocabulary of our design task and configuring them in certain ways, according to some system of actions, until we hit upon a design that satisfies our objectives. Of course, we may somehow alternate between these two strategies. We must be

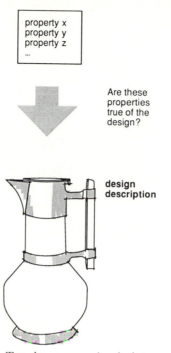

Figure 4.3   Top-down reasoning in interpreting designs.

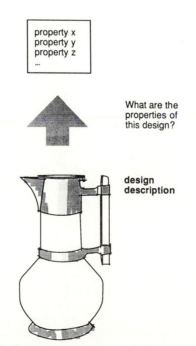

Figure 4.4   Bottom-up reasoning in interpreting designs.

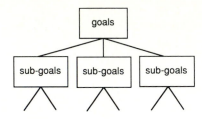

Figure 4.5   Goal decomposition.

able to represent control knowledge that accounts for these and other search strategies.

4.1.3   Failure Handling

An important component of the search operation is knowing what to do in the event of *failure*. In interpretation our reasoning may take us along a certain

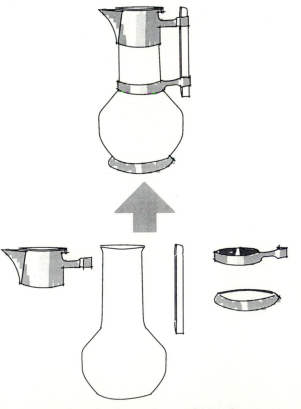

Figure 4.6   Bottom-up reasoning in producing design descriptions—starting with elements and configuring them to create a whole artifact.

path of inquiry that proves unfruitful. One way to handle this is simply to backtrack to some earlier point in the argument and follow another line of reasoning (Figure 4.7). But to what point do we backtrack, and then how do we know which path to follow next?

In producing a design we can imagine a designer developing a proposal in a certain direction. When some obstacle is encountered, or it seems that an important goal cannot be achieved, the designer may revert to earlier states in the development of the design and proceed along a new path. We must be able to model this aspect of knowledge about search.

4.1.4   Constraint Manipulation

A further refinement to the idea of goal-directed search is the distinction between constraints and objectives. A *constraint* is a statement about a design the truthfulness of which does not depend on any tradeoff with goals. Examples of constraints are the statements that the manufacturing cost of an artifact is to be less than $200, that the rentable floor area of an office development is to be 400 square meters, and that a building design must fit within the shape imposed by its site boundaries. The site-boundary constraint is true or false irrespective of whether or not the required rentable area is provided. On the other hand, an *objective* is a goal statement that constitutes something to aim for, such as the statement that the manufacturing cost of an artifact is to be as low as possible or that the design of an office development is to provide the greatest possible rentable floor area. There exists a duality between objectives and constraints; in many cases an objective can be stated in terms of constraints. For example, the objective "maximize sound insulation of a partition" can be stated as the constraint that the partition is to have a sound insulation factor greater than or equal to 50.

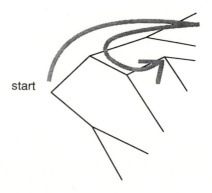

Figure 4.7   Backtracking.

Many such objectives may be evident in a design task, each making conflicting demands on the form of the final artifact. We must be able to model the knowledge by which a designer can map such objectives onto design decisions and can make tradeoffs between conflicting objectives. It is also necessary to model the knowledge by which designers appear able to impose provisional constraints on design tasks that can later be relaxed in the event of failure.

4.1.5   Non-Monotonic Reasoning

A further manifestation of this complexity in reasoning is how we manage to maintain temporary inconsistencies in reasoning steps in order to arrive at valid interpretations. Often logical argument is something we formulate after arriving at an interpretation or a design decision; sometimes it bears very little relation to how we actually arrive at the decision in the first place. People seem able to tolerate, even to require, temporary inconsistencies in reasoning. In terms of search we can see such *non-monotonic* reasoning as a process in which search proceeds temporarily, rather than failure and backtracking in the event of contradiction. Modeling the non-monotonic aspect of design reasoning presents great difficulties.

4.1.6   Multiple Knowledge Sources

Knowledge can emanate from many sources—in design, these sources can include "external" storage media in the form of texts, codes, and diagrams, as well as teams of experts working on a particular task. In accounting for control knowledge in design systems, we also must model the knowledge by which a designer organizes and gains access to knowledge from disparate sources. Another part of the control knowledge at a designer's disposal, therefore, concerns knowing where to find things out and knowing what sources of knowledge are relevant to the task at hand.

4.1.7   Multiple Abstractions

In Chapter 3 we considered the role of multiple representations in design. The current state of a design can be described in many ways. In the case of a VLSI chip, for example, there is a description of the design in terms of input and output behavior; there is the "floorplan," which is concerned with the gross physical structure of the layout; there is the logic diagram, in which gates are the basic unit; finally, there is the physical abstraction, the layout, in which the fine physical structure is described in terms of polygons (Figure 4.8).

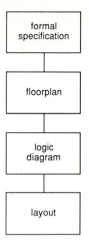

Figure 4.8   Multiple abstractions of a circuit design.

These representations are related, perhaps hierarchically, and we can consider design as a process in which there is free communication between types and levels of representational abstraction. Accomplishing design tasks appears to require the negotiation of tasks at various abstraction levels. How we organize and manipulate design knowledge in order to facilitate this movement is an issue of control. In all of this, we may say that control knowledge appears to be a manifestation of design knowledge at different abstraction levels. This theme will be pursued more vigorously in Chapter 7 in the context of generative knowledge.

4.1.8   Multiple Control Levels

Certain design tasks lend themselves to forms of hierarchical control exhibited by many organizational systems. As well as taking place at various levels of representational abstraction, design can be considered to take place at various levels of control. We are familiar with this concept in the context of group design projects. There may be a team leader, who exercises control over subordinates, each of whom makes decisions within the context of his or her own bounded subtasks. When subordinates encounter insurmountable difficulties in their subtasks, some overall strategy may be introduced by the team leader to improve interactions between the subordinates; complex interactions between levels may be involved. The management of the hierarchical decomposition of tasks is a complex issue of control knowledge.

Multiple control levels can also be manifested in the way design actions are treated. Design can be characterized as a process in which certain actions are brought to bear in manipulating and assembling components. We can consider a different abstraction of this task: as one in which there are "control actions" for manipulating and "assembling" design actions. This suggests a hierarchy of control abstractions in which the actions of one level become the objects of the controlling level above. Multiple control levels are addressed more fully in Chapter 7.

We will not attempt to address all of these issues in the ensuing discussion. They serve to indicate something of the complexity of the problem of effective control in design systems, and we expect a model of control in design systems to afford the promise of accounting for these issues, if not their realization.

So far we have discussed the issue of control in design primarily from the point of view of what it is we want to model. Let us return now to our basic model of computation, which views the representation of knowledge as concerned with objects and the relationships between them, and then to several basic models of control as they are understood in terms of automated reasoning systems. In the process we discuss how each can be said to accommodate certain aspects of design reasoning. This discussion leads to a view of control knowledge in design as concerned essentially with the propagation of information through many abstraction levels of representation and reasoning.

4.2   Control as Meta-knowledge

The type of reasoning with which we are concerned here is essentially that associated with deduction. The objective of this type of reasoning is to infer new facts from what is known—for example, deriving the properties of an artifact described in a computer database that are not explicitly stated, a process of interpretation. The control of those processes by which designs are produced is discussed in Chapters 6 and 7.

We have characterized *facts* as statements about the relationships between objects, and *knowledge* as concerned with the relationships between classes of facts. *Control* is essentially concerned with mappings between items of knowledge. It is convenient to discuss control issues in the context of rules. Rules, we have seen, are explicit statements mapping facts: if certain facts (the preconditions) are true, then certain other facts (the consequents) are also true.

We can say something similar about *meta-knowledge* statements, which relate knowledge statements to each other and to the facts to which they apply. There are two important ways in which we can define these relationships. In the first characterization, we concentrate on the relationships between knowledge statements and exploit how they are linked or chained together in order to derive new facts. These linkages can be formed from commonalities between

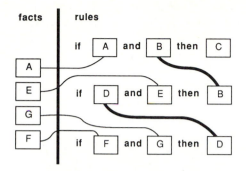

Figure 4.9   Rule chaining.

the preconditions and consequences of rules and between rules and facts that are already known (Figure 4.9).

In the second characterization of control, knowledge statements are independent entities, and any linking between them in terms of matches between preconditions and consequences is incidental. In this case the relationship between rules and facts is considered important. In this characterization, rules are regarded as actions that change the state of our understanding in some way (Figure 4.10).

The utility of the meta-knowledge idea is immediately apparent when we consider that it is possible to represent control knowledge in terms of rules—or possibly meta-rules. An example of meta-rules embodying the first model can be expressed as follows:

A fact can be proven **if** it matches with something that is already known to be true.

A fact can be proven **if** it is the consequent of some interpretive rule and all the preconditions of that rule can be proven.

These are meta-knowledge statements which take account of the relationships between rules. They also amount to rule definitions of a control strategy usually called *backward chaining* (Figure 4.11). In this model we effectively start by trying to establish whether a fact is true of our design (a goal statement). The fact is true if we can chain back through all the rules until we arrive at facts that are known explicitly to be true.

A different control strategy, called *forward chaining*, can also be defined informally by the following rules:

A fact can be inferred from an existing fact **if** they are the same.

A new fact can be inferred from an existing fact **if** the existing fact is part of the precondition of a rule and the other preconditions of the rule can be proven and the consequence of that rule leads to the derivation of the new fact.

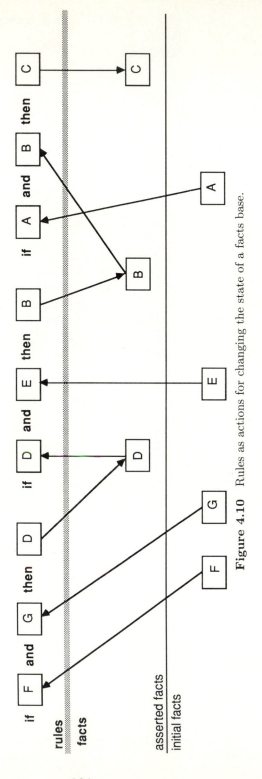

Figure 4.10   Rules as actions for changing the state of a facts base.

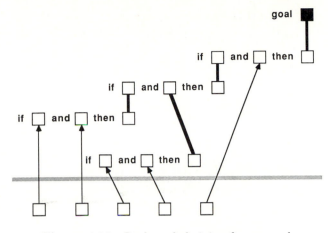

Figure 4.11 Backward chaining from a goal.

These rules are perhaps more difficult to understand. We effectively start by chaining through each of the rules from a particular known fact, within a set of known facts, and declare the new inferred facts made along the way. These control strategies are described more formally in the next section.

In the second characterization of control, knowledge statements can be regarded as a kind of "object," that is, as *actions* that change the state of our understanding. In the case of interpretation this generally involves some kind of incremental improvement in understanding. In this model the application of each rule brings about an accretion of information, as represented by the following statement:

> Some new facts can be determined **if** the preconditions of a rule match with what is known and the consequences of that rule can be asserted.

This essentially states that a new fact can be established if it can be asserted as the consequent of a rule, the preconditions of which have already been established as true (Figure 4.12). It results in a control strategy in which rules are effectively "fired" in turn to change the state of what is currently known. There is no rule chaining, but rules appear to "respond" to the facts base as the facts appear relevant.

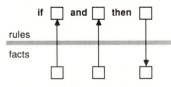

Figure 4.12 The application of an action to bring about an accretion of information.

In this discussion we have attempted to establish that a uniformity exists in the way we represent knowledge and the way we represent the control of that knowledge. In the following section we turn to a detailed discussion of control in the area of interpretation, considering various automated-reasoning strategies as meta-knowledge issues. We also consider their relevance to interpretation in design.

4.3   Control in Design Interpretation

Various automated-reasoning systems such as PROLOG (Clocksin and Mellish, 1981) and OPS5 (McDermott, 1982) incorporate built-in control mechanisms. Automated theorem provers such as PLANNER (Hewitt, 1972) and PROLOG enable us to represent statements in logic. They employ some kind of procedure, usually *resolution* (Robinson, 1965), to prove the consistency of these statements. Implicit in this procedure is some device for the exhaustive exploration of partial proofs. We will not attempt to describe these proof procedures here, only noting that a common method of implementing exhaustive search is to employ linear resolution and chronological backtracking. It is important to remember that the validity of logic statements stands independently of the methods of resolution employed. It serves our purpose here to employ logic as a language for describing control models. We assume that we have at our disposal some automated theorem prover that tests the truthfulness of a set of logic statements, and that it does so systematically, chronologically, and exhaustively, perhaps employing the idea of backtracking on failure (though other devices are explored later).

Here we consider three standard strategies of control applicable to interpretation in design—backward chaining, forward chaining, and goal-directed forward chaining—starting with the simplest and generally progressing toward the most sophisticated. We then consider the control of knowledge as production systems in which knowledge statements are actions that transform descriptions. We look at the control of production systems in terms of scheduling and focus of attention. We then explore control mechanisms that facilitate a kind of "expert reasoning," bringing us to a discussion of expert systems. We explore how these standard strategies, or models, can effectively be brought to bear on design interpretation tasks, indicating how the ideas must be extended to address some of the design issues raised in Section 4.1, such as non-monotonic reasoning.

We utilize a simple example to demonstrate how each of the different control strategies can be implemented: determining the meaning of a set of lines stored in a computer database. This task provides an example for the general task of inferring properties of a design from an object description. Because

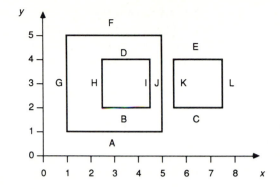

Figure 4.13 A geometrical representation of a facts base.

the interpretation of simple geometries is fairly well understood and belongs to a domain with which most people are familiar, we select this interpretation as a useful design task. The task is similar to many design domains—such as floorplanning in electronics—in fact, to any domain that lends itself to representation in a CAD system. The principles of control brought out by this example apply to all design domains.

As a starting point, we consider a set of facts representing the line segments depicted in Figure 4.13. We can describe this scene in predicate calculus notation in terms of line segments, labeling these predicates "Primitives":

Primitive(A, 1.0, 1.0, 5.0, 1.0)

Primitive(B, 2.5, 2.0, 4.5, 2.0)

Primitive(C, 5.5, 2.0, 7.5, 2.0)

Primitive(D, 2.5, 4.0, 4.5, 4.0)

Primitive(E, 5.5, 4.0, 7.5, 4.0)

Primitive(F, 1.0, 5.0, 5.0, 5.0)

Primitive(G, 1.0, 1.0, 1.0, 5.0)

Primitive(H, 2.5, 2.0, 2.5, 4.0)

Primitive(I, 4.5, 2.0, 4.5, 4.0)

Primitive(J, 5.0, 1.0, 5.0, 5.0)

Primitive(K, 5.5, 2.0, 5.5, 4.0)

Primitive(L, 7.5, 2.0, 7.5, 4.0)

The arguments of these predicates indicate the names of the line segments and the x and y coordinates of their two end points in turn. From here on, this set of facts is referred to collectively as the *facts base*.

4.3.1   Backward Chaining

The first strategy we consider is that of *backward chaining*. When we use backward chaining in interpreting designs, we begin with a statement that constitutes an interpretation of the design and attempt to determine whether it is true. This statement constitutes a goal; we determine its truth either by matching it against one of the facts in the design description (the facts base) or by matching it against the consequence of a rule in the knowledge base. The preconditions of the rules with which it matches constitute subgoals, the truth of which must also be established similarly. This process is undergone recursively until any subgoals required match the facts base. We do not consider here how this procedure can be formulated using procedural algorithms. (This matter is addressed in detail by Nilsson (1982).) Rather, we describe the process declaratively as meta-knowledge.

As a goal we consider the interpretive statement that the rectangle defined by line segments B, H, D, and I is inside the rectangle made up of segments G, F, J, and A. We can represent the knowledge that provides the mappings between design descriptions and interpretations (that is, between goals and subgoals) with the following rules:

If   there are two rectangles **and**
      the north edge of the first rectangle is below the north edge
         of the second rectangle **and**
      the south edge of the first rectangle is above the south edge
         of the second rectangle **and**
      the east edge of the first rectangle is to the left of the east edge
         of the second rectangle **and**
      the west edge of the first rectangle is to the right of the west edge
         of the second rectangle
then   the first rectangle is inside the second rectangle (Figure 4.14).

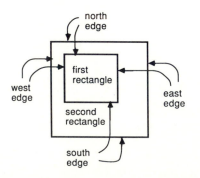

Figure 4.14   The geometry underlying a rule for determining whether a rectangle is inside another rectangle.

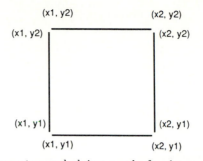

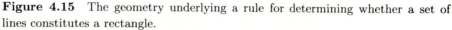

Figure 4.15 The geometry underlying a rule for determining whether a set of lines constitutes a rectangle.

If   there are four lines and two of the lines are parallel to the x axis
            sharing the same x values at their end points **and**
        there are lines parallel to the y axis with the same x values
            as each of these two end points **and**
        the y values of these two lines correspond with the y values
            of the lines parallel to the x axis
then   the composition of these four lines is a rectangle (Figure 4.15).

If   a primitive element has two end points x_1, y_1 and x_2, y_2
then   it is a line (Figure 4.16).

We can also represent these rules more formally as follows:

If   Rectangle(a, n_a, s_a, e_a, w_a) **and**
        Rectangle(b, n_b, s_b, e_b, w_b) **and**
        Not$(a = b)$ **and**
        $n_a =< n_b$ **and**
        $s_a >= s_b$ **and**
        $e_a =< e_b$ **and**
        $w_a >= w_b$
then   Inside(a, b).

If   Line(w_a, w, s, w, n) **and**
        $n > s$ **and**
        Line(e_a, e, s, e, n) **and**
        $e > w$ **and**
        Line(s_a, w, s, e, s) **and**
        Line(n_a, w, n, e, n)
then   Rectangle$($Components$(n_a, s_a, e_a, w_a), n, s, e, w)$.

If   Primitive(a, x_1, y_1, x_2, y_2)
then   Line(a, x_1, y_1, x_2, y_2).

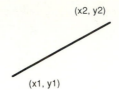

Figure 4.16　The geometry underlying a rule for determining whether a primitive element is a line.

These rules happen to be in a form in which chaining can be achieved relatively simply. This kind of rule, in which there there may be several preconditions but only a single consequent, is known as a *Horn clause*. The medium is very rich; any collection of logic statements can be represented as Horn clauses after some manipulation (Kowalski, 1979).

From now on we will refer to these rules collectively as the *knowledge base*. The goal can be stated as follows:

Goal(Inside(Components(H, D, I, B), Components(G, F, J, A)))

To test the consistency of this statement against the preceding knowledge base, we must be able to establish a match between goals and the consequents of rules, and between the arguments of goals, that is, instantiations of proposition arguments. The process involves a search through a space of rules and a space of instantiations. This control knowledge can be represented as meta-rules:

A goal is true **if** it matches a fact in the facts base.

A goal is true **if** there is a rule in the knowledge base the head of which matches the goal, and all the propositions in the consequent of the rule can be matched.

A list of propositions can be matched **if** the first proposition in that list is true as a goal and the rest of the propositions can be matched.

A single proposition can be matched **if** it is true as a goal.

We can represent this knowledge more formally as follows:

Goal(a) $\Leftarrow a$

Goal(a) $\Leftarrow$ (**if** b **then** a) $\land$ Match(b)

Match(a **and** b) $\Leftarrow$ Goal(a) $\land$ Match(b)

Match(a) $\Leftarrow$ Goal(a)

Readers are referred to Appendix A for an explanation of the symbols used in these statements. In this chapter we use logic to define control knowledge and the if-then rule form for knowledge pertaining to the interpretation task.

The preceding logic statements collectively constitute the meta-knowledge of our system.

An attempt is made to match the goal statement against the first clause in the meta-knowledge of our system. The goal does not match the facts base, so an attempt is made to match the second clause. The goal is true if there is a rule with the goal as a consequent and if the preconditions of the rule are true. An attempt is made to match the list of preconditions by taking each precondition in turn as a goal and repeating the process. Eventually, this backward chaining process results in a set of goals that match the facts base, and the initial goal is true (Figure 4.17).

We should also be able to satisfy a goal containing variables such as the following:

$$\exists u \exists v [\text{Goal}(\text{Inside}(u, v))]$$

(The $\exists$ symbol means "there exists.") When the initial goal contains variables, this chaining process should also result in the instantiation of values. Because several possible facts could be matched against each "Primitive" subgoal, several instantiations could be made for the variables in that subgoal. Not all values will be suitable for all necessary combinations of subgoals, so the derivation of a set of instantiations that makes the initial goal true will involve search, which requires backtracking.

The search process for the knowledge base just described is too unwieldy to show on a diagram. The idea is demonstrated in Figure 4.18 with a much simpler knowledge base. Here there is only one rule:

If $\text{Area}(x, a)$ and $a > 20$
then $\text{Big}(x)$.

An object is "Big" if its area is greater than 20 units. The facts base is

$\text{Area}(\text{P, 10})$

$\text{Area}(\text{Q, 25})$

The search for the goal $\text{Big}(x)$ involves an initial instantiation where x is P and a is 10. The subgoal $a > 20$ fails, making it necessary to backtrack to find another fact that matches $\text{Area}(x, a)$. The variable x is reinstantiated to Q and a is reinstantiated to 25. The value of a satisfies the subgoal $a > 20$. The system returns Q as the value for which the goal is true.

When there are a large number of facts and rules, this kind of search can be cumbersome. The advantage of this logic representation is that further control can be exercised over the order by which the space is searched. We can, for example, incorporate some kind of ordering heuristic to further control the search. We can also incorporate some kind of device for controlling communication with other systems. More is said about these issues in Section 7.2.

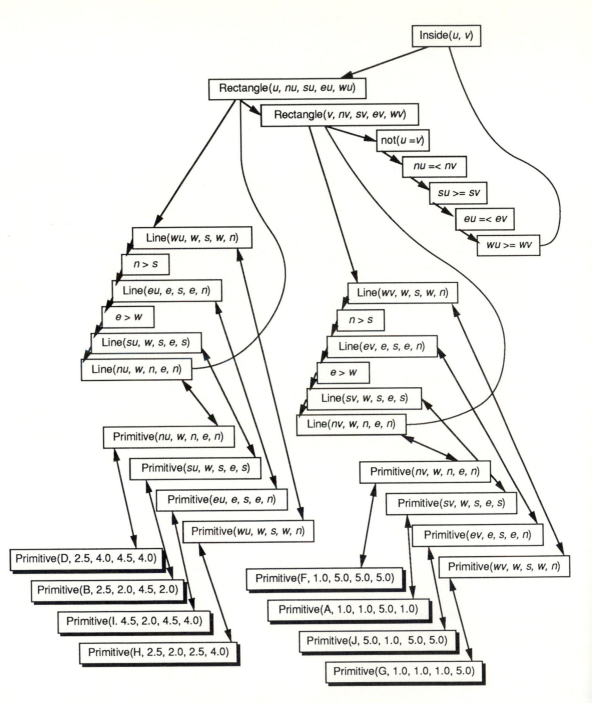

Figure 4.17   The backward chaining process.

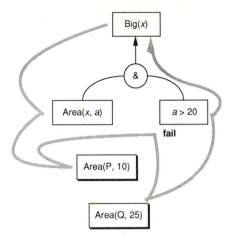

Figure 4.18 A simple example of backtracking involving the instantiation of variables. Backward chaining is indicated by the gray line.

4.3.2 Forward Chaining

Conceptually, forward chaining begins from a set of facts about the design that are already known and asks what can be inferred from these facts. We must be able to establish that a statement such as

Infer(Primitive(H, 2.5, 2.0, 2.5, 4.0), Rectangle(Components(H, D, I, B), 4.0, 2.0, 4.5, 2.5)))

is consistent with the facts base and the knowledge base—that is, that the line segment labeled H is instrumental in establishing the fact that there is a rectangle made up of components H, D, I, and B. More interestingly, we may want to prove the statement

$\exists x[\text{Infer}(\text{Primitive}(H, 2.5, 2.0, 2.5, 4.0), x)]$,

where x is a variable. This statement says that it is possible to infer something from the fact that there is a line primitive H. As a by-product, x must be instantiated to a proposition or propositions. There may be several ways in which the statement can be proven true and the instantiations of x will probably interest us most. (Automated theorem provers generally provide all the instantiations under which a goal statement can be said to be consistent.) In effect we require some meta-knowledge statements that enable us to find out all the facts that can be inferred from a given fact.

It is possible to infer a fact from itself.

It is possible to infer a new fact from an existing fact **if** there is a rule in the knowledge base, the existing fact is a member of the list of preconditions

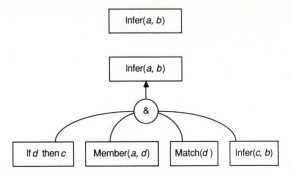

Figure 4.19   The logic of forward chaining.

of that rule, all the preconditions of the rule can be matched, and it is possible to infer the new fact from the consequent of that rule.

This knowledge can be represented more formally as follows (Figure 4.19):

$\mathrm{Infer}(a, a)$

$\mathrm{Infer}(a, b) \Leftarrow (\textbf{if } d \textbf{ then } c) \wedge \mathrm{Member}(a, d) \wedge \mathrm{Match}(d) \wedge \mathrm{Infer}(c, b).$

The Match predicate can be proven according to its definition in the previous section. The Member predicate must also be defined (Figure 4.20).

A proposition is a member of a list of propositions if there is only one proposition in the list and they are the same.

$\mathrm{Member}(u, u)$

A proposition is a member of a list of propositions if it is the same as the first proposition in the list.

$\mathrm{Member}(u, u, \textbf{and } v)$

A proposition is a member of a list of propositions if it is a member of the tail of a list—that is, the rest of the list after the first element is removed.

$\mathrm{Member}(u, w, \textbf{and } v) \Leftarrow \mathrm{Member}(u, v)$

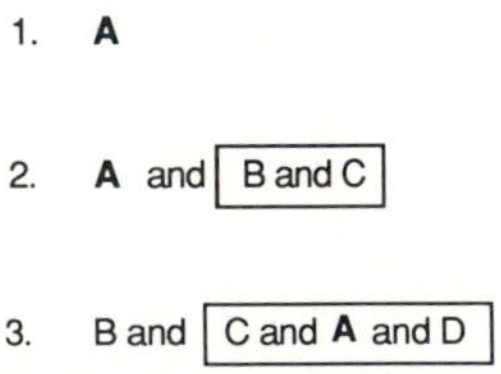

1.   **A**

2.   **A** and $\boxed{\text{B and C}}$

3.   B and $\boxed{\text{C and } \textbf{A} \text{ and D}}$

Figure 4.20   Cases in which it can be said that A is a member of a list: (1) the list contains only one element, which is A, (2) the list starts with A, and (3) A is in the tail of the list.

Presented with the goal

$$\exists x[\text{Infer}(\text{Primitive}(\text{H, 2.5, 2.0, 2.5, 4.0}), x)],$$

the system will match the statement $\text{Infer}(a, a)$ and produce a value of Primitive(H, 2.5, 2.0, 2.5, 4.0) for x—that is, a fact can be inferred from itself. Of much greater interest are the other values of x that will be generated. According to the second statement about Infer the system will find an if-then rule with Primitive(H, 2.5, 2.0, 2.5, 4.0) in its preconditions. It will check all those preconditions to see whether they are true according to the Match predicate. Then the system will Infer what it can from the consequent of the if-then rule. Because a fact can be inferred from itself, that predicate will be true. The value of its second argument will be the consequent of the rule, which is declared as a proposition that can be inferred from Primitive(H, 2.5, 2.0, 2.5, 4.0). Because the Infer predicate can be proven according to two statements, one of them is recursive, further inferences can be made. The overall effect is to chain upward through the if-then rules declaring their consequents as possible values of x. The instantiations of x are therefore as follows:

Primitive(H, 2.5, 2.0, 2.5, 4.0)

Line(H, 2.5, 2.0, 2.5, 4.0)

Rectangle(Components(H, D, I, B), 4.0, 2.0, 4.5, 2.5)

Inside(Components(H, D, I, B), Components(G, F, J, A))

Figure 4.21 illustrates the outcome of forward chaining, the traversal through a search tree proceeding from the bottom to the top.

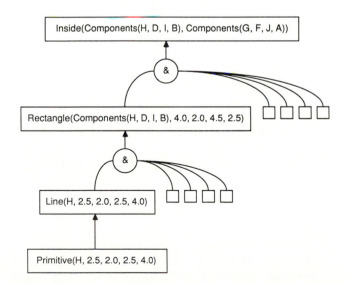

Figure 4.21 Forward chaining through an inference network.

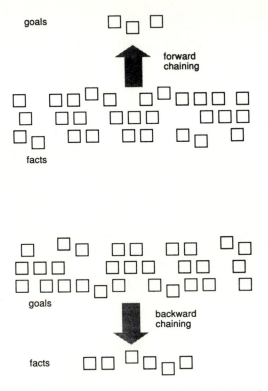

Figure 4.22   Systems that favor forward and backward chaining.

A forward-chaining approach to reasoning is valuable because the interpretation required need not be anticipated from a facts base and a knowledge base. This has some appeal in the context of design. Unexpected interpretations arising out of partial and incomplete designs can stimulate the exploration of new ideas. This reasoning approach carries with it the problem, however, that an excessively large space of interpretations may be generated, including many of no interest. From the point of view of efficiency, we would expect forward chaining to be more favorable for a facts base and a knowledge base for which there are many facts and few possible interpretations, or goals. A system for which there are few facts and many possible goals would favor the backward-chaining approach (Figure 4.22).

4.3.3   Goal-Directed Forward Chaining

We may consider these two control strategies of backward and forward rule chaining as standard methods of determining new facts from known facts, at least in terms of our understanding of meta-knowledge as concerned with the relationships between rules. Further refinements can be added to these

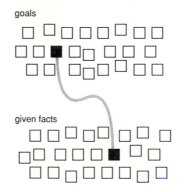

Figure 4.23   Finding a path to determine whether two propositions are related.

models—for example, what we might call *goal-directed forward chaining*. The forward-chaining model, as just described, may involve the expenditure of a great deal of effort to establish that a particular given fact contributes to the proof of a new fact. Unnecessary effort may go into establishing the truth of propositions that turn out to be unrelated to any goal proposition. It would appear to be desirable to determine that two facts are indeed connected before attempting to prove all the propositions along the path joining them (Figure 4.23). This involves finding a path from a fact to a goal and then attempting to prove the statements along that path—that is, for a path connecting two propositions before making instantiations.

We need not go into the details of how this process can be described fully in logic except to demonstrate that it is relatively straightforward to discover all the propositions that lie along a path connecting two propositions in an inference network.

> The path joining a goal to itself is the same as the list of all propositions already tried.

> There is a path linking a proposition to another proposition **if** there is a rule the precondition of which has the first proposition as a member, the proposition is not contained within the list of propositions already tried, and there is a path connecting the consequent of the rule to the second proposition.

We can restate this meta-knowledge more formally (Figure 4.24):

$\text{Path}(a, a, b, b)$

$\text{Path}(a, b, c, d) \Leftarrow (\textbf{if } f \textbf{ then } e) \wedge \text{Member}(a, f) \wedge \text{Not}(\text{Member}(e, c))$
$\wedge \text{Path}(e, b, e \textbf{ and } c, d)$

The first argument of Path is the fact from which the journey is to commence, while the second argument is the fact at the destination. The third argument is a list of propositions that are already on the path and that must not be

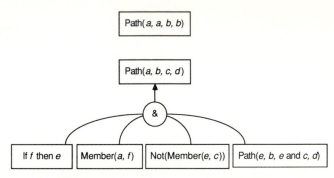

Figure 4.24   The logic of path finding.

considered again (initially the empty list), and the final argument is a list of all the propositions along the path.

With this knowledge we can attempt to test the consistency of a goal such as

$$\exists u \exists v \exists x \ [\text{Path}(\text{Primitive}(H, 2.5, 2.0, 2.5, 4.0), \text{Inside}(u, v), \text{Nil}, x)].$$

We would expect this goal to result in the instantiation of x to the following list of intermediate propositions:

Inside(Components(H, _, _, _), _) **and**

Rectangle(Components(H, _, _, _), 4.0, 2.0, _, 2.5)) **and**

Line(H, 2.5, 2.0, 2.5, 4.0) **and**

Primitive(H, 2.5, 2.0, 2.5, 4.0) **and** Nil

This list shows the path connecting the two propositions in the goal statement, in reverse order. The underscore symbol (_) indicates variables that are not instantiated. Using the meta-knowledge of Section 4.3.1, each proposition must be proven for these variables to be instantiated. In a problem with many if-then rules, several paths may connect facts to goals. Each of these paths would be generated by the preceding meta-knowledge. The effect of this path-finding process is shown notionally in Figure 4.25. In this example two paths connect a declared fact and the goal. Finding such paths before attempting to prove the propositions along them has implications for efficiency—that is, path finding is relatively economical compared with "proving" the path. In certain cases we would expect path finding to be more efficient than backward or forward search, in which it is attempted to prove propositions during the search process. Of course, we may also use the path-finding knowledge to discover all the facts that bear on a particular goal

$$\exists y \exists u \exists v \exists x \ [\text{Path}(y, \text{Inside}(u, v), \text{Nil}, x)],$$

where y will be instantiated to all the propositions that are connected to the goal Inside(u, v).

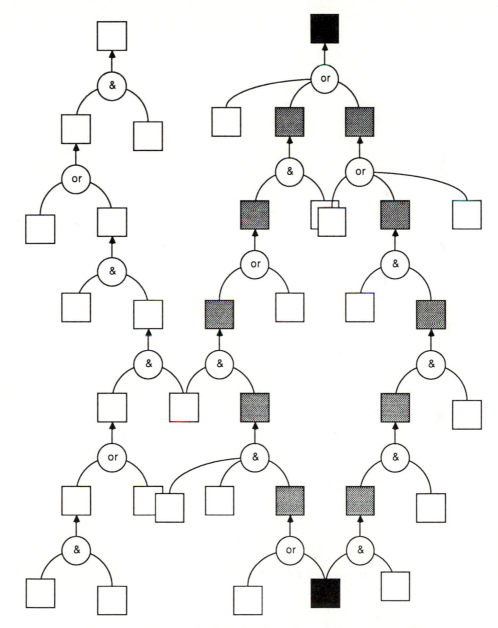

Figure 4.25   Multiple paths through an inference network.

Informally, we can characterize these path-finding strategies as a means of determining the relevance of a particular fact to an interpretation before determining the validity of the interpretation. This is akin to deciding on admissible evidence in a court case before arguing from the evidence about the charge. In our simple interpretation task the process is one of reasoning that the properties of lines are relevant to determining whether one shape is inside another. A similar process could be applied to identifying the relevant factors in determining the bending moment of a beam, the thermal properties of a building envelope, or the aerodynamic properties of an aircraft.

4.4   The Control of Actions

In interpretation we are determining new facts from existing facts about a design. We have some kind of representation of a design before us, perhaps in the form of a drawing. All that exists on paper is a configuration of lines, points, and labels. How do we model what happens when we try to bring our knowledge to bear in interpreting that representation? Initially there may be some kind of mapping of these configurations onto an understanding of geometrical entities, such as rectangles, circles, axes of symmetry, and grid lines. From this kind of understanding comes an appreciation that certain entities are rooms, walls, machine parts, or columns, and that these entities have certain relationships with each other. From this understanding may come a realization that certain parts of the design conform to the canons of good practice, while others appear unsatisfactory in some way. What this suggests is a kind of "bottom-up" process of interpretation. The drawing appears to be "telling" us what it can about itself in the light of our knowledge about good design (Figure 4.26).

The system is guided by changes to a store of facts rather than an attempt to satisfy some goals. As the opportunity arises, an item of knowledge—such as a rule—fires. In this system, an alternative control structure to the idea of rule chaining described in the previous sections is used. The meta-knowledge is generally concerned with the relationship between each knowledge statement and the facts base.

Section 4.4.1 may be difficult for readers who are unfamiliar with formal logic. Such readers may wish only to skim through the material before turning to Section 4.4.2.

4.4.1   Rules as Actions

In this model we start with a facts base and attempt to match the preconditions of each rule with it. As a rule becomes applicable, we instantiate its preconditions and assert the consequent into the facts base. We apply each

conformity

relationships

objects

geometrical
entities

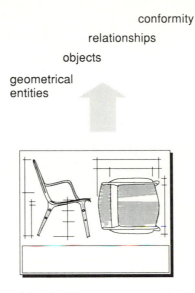

Figure 4.26   A kind of bottom-up reasoning in interpretation.

rule exhaustively (that is, with all possible instantiations) in turn, then cycle through the process again. The facts base grows in size with each rule firing. (As indicated later, there are several methods of cycling through rules, but this characterization conveys the general idea.)

The idea of making assertions into facts bases does not fit comfortably with the rigors of formal logic. It would be more correct to consider the facts base as passing through various states and to label facts according to the states to which they apply. It would then be possible to chain through the rules in the knowledge base to establish whether a particular fact is true in a particular state. This method must be considered in some detail because the idea of actions is also important in the context of control in generative systems. We also want to establish that the control of actions fits firmly within the idea of meta-knowledge.

We could, for example, consider the facts base described earlier as the initial facts base. We can tag the facts in terms of the states to which they belong. The facts base can therefore be expressed in the following way:

Fact(Primitive(A, 1.0, 1.0, 5.0, 1.0), Start)

Fact(Primitive(B, 2.5, 2.0, 4.5, 2.0), Start)

Fact(Primitive(C, 5.5, 2.0, 7.5, 2.0), Start)

Fact(Primitive(D, 2.5, 4.0, 4.5, 4.0), Start)

Fact(Primitive(E, 5.5, 4.0, 7.5, 4.0), Start)

Fact(Primitive(F, 1.0, 5.0, 5.0, 5.0), Start)

Fact(Primitive(G, 1.0, 1.0, 1.0, 5.0), Start)

Fact(Primitive(H, 2.5, 2.0, 2.5, 4.0), Start)

Fact(Primitive(I, 4.5, 2.0, 4.5, 4.0), Start)

Fact(Primitive(J, 5.0, 1.0, 5.0, 5.0), Start)

Fact(Primitive(K, 5.5, 2.0, 5.5, 4.0), Start)

Fact(Primitive(L, 7.5, 2.0, 7.5, 4.0), Start)

It would also be useful to tag the rules in some way. We can re-express the if-then rules of Section 4.3.1 as the following table:

Action	Preconditions	Consequents
1	$\text{Rectangle}(a, n_a, s_a, e_a, w_a)$ $\text{Rectangle}(b, n_b, s_b, e_b, w_b)$ $\text{Not}(a = b), n_a =< n_b, s_a >= s_b$ $e_a =< e_b, w_a >= w_b$	$\text{Inside}(a, b),$
2	$\text{Line}(w_a, w, s, w, n)$ $n > s, \text{Line}(e_a, e, s, e, n)$ $e > w, \text{Line}(s_a, w, s, e, s)$ $\text{Line}(n_a, w, n, e, n)$	$\text{Rectangle}(\text{Composition}$ $(n_a, s_a, e_a, w_a),$ $n, s, e, w),$
3	$\text{Primitive}(a, x_1, y_1, x_2, y_2)$	$\text{Line}(a, x_1, y_1, x_2, y_2),$

We can express these actions as facts in predicate calculus. The general form might be

Action(*number, consequent, preconditions*),

where preconditions are lists of facts joined by "and."

Action(1, Inside(a, b),
  Rectangle(a, n_a, s_a, e_a, w_a) **and**
  Rectangle(b, n_b, s_b, e_b, w_b) **and**
  Not($a = b$) **and** $n_a =< n_b$ **and** $s_a >= s_b$ **and**
  $e_a =< e_b$ **and** $w_a >= w_b$) **and** Nil).

Action(2, Rectangle(Composition(n_a, s_a, e_a, w_a), n, s, e, w),
  Line(w_a, w, s, w, n) **and** $n > s$ **and**
  Line(e_a, e, s, e, n) **and** $e > w$ **and**
  Line(s_a, w, s, e, s) **and** Line(n_a, w, n, e, n) **and** Nil).

Action(3, Line(a, x_1, y_1, x_2, y_2),
  Primitive(a, x_1, y_1, x_2, y_2) **and** Nil).

We can define a state by a list of actions that produces that state. A statement that a fact is true in a particular state may therefore appear as

Fact(Rectangle(Composition(H, D, I, B), 4.0, 2.0, 4.5, 2.5),
   (2 and 3 and 3 and 3 and 3 and Start)),

indicating that the rectangle fact is true in the state produced by the repeated firing of rule 3 and then of rule 2. But we require a method of generalizing this type of information as meta-rules. We must establish whether any fact is true in a particular state. There are five conditions under which a fact can be said to hold true in a given state.

1. A fact holds **if** it is the Nil fact.
2. A fact holds **if** it is a fact that is true independently of any state—for example, that 12 is greater than 4.
3. A fact holds **if** there is some explicit statement in the facts base that the fact is true in that state—for example, that Primitive(G, 1.0, 1.0, 1.0, 5.0) is true in the initial state.
4. A fact holds **if** the fact is a consequent of the last action that contributed to that state.
5. A fact holds **if** it held before the last action contributing to the current state.

We can put this more formally as follows:

Holds(Nil)

Holds(a, u) $\Leftarrow a$

Holds(a, b) $\Leftarrow$ Fact(a, b)

Holds(a, b **and** u) $\Leftarrow$ Action(b, a, u)

Holds(a, b **and** c) $\Leftarrow$ Holds(a, c)

We have defined this system as *monotonic*. If a fact holds in one state, it holds in all states. We must also establish whether a state is a possible state—that is, whether actions can logically follow from one another to produce a valid state:

A state is possible **if** it is the initial state.

A state is possible **if** the sequence of actions that produced the last state is possible and the last action is consistent with the previous sequence of actions.

This can be restated as follows:

Possible(Start)

Possible(a **and** b) $\Leftarrow$ Possible(b) $\wedge$ Consistent(a, b)

This definition requires that we can ascertain whether an action is consistent with the actions that have gone before it.

An action is consistent with a list of actions **if** the action exists and the preconditions of that action all hold in the state produced by the list of actions.

Consistent$(a, b) \Leftarrow$ Action$(a, c, u) \wedge$ AllHold(c, b)

We also require a definition about whether a list of facts holds:

A single fact holds in a certain state **if** it holds in that state.

A list of facts holds in a certain state **if** the first fact holds in that state and the rest of the facts all hold.

We can restate this as follows:

AllHold$(a, b) \Leftarrow$ Holds(a, b)

AllHold$(a \text{ \bf and } b, c) \Leftarrow$ Holds$(a, c) \wedge$ AllHold(b, c)

Armed with these meta-knowledge statements, we can discover valid states and establish whether a fact is true in any given state. We can therefore establish the consistency of such statements as

$\exists x$[Holds(Inside(Rectangle(Composition(H, D, I, B), Composition(G, F, J, A)), x), Possible(x)].

The system will establish not only the truth of a goal statement, but also the state in which it is true—that is, the sequence of actions that produces that state. This process effectively simulates the backward-chaining model described earlier, with the added benefit that, as well as establishing the truth of a goal, we can derive the sequence of actions (or if-then rules) that enable us to satisfy that goal.

The major benefit of this meta-knowledge formulation, however, is that we can derive new facts in a rule-driven fashion. We can ask what the state of the facts base looks like after the firing of a particular rule or a sequence of rules. For example, in testing the consistency of the statement

$\exists x$[AllHold$(x, 2 \text{ \bf and } 1 \text{ \bf and } \text{Start})$]

with the meta-knowledge statements, we find that x is instantiated to a list of facts that are true in that state.

This entire system of meta-knowledge representation, however, is fairly cumbersome and inefficient. If rules are actions, then why not simply change the facts base as the rules are fired? The reason is that in doing so we effectively lose our ability to reason formally about facts and rules, and the idea of proof is very difficult to realize—in other words, we abandon the strictures of formal logic. We may also lose a mechanism for defining what sequence of actions produces any particular state.

We can introduce a shortcut to the preceding formal representation of actions by incorporating certain *extra-logic* statements that facilitate modifications to a facts base. The Assert function is such a predicate. The meta-knowledge for firing the actions can be expressed simply as follows:

> An action can be fired **if** there exists an action of that name, the preconditions of the action are true in the facts base, and the consequents of the action can be asserted.

> $\text{Fire}(a) \Leftarrow \text{Action}(a, b, c) \wedge \text{Test}(c) \wedge Assert(b)$

We require some means of testing whether a list of facts is true in the facts base:

> A list of facts satisfies the test **if** the first fact exists in the facts base and the rest of the facts can be tested.

> A single fact satisfies the test **if** it exists.

> $\text{Test}(a \textbf{ and } b) \Leftarrow a \wedge \text{Test}(b)$

> $\text{Test}(a) \Leftarrow a$

It should be apparent that this type of control system includes no chaining between items in the knowledge base. It could be said that rules "communicate" only with the facts base, not with each other. We can operate the system by simply presenting the statement

> $\text{Fire}(1)$

to be tested for consistency with the system. As a by-product of the "proof" procedure, we would expect the facts base to be changed. We will not concern ourselves here with the refinements that must be added to this system, such as the requirement that there be some mechanism to ensure that the same facts are not asserted repeatedly. Having established that this kind of control can indeed be described in terms of meta-knowledge, let us turn from a discussion of logic to concentrate on the operational aspects of the control of systems of actions—or *production systems*, as they are usually called.

4.4.2   The Control of Production Systems

The production-system formalism is a way of organizing a computational system into particular functional components. A major advantage of a production system is that it is simple to implement in a computer system. Also, although we are able to establish its basis in formal logic, it does not require automated theorem proving to function. It is relatively straightforward to implement production systems in both procedural and symbolic computer-

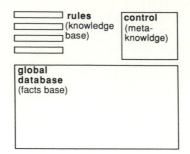

Figure 4.27   The components of a production system.

programming languages, particularly languages such as LISP and PROLOG. There is also a substantial body of theory associated with production systems, borrowed substantially from formal language theory, that we can use.

The theory of production systems is attributed to the work of Post (1943). Davis and King (1977), Waterman and Hayes-Roth (1978), and Gips and Stiny (1980) provide summaries of the underlying structures common to most production-system formalisms. These structures can be described in various ways, but usually they are seen as consisting of three basic components, which map conveniently onto the components of knowledge-based systems discussed so far (Figure 4.27).

1. The domain on which the system is to work—the "global database" or *facts base*.

2. Production rules, or actions, for transforming the domain from one state into another—the *knowledge base*.

3. A control strategy—the *meta-knowledge*.

The task of a production system is to transform the facts base from some initial state to an end state, with any number of intermediate steps along the way. The application of the actions transforms one state into another state.

The major control issue we wish to consider here concerns the selection of actions. Given any state of the facts base, several actions will be available for transforming that state. Typically there are many choices and hence many paths through the space of states. Thus it is necessary to adopt some strategy for choosing which actions to execute in order to achieve a particular state.

We can consider several simple control strategies for cycling through actions in production systems. Two kinds of choices must be made at each cycle: the choice of action and the choice of instantiations. A cycle generally includes the selection of an appropriate action, the instantiation of the preconditions of the action, and the firing of the action (Figure 4.28). There are at least two methods of implementing this kind of cycle. In the first, each action is applied exhaustively before the next action is considered (Figure 4.29).

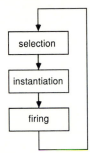

Figure 4.28   The three major elements in the control of a production system.

Method 1

1. Select first action.
2. Instantiate the preconditions of the action.
3. Fire the action.
4. Find another instantiation of the preconditions of the same action.
5. Fire the action and repeat for all possible instantiations.
6. Select the second action and continue in the same manner, firing for all possible instantiations.
7. Repeat for all actions.
8. Proceed through all the actions again and repeat until no more new instantiations can be made.

In the second method, each action is applied in turn only once. Then the operation is repeated for different instantiations of each action (Figure 4.30).

Method 2

1. Select first action.
2. Instantiate the preconditions of the action.
3. Fire the action.
4. Select the next action and instantiate its preconditions.
5. Fire the action and repeat for all actions.
6. Return to the first action and find a new set of instantiations for its preconditions.
7. Fire the action and repeat for all actions.
8. Continuously cycle through all the actions in this manner until no more instantiations can be made.

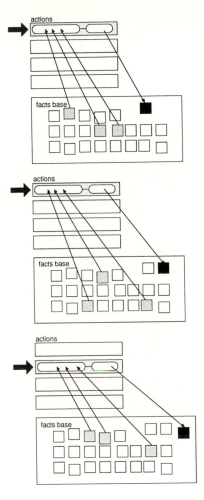

Figure 4.29   Controlling a production system by applying each action exhaustively.

Each of these methods can be demonstrated with the actions of Section 4.4.1. Considering Method 1, the initial state of the facts base is as follows:

Primitive(A, 1.0, 1.0, 5.0, 1.0)
Primitive(B, 2.5, 2.0, 4.5, 2.0)
Primitive(C, 5.5, 2.0, 7.5, 2.0)
Primitive(D, 2.5, 4.0, 4.5, 4.0)
Primitive(E, 5.5, 4.0, 7.5, 4.0)
Primitive(F, 1.0, 5.0, 5.0, 5.0)
Primitive(G, 1.0, 1.0, 1.0, 5.0)

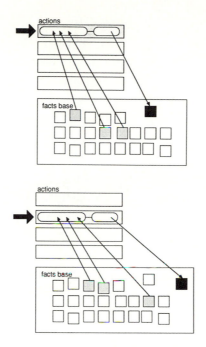

Figure 4.30   Controlling a production system by applying each action in turn once only.

Primitive(H, 2.5, 2.0, 2.5, 4.0)

Primitive(I, 4.5, 2.0, 4.5, 4.0)

Primitive(J, 5.0, 1.0, 5.0, 5.0)

Primitive(K, 5.5, 2.0, 5.5, 4.0)

Primitive(L, 7.5, 2.0, 7.5, 4.0)

Action 3 is the only one whose preconditions match the facts base. After firing action 3 for every possible instantiation of its preconditions, the following facts are added to the facts base:

Line(A, 1.0, 1.0, 5.0, 1.0)

Line(B, 2.5, 2.0, 4.5, 2.0)

Line(C, 5.5, 2.0, 7.5, 2.0)

Line(D, 2.5, 4.0, 4.5, 4.0)

Line(E, 5.5, 4.0, 7.5, 4.0)

Line(F, 1.0, 5.0, 5.0, 5.0)

Line(G, 1.0, 1.0, 1.0, 5.0)

Line(H, 2.5, 2.0, 2.5, 4.0)

Line(I, 4.5, 2.0, 4.5, 4.0)

Line(J, 5.0, 1.0, 5.0, 5.0)

Line(K, 5.5, 2.0, 5.5, 4.0)

Line(L, 7.5, 2.0, 7.5, 4.0)

Action 2 is the next action whose preconditions match the facts base. After firing action 2 for every possible instantiation of its preconditions, the following facts are added to the facts base:

Rectangle(Composition(D, B, I, H), 4.0, 2.0, 4.5, 2.5)

Rectangle(Composition(E, C, L, K), 4.0, 2.0, 7.5, 5.5)

Rectangle(Composition(F, A, J, G), 5.0, 1.0, 5.0, 1.0)

In matching the preconditions of the action against the facts base, of course, there is no direct match with the inequalities, such as $n_a =< n_b$. These must be evaluated from instantiations of n_a and n_b.

The next applicable action is action 1. After it is fired for the only possible instantiation of its preconditions, the following fact is added to the facts base:

Inside(Composition(D, B, I, H), Composition(F, A, J, G)).

In Method 1, actions are considered as operators that facilitate global changes to a facts base—a kind of "breadth-first" search strategy. Each action is implemented to generate the maximum number of facts. This is shown notionally in Figure 4.31. The initial state of the facts base is shown in Figure 4.31a as a "layer" of facts. After the exhaustive firing of action 3, a new set of facts is derived; these are shown in Figure 4.31b as rectangles, joined by arrows to the facts from which they were derived by the action. In Figure 4.31c we see the new facts derived by action 2. Eventually we get to the state shown in Figure 4.31d, produced by action 1.

With Method 2, action 3 is first fired to produce the following fact:

Line(A, 1.0, 1.0, 5.0, 1.0)

Then an attempt is made to match the other actions. No other actions are applicable, so action 3 fires again. Eventually sufficient facts about Lines are asserted for action 2 to be fired. Depending on the ordering of facts in the initial facts base, it may not be necessary to fire each action for every instantiation before other actions become applicable. This approach can be demonstrated most effectively by assuming that new facts are derived from existing facts according to the order in which they appear in the facts base. The initial

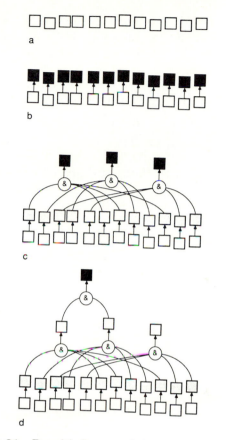

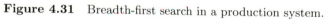

Figure 4.31   Breadth-first search in a production system.

facts can be reordered as follows:

Primitive(D, 2.5, 4.0, 4.5, 4.0)

Primitive(B, 2.5, 2.0, 4.5, 2.0)

Primitive(I, 4.5, 2.0, 4.5, 4.0)

Primitive(H, 2.5, 2.0, 2.5, 4.0)

Primitive(F, 1.0, 5.0, 5.0, 5.0)

Primitive(A, 1.0, 1.0, 5.0, 1.0)

Primitive(J, 5.0, 1.0, 5.0, 5.0)

Primitive(G, 1.0, 1.0, 1.0, 5.0)

Primitive(E, 5.5, 4.0, 7.5, 4.0)

Primitive(C, 5.5, 2.0, 7.5, 2.0)

Primitive(L, 7.5, 2.0, 7.5, 4.0)

Primitive(K, 5.5, 2.0, 5.5, 4.0)

The only action applicable, action 3, can be fired four times before another action becomes applicable. The repeated firing of action 3 will result in the addition of the facts

Line(D, 2.5, 4.0, 4.5, 4.0)
Line (B, 2.5, 2.0, 4.5, 2.0)
Line (I, 4.5, 2.0, 4.5, 4.0)
Line (H, 2.5, 2.0, 2.5, 4.0)

followed by the firing of action 2 to produce the fact

Rectangle(Composition(D, B, I, H), 4.0, 2.0, 4.5, 2.5).

Subsequent firings of action 3 produce the new facts

Line(F, 1.0, 5.0, 5.0, 5.0)
Line (A, 1.0, 1.0, 5.0, 1.0)
Line (J, 5.0, 1.0, 5.0, 5.0)
Line (G, 1.0, 1.0, 1.0, 5.0).

Action 2 produces the new fact

Rectangle(Composition(F, A, J, G), 5.0, 1.0, 5.0, 1.0).

Action 1 then produces the fact

Inside(Composition(D, B, I, H), Composition(F, A, J, G)).

It is interesting that in Method 2 it is not necessary to make inferences from every fact in the initial facts base before some useful new facts are derived, which suggests a "depth-first" search strategy. This approach, illustrated in Figure 4.32, is akin to focusing attention on a particular aspect of a design and deriving everything possible from that aspect. In this example the focusing depends entirely on the order in which facts are stored in the facts base. In Section 4.4.4 we consider how focusing may be accomplished by other control structures.

The disadvantage of this formulation of production systems, as in forward chaining, is that facts that are of little or no interest may be asserted. An analogy is making inferences from everything we know, which tends to be an inefficient way to reason. If we are interested in discovering only certain facts about a design, then we must introduce further sophistication into the control of actions. For example, we may assert facts temporarily, then retract them when they turn out to be irrelevant to the establishment of the goal facts. This suggests that the system might be formulated to undertake exhaustive search.

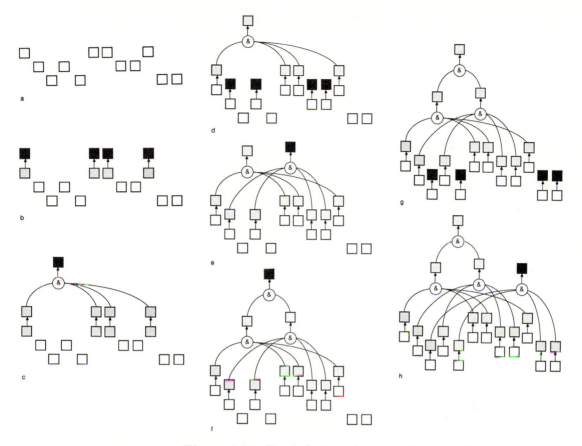

Figure 4.32 Depth-first search in a production system.

The idea of exhaustive search detracts from the simplicity of production systems because it is necessary to be able to "undo" the effect of actions when they lead to undesirable states. In this type of system, the effective selection of actions assumes greater importance than in rule-chaining approaches because of the computational overhead involved in manipulating facts bases and keeping track of actions. We consider therefore two methods by which the control of production systems can be enhanced: the idea of scheduling and focus of attention.

4.4.3 Scheduling

At any moment in the state of a facts base, several actions may be applicable for firing. Each action may also be applicable to different facts in the facts base. We can call each eligible combination of action firing and instantiation an *event*. A *schedule* is an ordered list of such events in which the event at the top of the list is to be activated first. Events are fired according to their place on the schedule.

In a scheduling system, some determination must first be made as to which events are eligible to be placed on the schedule. This can be decided by the process of triggering. An action is *triggered* when its preconditions can be matched against the facts base; the effects of the action are stored as a temporary record. The triggering of all rules for all possible instantiations will result in a set of records that can be compared. The events responsible for the records considered most desirable in some sense gain a higher place on the schedule. There must be some knowledge by which selections can be made between records and by which events can be ordered. This is often done on the basis of heuristics. Events are then fired according to their place on the schedule.

Of course, the schedule is dynamic. With each firing, the facts base changes and other rules may become eligible. An action may also cease to be eligible once another event ahead of it in the schedule has been activated. These possibilities imply that the schedule must undergo frequent revision.

What is the basis of selection between competing events? The scheduling approach assumes that knowledge is available by which choices between competing actions can be made. This suggests that the criteria for selecting between events should embody some idea of goals and the characteristics of states that are intermediate to the production of particular goal states. Unfortunately, the criteria for evaluating intermediate states are generally different from those used for evaluating goal states. The process becomes feasible, however, if the schedule does not have to be "right." If an "incorrect" path is selected through possible states of the facts base, then the only sacrifice is efficiency. The goal state will eventually be reached. The task of the scheduler is to direct the search.

Some simple scheduling "rules" might be formulated as follows.

1. Place events that produce the largest number of new facts higher up in the schedule.

2. Place events that produce the smallest number of new facts from the largest number of existing facts higher up in the schedule.

3. Place events that are implemented by the most recently fired action higher up in the schedule.

4. Place events that are implemented by the least recently fired rule higher up in the schedule.

5. Place events affecting the most recently derived facts higher up in the schedule.

6. Place events affecting the least recently derived facts higher up in the schedule.

Because these rules represent rival strategies, a scheduler will probably incorporate only some of them. When several scheduling operators are applicable, they may be designed to assign weighting factors to events on the schedule. Events may be placed on the schedule according to cumulative scores.

These scheduling rules imply different search strategies. For the domain discussed here, the use of scheduling rule 5 would tend to produce a depth-first approach to the derivation of new facts. The use of scheduling rule 6 would tend to produce a breadth-first search. The scheduling approach is powerful because it suggests that the search process can be changed by adopting different strategies. There may be some other control device working above the scheduler that brings different rules into play to change the search strategy— for example, if the search is taking too long or there is some kind of logical impasse. The scheduling idea is explored again in Chapter 7 in the context of the control of design generation.

OPS5 (Brownston et al., 1985) is the archetypal tool employing the forward chaining concept to create expert systems. It has been used, for example, to create the XCON expert system for configuring VAX computer systems (McDermott, 1982). OPS5 is one of the family of OPS languages in which the elements are production rules, working memory elements, and actions. The inference engine operates on the recognize-select-act cycle using three algorithms for match rules, select rules, and execute rules. Match rules match the condition elements on the left-hand side of the rules against working memory, select rules choose one dominant rule, and execute rules fire the selected rule by executing the actions of the right-hand side sequentially. For OPS5 to be implemented, there must exist at least one rule whose left-hand side can be matched—that is, whose conditions can be instantiated to existing working memory elements. As rules are fired, existing working memory elements and/or rules may be deleted, and new working memory elements and/or rules added. OPS5 proceeds until no more rules can be matched.

If the match step produces a set containing more than one rule, then one rule must be selected to be fired. OPS5 has two possible selection strategies available, both based on the recency of time tags attached to the working memory element of the left-hand side of the rules, the most "recent" rule being selected.

4.4.4 Focus of Attention

Scheduling suggests that it is possible to control where certain actions direct their attention—that is, on which parts of the facts base they operate to produce new facts. This idea can be exploited explicitly by partitioning the facts base in certain ways and by using heuristics to decide the part of the facts base to which actions should be applied to produce new facts.

At least two methods are available to partition the facts base describing the simple world of rectangles discussed earlier. One is to divide facts up

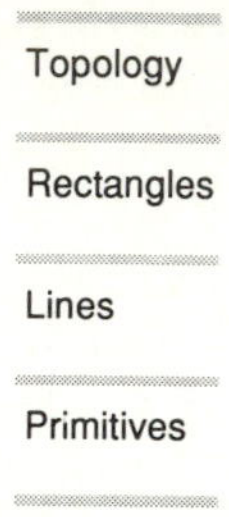

Topology

Rectangles

Lines

Primitives

Figure 4.33   Facts partitioned into a derivational hierarchy.

according to a derivational hierarchy (Figure 4.33). Thus, in our example, facts about Primitives would be at one level in the hierarchy, facts about Lines at another, facts about Rectangles above that, and facts about topology such as Inside at another level. Another way of partitioning facts is spatially (Figure 4.34). We could divide the coordinate space into zones, perhaps based on a spatial grid, and tag facts according to the zones to which they relate. We could adopt several overlapping systems of partitioning. We might then employ control rules that permit only the triggering and firing of actions applicable to certain partitions of the facts base. Examples of control rules applicable to these systems of partitioning are as follows.

1. Fire actions applicable to facts in the highest level in the derivational hierarchy.

2. Fire actions applicable to facts in the lowest level in the derivational hierarchy.

3. Fire actions applicable to the facts relating to the center-most zones of the spatial partitioning system.

Different search strategies can be modeled based on these scheduling rules. Control rule 1 simulates the depth-first search approach, while rule 2 produces the behavior of breadth-first search. Rule 3 causes the system to focus its

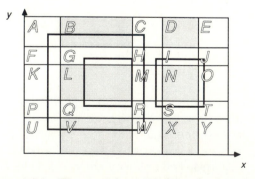

Figure 4.34   The spatial partitioning of a facts base.

attention on deriving facts pertinent to the central area of the layout before producing facts about the outer zones of the layout. There would appear to be little advantage in adopting such an approach for this simple example, but this heuristic might be useful for large facts bases—for example, when some interpretation is required of part of a design described in a conventional CAD system, such as determining the properties of a particular beam in a structural layout. In investigating what other members impinge on that particular beam, we could adopt a strategy that focuses on the elements in close proximity to that beam—sharing the same spatial zone. Only after taking account of neighboring structures will we consider elements in zones further afield. By the judicious use of heuristics for focusing attention, we can introduce powerful efficiencies into the process of deriving new facts from existing facts about a design. Further applications are discussed in Chapter 7.

4.4.5 Reasoning About Actions

Where several actions are competing to be implemented, knowledge about actions can be brought to bear in their selection. This assumes that something is known about the applicability of each action in producing a particular state in the facts base or that this knowledge can be derived from the actions. An example of a control rule about actions pertinent to the problem domain discussed earlier is as follows:

> **If**   you are interested in determining facts about area
> **then**   consider actions about dimensions before actions about orientation.

Such rules can be incorporated into an interpretation system by partitioning, or organizing, the actions in some way. Because it may then be necessary to include new knowledge into a framework, this works against the objectives of modularization and the independence of knowledge statements. Another strategy is to provide some facility by which certain features of actions can be identified, allowing the system to detect which actions are about dimensions and which are about orientation. A still more interesting approach is to have rules embodying some more general principle, such as

> Implement specific actions before general actions.

The system might use such domain-independent knowledge as a basis of selection between competing actions. The translation of general rules to affect specific actions, however, poses problems.

Combinations of these control strategies—scheduling, focus of attention, and reasoning about actions—may also be applicable in certain cases. In Chapter 7 we consider such combinations, incorporated into blackboard systems.

4.5   Expert Systems

Many of the control issues we have been considering can be embedded in computer programs called *expert systems*. As defined in Section 1.3.6, an expert system is a computer program dealing with knowledge in a specialized domain requiring expertise to provide solutions and give advice. Expertise is knowledge acquired over time and with experience, and is usually heuristic in nature. An expert system represents an attempt to emulate the performance of experts, without necessarily modeling their reasoning processes.

The main components of an expert system can be identified as an interface facility, a facts base, a knowledge base, an inference mechanism, an explanation facility, and a knowledge-acquisition facility (Figure 4.35). Not all existing expert systems have all of these facilities. The knowledge-acquisition facility enables knowledge about a domain to be incorporated, often in the form of rules. Additionally, an expert system may exhibit the capability to learn from examples of representative problems, but this process is too complex to warrant discussion here. The explanation facility, one of the main attributes of an expert system, enables the expert system to explain or justify its conclusions.

Expert system *shells* are expert systems with empty facts bases and empty knowledge bases (Figure 4.36). Ideally, by taking care of the control and support of information, an expert system shell serves as a domain-independent tool enabling the experts, or "knowledge engineers," to concentrate solely on providing and organizing the knowledge itself. In theory an expert system shell can be used with knowledge from widely different domains—that is, with different sources of knowledge. In practice the way a shell is structured has some

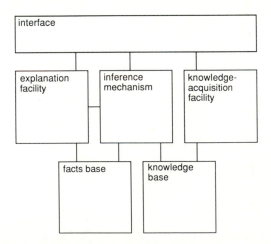

Figure 4.35   The components of an expert system.

Expert system shell

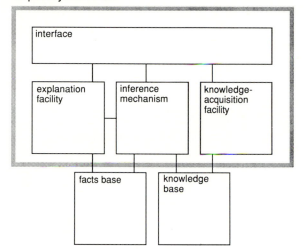

Figure 4.36   The components of an expert system shell.

bearing on the suitability of a particular shell to a particular domain. A major advantage of separating out knowledge pertaining to control (that is, the shell) from domain knowledge is that a high degree of human interaction can be sustained. A user of the system can respond to and make demands of the knowledge in various ways. This process includes employing the knowledge base to explain its own reasoning as well as to answer queries.

Existing expert systems and shells employ a variety of systems of knowledge representation, such as rules, frames, semantic networks, object-oriented programs (Section 4.7), or combinations of these (hybrid systems). These systems and shells also provide a variety of control mechanisms. They have been implemented to carry out a wide range of tasks. Comprehensive reviews of expert systems and shells have been made by Hayes-Roth, Waterman, and Lenat (1983), Buchanan and Shortliffe (1984), Allwood and Stewart (1984), Rychener (1985), and Waterman (1986). In addition, expert systems (shells and knowledge) of interest to designers include HI-RISE (Maher and Fenves, 1985), SICAD (Lopez, Elam, and Christopherson, 1985), EDISON (Dyer, Flowers, and Hodges 1986), DOMINIC (Howe et al., 1986), PROMPT (Murthy and Addanki, 1987), XCON (McDermott, 1982), RETWALL (Hutchinson, Rosenman, and Gero, 1987), PREDIKT (Oxman and Gero, 1987), and OPTIMA (Balachandran and Gero, 1987b). These systems carry out a wide variety of processes, including diagnosis, classification, interpretation, planning, selection, and configuration.

In the following sections we discuss expert system reasoning in terms of question answering and reasoning about objects, attributes, values, and classes.

```
┌─────────────────────────────────────────────────┐
│                                                 │
│   3.7.2 SUNLIGHT                                │
│                                                 │
│   PRINCIPAL OBJECTIVES                          │
│                                                 │
│   Buildings should be sited to ensure that      │
│   unobstructed sunlight is received on a        │
│   reasonable portion of the allotment, or the   │
│   face of the building, throughout the year.    │
│                                                 │
│   GENERAL REQUIREMENTS TO SATISFY OBJECTIVES    │
│                                                 │
│   The proposed layout of lots and the location  │
│   and form of dwelling units should be such     │
│   that:                                         │
│   — sufficient winter sunshine is received in   │
│     each dwelling;                              │
│   — there is adequate shading to protect the    │
│     dwelling from excessive summer sun,         │
│     particularly from north-west to west;       │
│   — no structure unduly restricts sunshine      │
│     available to an adjoining dwelling.         │
│                                                 │
└─────────────────────────────────────────────────┘
```

Figure 4.37   A clause from the Model Cluster Code (Cluster Titles Committee, Victoria, Australia, 1979).

4.5.1   Expert System Reasoning

Here we look in more detail at the reasoning processes carried out by expert systems. As an example, we take a very simple system concerned with building codes. Figure 4.37 shows a clause from a provisional design code in Australia in the context of building design known as the Model Cluster Code (Cluster Titles Committee, 1979). The intention of this code is to evaluate certain types of residential development. It can be seen that such a clause lends itself to the if-then construct of production rules described in Chapter 3.

Statements of this kind can be represented in an inference network, where the consequents of one rule may become the antecedents of another. The inference network for a section of the code appears in Figure 4.38. The hierarchical nature of the system is immediately apparent from the diagram. The knowledge can be processed in several ways, including forward and backward chaining. For a realistic set of inference rules, the number of facts that can be derived is likely to be very large. There is therefore more work involved in asking a design, "What attributes can be inferred from its description?" than from asking, "Does the design exhibit this particular performance?" The former question suggests a forward-chaining approach; from a design description, an attempt is made to infer as much as the rules will allow. The latter question implies a goal-directed, or backward-chaining, approach that begins with various statements about the design's intended performance and tries to discover whether the artifact possesses those attributes. Let us explore the goal-directed approach here.

The issue of control is particularly important when all the facts from which inferences are to be made are not represented, either explicitly or implicitly.

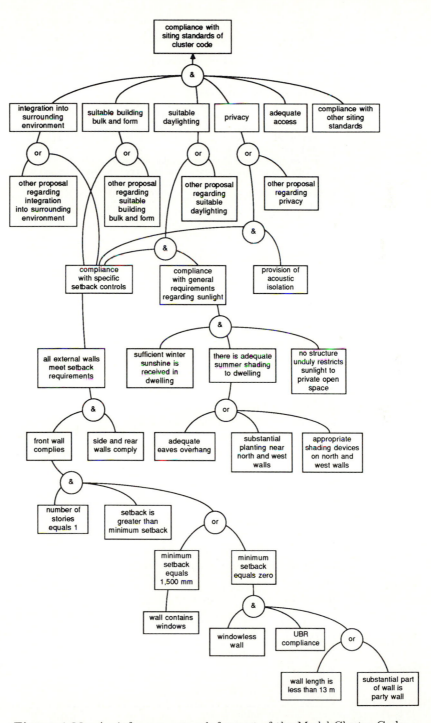

Figure 4.38 An inference network for part of the Model Cluster Code.

```
>> prove: compliance with siting standards of cluster code.

Enter value of number_of_storeys ... 1.
wall contains windows? (yes or no) ... no.
windowless wall (yes or no) ... yes.
Enter value of wall_length ... 9.
ubr compliance? (yes or no) ... yes.
Enter value of setback ... 1200.
side and rear walls comply? (yes or no) ... yes.
sufficient winter sunshine is received in dwelling? (yes or no) ... yes.
adequate eaves overhang? (yes or no) ... no.
substantial planting near north and west walls? (yes or no) ... yes.
no structure unduly restricts sunlight to private open spaces? (yes or no) ...
yes.
provision of acoustic isolation? (yes or no) ... yes.
adequate access? (yes or no) ... yes.
compliance with other siting standards? (yes or no) ... yes.
compliance with siting standards of cluster code
*TRUE*

>> status facts.

The following facts are true ...

    compliance with siting standards of cluster code
    compliance with other siting standards
    adequate access
    privacy
    provision of acoustic isolation
    suitable daylighting
    compliance with general requirements regarding sunlight
    no structure unduly restricts sunlight to private open space
    there is adequate summer shading to dwelling
    substantial planting near north and west walls
    sufficient winter sunshine is received in dwelling
    suitable building bulk and form
    integration into surrounding environment
    compliance with specific setback controls
    all external walls meet setback requirements
    side and rear walls comply
    front wall complies
    setback > min_setback
    min_setback equals 0
    ubr compliance
    wall_length < 13
    windowless wall
    number_of_storeys = 1

The following facts are not true ...

    adequate eaves overhang
    wall contains windows

>> status values.

The following assignments have been made ...
    setback is 1200
    min_setback is 0
    wall_length is 9
    number_of_storeys is 1
```

Figure 4.39 A dialogue with an expert system based on the rules of Figure 4.38.

In an interpretation system we would expect a proposition to be untrue if any or all the preconditions of that proposition were not true. A proposition is not true if there is an explicit statement that a fact is not true or if there are no explicit or inferable statements existing in the facts base stating that the fact is true. A human expert is unconstrained by this consideration. If a particular fact is not known to be true, there are usually useful ways to determine the validity of statements. Perhaps the simplest mechanism available to a human expert, and the one we can model most easily, is to ask questions.

Even when there are no statements constituting a building description, a "partial logic system" is of value. The "leaf nodes" of the tree of Figure 4.39, which correspond to requests for facts about the design, can be handled interactively by means of prompts to the user. These nodes can also be regarded as entry points to other interpretation systems that interpret computer databases, or to other building-code systems. When incorporated into a general-purpose inference system, a dialogue such as the one illustrated in Figure 4.39 is produced.

Although this is a very simple example, it demonstrates some of the features of an expert system. Such a system is intended to be consultative; an ideal system communicates by natural language. Communication also concerns the issue of deciding at which level in the interpretation hierarchy the consultation should take place. If the system is to communicate with a novice, someone who knows little about the domain of expertise represented by the system, then there may be justification in making the leaf nodes the propositions about which questions are asked.

For experts, however, questions about propositions higher up in the network may eliminate unnecessary search. A sample session with the system is presented in Figure 4.40, in which the response "how" provides an instruction to the inference system that it is to proceed to a lower level of inquiry. A simple explanation facility is also provided by use of the "why" response, which causes the lines of reasoning so far employed to be traced, thus explaining why a particular question was asked.

Let us now consider the meta-knowledge for generating requests from an expert system. We consider an even simpler example than the last one: two rules for interpreting information about shape in order to determine the category of a particular structural engineering component. The rules are not intended to be universal, of course, but rather are pertinent to a limited domain (Figures 4.41–4.43).

If   shape is cylinder **and**
       extends floor to ceiling
then column.

If   plan is circle **and**
       elevation is rectangle
then shape is cylinder.

```
>> prove: compliance with general requirements regarding sunlight.

sufficient winter sunshine is received in dwelling? (yes, no, how or why) ...
yes.
there is adequate summer shading to dwelling? (yes, no, how or why) ... why.

        i.e. why do you want to know about -
             there is adequate summer shading to dwelling?

BECAUSE

compliance with general requirements regarding sunlight - is proved by ...

        sufficient winter sunshine is received in dwelling
        there is adequate summer shading to dwelling
        no structure unduly restricts sunlight to private open space

there is adequate summer shading to dwelling? (yes, no, how or why) ... how.

        i.e. How do I prove -
             there is adequate summer shading to dwelling?

there is adequate summer shading to dwelling - is proved by ...
        adequate eaves overhang
        OR
        substantial planting near north and west walls
        OR
        appropriate shading devices on north and west walls

adequate eaves overhang (yes, no, how or why)? ... yes.

there is adequate summer shading to dwelling
*TRUE*

no structure unduly restricts sunlight to private open space? (yes, no, how or
why) ... no.

compliance with general requirements regarding sunlight
not true
```

Figure 4.40   A dialogue exploiting different levels in the inference network of Figure 4.38.

We assume that rules have only a single consequent. If the propositions connected by **and**s are true, then the consequent is true. The logic of a simple interpreter to handle these if-then rules can be expressed as follows (Figure 4.44):

$$\text{Prove}(a) \Leftarrow (\textbf{if } b \textbf{ then } a) \wedge \text{Match}(b)$$
$$\text{Prove}(a) \Leftarrow Write(a) \wedge Write(?) \wedge Read(\text{Yes})$$

This logic states that the proposition a can be inferred to be true if there is a rule with a as its consequent and the antecedents of the rule can be matched. If there is no rule with a as its consequent, then prompt for verification. We assume here that $Write$ and $Read$ are extra-logical functions that facilitate

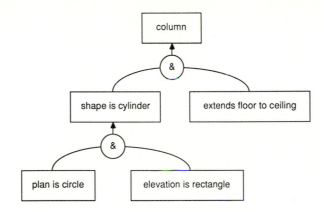

Figure 4.41 An inference network for determining whether an object is a column.

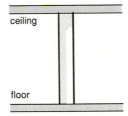

Figure 4.42 A column extends from floor to ceiling.

communication with the outside world. The prompt for a proposition is successful if it can be written out followed by a question mark and the response is the term Yes. We also assume that a theorem prover to process these statements would attempt to prove theorems by inspecting these statements in the order written here and by proving the preconditions of a rule in a left-to-right order.

The knowledge for determining whether a list of propositions matches is

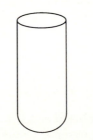

Figure 4.43 The shape is cylindrical.

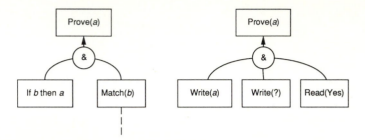

Figure 4.44   The logic of a simple expert system shell.

similar to the definition given in Section 4.3.1:

$$\text{Match}(a \textbf{ and } b) \Leftarrow Prove(a) \wedge \text{Match}(b)$$
$$\text{Match}(a) \Leftarrow \text{Prove}(a)$$

A list of preconditions is Matched if the first element of the list can be proven and the rest of the list can be matched, or if there is only one precondition and it can be proven.

In attempting to prove the following theorem,

$$\text{Prove}(\text{Column}),$$

we would expect the following dialogue, where the response of a human operator is in bold type:

Plan is circle? **Yes**.

Elevation is rectangle? **Yes**.

Extends floor to ceiling? **Yes**.

As a consequence the theorem, Prove(Column), would be proven true. The formulation requires further enhancements to handle a No response elegantly. Rules can also contain variables, but further mechanisms are needed to handle prompts for values. The way in which the tree of propositions formed by the if-then rules is searched depends on the theorem prover. If we assume a theorem prover that uses linear resolution, then the preconditions of the interpretive rule are processed from left to right. It is also possible to introduce various control devices for directing attention down the inference tree, such as changing the order in which propositions are handled in the Match definition. We could therefore replace the first logical statement with the following,

$$\text{Prove}(a) \Leftarrow (\textbf{if } b \textbf{ then } a) \wedge \text{Reorder}(b, c) \wedge \text{Match}(c),$$

where Reorder is a function that reorders the preconditions of the interpretive rule according to some further control knowledge (Figure 4.45). Another point about the logic of this rule interpreter is that, assuming the theorem prover

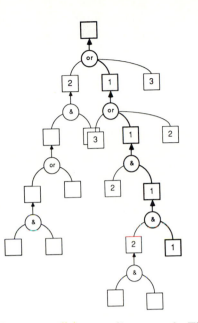

Figure 4.45   Reordering preconditions to direct search. The first path is shown.

works from the top down and from left to right, questions are asked only at the branch ends of the tree. Further control devices are needed to facilitate requests for information at intermediate nodes in the decision tree. In this example, swapping the order of the first two statements in the interpreter would ensure that the system asks the user to verify each proposition before attempting to prove it by chaining through the rules:

$$\text{Prove}(a) \Leftarrow Write(a) \wedge Write(?) \wedge Read(\text{Yes})$$
$$\text{Prove}(a) \Leftarrow (\textbf{if } b \textbf{ then } a) \wedge \text{Match}(b)$$

These and other issues have been further explored by Clark and McCabe (1982), Hammond (1982), Hammond and Sergot (1983), Mizoguchi (1983), and Rosenman and Gero (1985), all in the context of working expert systems in PROLOG.

4.5.2   Reasoning about Objects, Attributes, Values, and Classes

Ideally, an expert system should be able to reason from statements that resemble sentences in natural language, or even in graphical languages. This requires facilities in natural-language understanding that are beyond the capabilities of most expert system shells. Because of the formality and uniformity of rules, it is simpler to limit ourselves to them. Even with this limitation, an expert system shell should be able to interpret rules and employ some understanding of their meaning to conduct its reasoning. This can be achieved quite simply if the propositions within rules can be structured in some way. Let us investigate this structuring with reference to some design-code knowledge.

To illustrate, we consider a prototypical expert system based on the Australian Model Uniform Building Code (1982), which is typical of many design codes in building, engineering, and other fields. The operations of the system are based on the control structure described in Section 4.5.1. One part of the code deals with room sizes and heights. Four rules typical of the knowledge in the code are given here.

> **If**   room **is a** bedroom **or** living room **or** kitchen **or**
> dining room **or** sunroom
> **then**   type **of** room **is** habitable room.

> **If**   building **is a** dwelling house **or** flat **and**
> type **of** room **is** habitable room **and**
> **not**(room **is a** kitchen) **and**
> number **of** habitable rooms contained **is equal to** 1 **and**
> **not**(increase in floor area of habitable room requirement)
> **then**   required minimum floor area **of** room **is** 18.5.

> **If**   building classification **is** classI **or** classII **or** classIII **or** classIV **and**
> room **is a** bathroom **and**
> the bathroom is provided with a separate shower **and**
> **not**(increase in bathroom floor area requirement)
> **then**   required minimum floor area **of** room **is** 2.8.

> **If**   building classification **is** classI **or** classII **or** classIII **or** classIV **and**
> type **of** room **is** habitable room
> **then**   required minimum height **of** room **is** 2,400.

The form of these rules includes features in addition to the simple if-then structure used in Section 4.5.1. As well as accommodating the representation of disjunctions (**or**) and negation (**not**), the rules facilitate the structuring of information in terms of objects, attributes, values, and classes (or types). One of the general forms for such propositions is as follows:

> Attribute **of** Object **is** Value

The words **of** and **is** are effectively operators. We can introduce a new operator **is a** to designate the type of an object:

> Object **is a** Type

Or we can make statements about classes by considering Type as a particular attribute:

Type **of** Object **is** Value

There are also operators that can be evaluated, such as Boolean operators:

$Value_1$ **is equal to** $Value_2$

In this scheme, Object can be any atom and Value and Type can be atoms or numbers. The advantage of this representation is that it enables the expert system shell to reason about objects, attributes, values, and types. There are four ways in which we can take advantage of this structuring.

1. The structuring can help the expert system shell in the way it directs questions to the user or interrogates a facts base. As an illustration, consider the second rule just given. In attempting to prove the proposition "room type **is** habitable room," the system can ask,

   What is type of room?

   or

   Type of room is (please supply value).

   This form of questioning is quite straightforward to implement when there is an explicit structure to the proposition that can be easily parsed. This is a much more natural way of communication than the question

   Type of room is habitable room (answer yes or no).

2. With this structuring of propositions, the expert system shell can take the "value" supplied by the user—in this case habitable room—and store it for later use. If the system must ask about the room type further along in its reasoning (for example, to prove the proposition "type of room is storage") then the query has already been answered.

3. The shell can conduct some reasoning on the knowledge base before using it to investigate designs. It can look at the rules and extract information about objects, attributes, values, and types to construct lists that will be useful in a dialogue with the human user. The expert system shell effectively parses the rules to discover candidate attributes, values, and types. This enables questioning such as the following:

   The type of room is?
   (possible values are habitable room and non-habitable room).

4. We can present inquiries to an expert system in terms of requests for information about objects, attributes, values, and types. Thus we can tell the expert system to

   Find required minimum floor area of room.

This will involve the expert system in identifying all rules with the object attribute required—minimum floor area of room—as a consequent and in discovering the value of the attribute by reasoning through the rules.

There are many ways in which propositions within rules can be structured, only one of which was presented here. A wide range is demonstrated among the various commercial expert system shells available (Waterman, 1986). Various structures have been proposed for representing information in a uniform manner. One notable approach is Sowa's (1984) conceptual structures.

4.6   Reasoning about Objects

Much of our previous discussion about design makes use of compositional hierarchies. Design elements in computer systems constitute objects. Certain interpretive knowledge is concerned with objects and their properties; generative knowledge can also be represented in terms of actions that operate on objects. In this section we explore the idea that certain design knowledge can be organized in such a way that we focus on the representation of objects, their properties, and the knowledge associated with them. This *object-centered* approach to knowledge representation contrasts with the action-centered approach underlying most of the discussion in the preceding sections.

The object-centered approach can be demonstrated in the way we might organize knowledge in an expert system. A further example of expert system rules belonging to the domain discussed in Section 4.5.2 are as follows:

R1 **If**   room **is a** bathroom **and**
          not (increase in area requirement)
  then   required minimum floor area in square meters **is** 2.2.

R2 **If**   room **is a** bathroom **and**
          increase in area requirement
  then   required minimum floor area in square meters **is** 2.8.

R3 **If**   bathroom contains a separate shower
  then   increase in area requirement.

We can identify "room" as an object. Although they are not physical design elements in the same way as "room", we could also identify "increase in area requirement" and "required minimum floor area in square meters" as objects. We can itemize these objects as the first argument of an Object predicate. The second argument is a list of rules that contains the object as antecedents. The third argument is a list of rules that contain the object as a consequent.

We can then extract the following set of objects with their list of rules as follows:

Object(room, [R1, R2], [ ])

Object(increase in area requirement, [R1, R2], [R3])

Object(required minimum floor area in square meters, [ ], [R1, R2])

Object(bathroom contains a separate shower, [R3], [ ])

This representation facilitates reasoning about objects. Thus, if we require information about "required minimum floor area in square meters," the knowledge associated with determining this object's properties is found in rules R1 and R2. This is also useful for data-driven systems. Given the fact "room is a bathroom," we can find that in order to infer some new facts we must use rules R1 and R2. We can also represent facts that contain lists of possible attributes and values of objects. There are operational advantages in this object-centered approach in terms of speed and the way queries are presented to a user of the system.

We can extend this idea more formally in object-oriented systems (Stefik and Bobrow, 1986). Such systems organize information so that objects have associated with them knowledge about how they behave. Objects can "respond" to messages. They may also "transmit" messages. For example, if we imagine a robot reigning over a world of toy building blocks, we may command the robot to pick up blocks, move them, and replace them. We pass instructions to the robot in terms of actions to be performed on objects (Figure 4.46). This is an action-oriented system. In an object-oriented system, however, we pass messages to the blocks themselves. Thus we may tell block A to move on top of block C. Associated with block A is the knowledge that before it can move, it must be clear of obstacles. If block B is sitting on block A, then A will transmit a message that the object on top of it is to move somewhere else. Block B will respond accordingly. This knowledge associated with objects is often termed a *method*.

Objects are organized into classes according to their properties, simplifying the association of knowledge with objects. For example, if we know that A, B, and C are blocks, then each can inherit the methods and message-passing capabilities of a generally defined blocks class. This idea is most useful in domains with wide disparities among object types. Thus we may define blocks, wheels, and pulleys as object types, with different properties, and define certain elements as instances of these object types as the design develops.

SMALLTALK (Goldberg and Kay, 1976) and LOOPS (Bobrow and Stefik, 1981) are object-oriented programming languages in which objects and classes of objects can be defined and manipulated in this way. The most visible application of object-oriented techniques is in operating-system design, in which files are typed and treated as objects with associated methods. A user communicates with the operating system by sending messages to individual files that respond according to their associated methods. When combined with

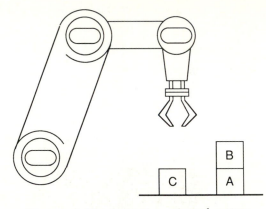

Figure 4.46   An action-oriented system.

graphics, as in the Apple Macintosh operating system, such systems can be both very powerful and very simple to use.

In an object-oriented system, the objects seem to "know" the behavior expected of them. This idea has advantages in the area of simulation (Stefik and Bobrow, 1986). The idea of objects behaving according to the knowledge associated with them is also useful in a design domain. For example, it is sometimes helpful to assume that a plug "knows" that it cannot go into a power outlet upside down (Figure 4.47), or that a roof "knows" what will happen to it in the event of gale-force winds. The message-passing analogy can also be helpful. For example, structural components in a building "pass messages" to each other in the form of loads (Figure 4.48).

The major advantage of the object-oriented approach is that it absolves a controller from having to contain knowledge about all there is to know about objects. The knowledge about behavior resides with the objects, simplifying the control of the system, as well as the task of adding new elements to a design system in a modular fashion.

The object-oriented approach is adopted in many knowledge-based systems. The control of an object-oriented system can be represented in standard

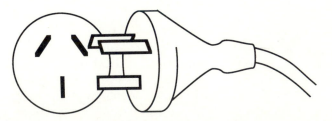

Figure 4.47   A power plug is constrained to behave in a particular way in relation to a power outlet.

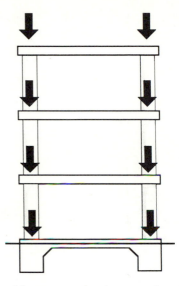

Figure 4.48   Message passing in an engineering structure.

logic. We refer here to Stabler (1986), who describes a logic programming approach to object-oriented programming. It is not necessary to follow predicate calculus in defining object-oriented systems, but for the sake of uniformity with the other systems described in this book let us resort to logic to describe the basic idea.

We can describe objects as facts. The general definition of an object would then be

Object(*name, method*),

where *name* is the identifier of the object and method is some knowledge attached to the object for computing something. The *name* may also incorporate arguments. These arguments can be the *variables* of the system. We want to be able to communicate with an object-oriented system by sending messages in the following general form:

Send(*name, message*).

An example of a simple object is a regular polygon with number of sides n and length of sides l.

Object(RegularPolygon(n, l), **if** p **is** $n \times l$ **then** Perimeter(p))

This object is defined by using a rule for determining the perimeter length of the polygon (Figure 4.49). We wish to be able to send messages to such an object. The control of a simple message-passing system can be represented

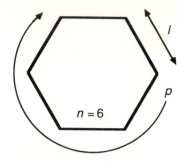

Figure 4.49   A regular polygon as an object.

with the following statement:

$$\text{Send}(object, message) \Leftarrow \text{Object}(object, method)$$
$$\wedge\ \text{CheckRule}(message, method)$$

This says that a message can be sent to an object if the object has a method and that method passes the CheckRule test. CheckRule is true if its second argument is a rule whose precondition is true.

$$\text{CheckRule}(b, \textbf{if } a \textbf{ then } b) \Leftarrow a$$

We can present such a system with a goal:

$$\exists p[\text{Send}(\text{RegularPolygon}(4, 5), \text{Perimeter}(p))]$$

This message effectively "asks" for the perimeter of a four-sided regular polygon with sides of length 5. This message prompts the method for computing the perimeter to fire, instantiating the variable p to 20:

$$p = 20$$

There may be several methods, some of them facts or procedures. In this case, the second argument of Object would have to be a list, and the control of the system would be slightly more complex than just indicated.

We also require some ways to represent inheritance between objects and classes of objects. This can be demonstrated by considering a square as a subclass of objects that are regular polygons. The following fact denotes a square object. In this example the method is simply a description, not a rule:

$$\text{Object}(\text{Square}(length\_of\_side),$$
$$\text{Description}(\text{"four-sided-polygon with equal sides"}))$$

We need relations that specify the lines of inheritance. We can state these relations as simple facts linking classes and objects. For example, a square is a regular polygon with four sides, and a pentagon is a regular polygon with

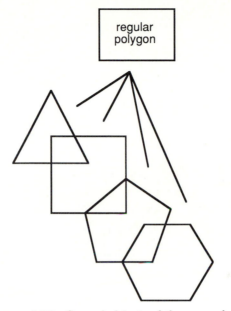

Figure 4.50   Several objects of the same class.

five sides (Figure 4.50):

IsA(Square(l), RegularPolygon(4, l))

IsA(Pentagon(l), RegularPolygon(5, l))

We must rewrite the message-processing control knowledge given earlier to facilitate chaining through the more general methods associated with the object, allowing us to see whether one of the parent methods applies. The message

$\exists l \exists x [\text{Send}(\text{Square}(l), \text{Description}(x))]$

returns a value of

$x =$ "four-sided polygon with equal sides"

The goal

$\exists p [\text{Send}(\text{Square}(5), \text{Perimeter}(p))]$

produces a value of

$p = 20$

The advantage of this approach is that new objects can fit into a hierarchy of objects already defined; they will inherit the methods of their parents except for those messages for which they have their own special methods that override those of the parents.

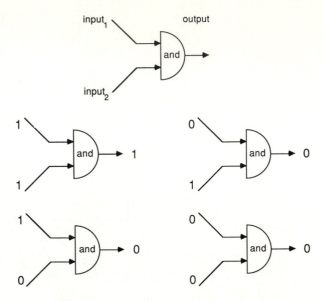

Figure 4.51   An "and" gate object.

Stabler (1986) also presents another simple example applicable to the simulation, design, or analysis of logic circuits. We can define an "and gate" as an object with two associated methods (Figure 4.51):

$$\text{Object}(\text{AndGate}(input_1, input_2),$$
$$[\textbf{if } input_1 = 1 \textbf{ and } input_2 = 1 \textbf{ then } \text{Output}(1),$$
$$\textbf{if } input_1 = 0 \textbf{ or } input_2 = 0 \textbf{ then } \text{Output}(0)])$$

In this example the methods are represented as rules, two of which apply to the AndGate object. Similarly, we can define an object that is an "or gate"

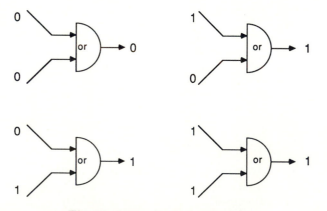

Figure 4.52   An "or" gate object.

Figure 4.53 A "not" gate object.

(Figure 4.52) and one that is a "not gate" (Figure 4.53):

Object(OrGate($input_1, input_2$),
   [**if** $input_1 = 0$ **and** $input_2 = 0$ **then** Output(0),
   if $input_1 = 1$ **or** $input_2 = 1$ **then** Output(1)])

Object(NotGate($input$),
   [**if** $input = 1$ **then** Output(0),
   if $input = 0$ **then** Output(1)])

We can then define circuits in terms of these basic gates. For example, with the appropriate method, the circuit of Figure 4.53 can be represented as an object.

Object(Circuit$_1$($input_1, input_2$),
  [**if** Send(NotGate($input_1$), Output(not_1)) **and**
     Send(NotGate($input_2$), Output(not_2)) **and**
     Send(OrGate(not_1, not_2), Output(out))
  then Output(out)])

Again the method is represented as a rule. In this case the rule preconditions are messages to be sent. If the messages can be sent, then the Output will be *out*. If we send the following message to the system,

$$\exists out[\text{Send}(\text{Circuit}_1(1, 0), \text{Output}(out))],$$

then *out* is instantiated as follows:

$$out = 1$$

The following message,

$$\exists out\ [\text{Send}(\text{Circuit}_1\ (1, 1), \text{Output}(out))],$$

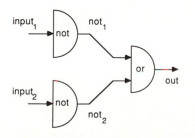

Figure 4.54 A simple circuit.

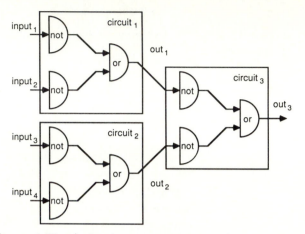

Figure 4.55   A circuit as a composition of smaller circuits.

returns

$$out = 0$$

Circuit$_1$ is an object that can be used to build more complex objects. The following fact describes the circuit of Figure 4.55:

Object(Circuit$_1$($input_1$, $input_2$, $input_3$),
    [**if**   Send(Circuit$_1$($input_1$, $input_2$), Output(out_1)) **and**
           Send(Circuit$_1$($input_3$, $input_4$), Output(out_2)) **and**
           Send(Circuit$_1$(out_1, out_2), Output(out_3))
  then   Output(out_3)]])

We need not pursue the example further here. There is some intuitive appeal in organizing information in this way to give priority to objects, and certainly an advantage in terms of modularity. There may also be benefits in terms of efficiency, though these are harder to justify.

Several of the example systems that we describe in Chapter 6 make use of object-oriented approaches. Objects need not only be physical design elements. We may talk of constraints, goals, and tasks as objects. The PRIDE system (Mittal, Dym, and Morjaria, 1986), described in Section 7.4, is an interesting application of the object-oriented approach. Knowledge is attached to the various physical elements (the mechanical parts of the paper-feed system). Non-physical objects—goals, constraints, methods, advice, and calculations—are also defined in the system. The system is written in the LOOPS language (Stefik and Bobrow, 1981), which provides a general object-oriented programming environment.

Of course, an object-oriented program need not follow the predicate calculus formalism just described. More usually a frame-based representation is employed. It is worth showing a frame used in PRIDE:

Slot	*Value*
Type	Simple task
Name	Goal 5
Descriptor	Decide number and location of roll stations
Status	Initial
Antecedent goals	Design paper path
Input parameters	Paper path, length of paper path
Output parameters	Number of roll stations, location of all rolls
Design methods	Subtasks
	Decide minimum number of roll stations
	Decide abstract placing
	...
Constraints	First station <= 100mm
	Distance between adjacent stations
	<= 160mm
	...

Objects are represented by frames as instances of various classes with associated design methods. In this example the object is actually a goal, not a design element. There are hierarchies of goals and methods by which goals can be satisfied.

4.7 Non-Monotonic Reasoning

The systems of reasoning discussed so far all require a consistent facts base and knowledge base. If we attempt to prove something that is inconsistent with the system, then it will register a failure in some way. Such systems are called *monotonic* because introducing new statements does not invalidate theorems that have already been established (McDermott and Doyle, 1980). A statement is true for the life of the system.

In Section 4.4.1 we considered how some facts may be true in particular states and how we can reason about actions for transforming states. There we considered only the addition of facts by actions, but we may also be interested in actions that delete facts, as in the case of what we might call *design actions*—"make kitchen bigger," "connect the manifold to the carburetor," "double the value of the resistor." (We discuss the control of these kinds of actions in Chapter 7.) Adding and retracting facts is a form of *non-monotonicity*. In Section 4.4.1 we described how consistency is maintained by labeling each fact with the state in which it is true.

Here we explore other interesting manifestations of non-monotonicity. Design processes do not always appear wholly logical or consistent. Designers seem to tolerate inconsistencies in reasoning, even when interpreting a design description. Designers would appear not to resolve all inconsistency problems as they arise, but may sustain many lines of reasoning, deferring a resolution of the problem until the moment when it is necessary to make a decision. These problems can be addressed by reasoning with weighted evidence and by "fuzzy reasoning." Here, however, we highlight the issue in the framework of formal logical reasoning. Readers interested in more detail are directed to Doyle (1981) and de Kleer (1986).

The issue of non-monotonicity can be demonstrated most clearly in cases when there is some incompleteness in our knowledge base. We rarely have sufficient resources to ensure that our knowledge is correct in every detail. For example, it is within the nature of generalizations to sometimes gloss over exceptions—for example, the statement "All animals with beaks or bills are birds." For most purposes this serves as a useful means of identifying birds, until someone introduces us to a platypus, a marsupial with an apparent bill (Figure 4.56). There are several ways to handle this apparent contradiction. One is by denying some feature of the observation statement: the creature does not have a bill, or it is not an animal. We could also modify the original statement: denying that all animals with beaks or bills are birds, or amending it to "All animals with beaks or bills are birds, except animals that are platypuses." Clearly the resolution of the problem does not always lie in rejecting the last statement made. This example suggests that we should address the problem of inconsistency more creatively than possible using (first-order) logic.

This problem of incomplete knowledge is important in the interpretation of designs, both in identifying objects and in anticipating the properties of things. For example, if a building has a door to the outside, has a kitchen,

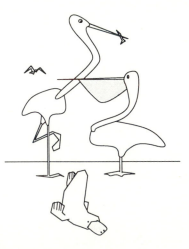

Figure 4.56   The difficulty of identifying all animals with beaks or bills as birds.

and is used for living in, then it can be identified as a house. We may also formulate our knowledge for anticipating the properties of a design: if we know that our building belongs to the house class, then we expect it to have a door to the outside, to have a kitchen, and to be used for living in. In either case our knowledge is likely to be incomplete. We can readily think of exceptions to any general rule: a dwelling with a kitchen and a door to the outside may be a motel, or a building may be a house but not be used for living in, such as a promotional display house or a historic building. Rather than suffer the frustrations of trying to account for every possible exception in our knowledge base, we may reason with "default assumptions," which may suffice until contrary evidence is found.

Let us consider the issue of non-monotonicity with a slightly more interesting body of knowledge. We can consider three items of knowledge pertaining to the thermal properties of a building (Figure 4.57). First, if most of the windows are on the side facing the equator, then the building will exhibit a good thermal performance. Second, if there are no awnings over window openings, then there will be excessive solar gain. Third, a design cannot have both good thermal performance and excessive solar gain. After a little reflection the deficiencies of this knowledge are fairly obvious. We can easily see how it could be reformulated to solve consistency problems, but this is not usually the case in large knowledge bases. Inconsistencies and incompleteness may be effectively concealed.

If we decide that most of the windows are on the side of the building facing the equator (north for the southern hemisphere), then we must conclude that the building will exhibit good thermal performance. But if we add the fact that there are no awnings over window openings, this produces a contradiction. The contradiction can be made clearer by formalizing the statements in logic:

$$\text{ThermalPerformance}(x, \text{Good}) \Leftarrow \text{WindowConcentration}(x, \text{North}) \qquad (1)$$

$$\text{SolarGain}(x, \text{Excessive}) \Leftarrow \text{WindowAwnings}(x, \text{None}) \qquad (2)$$

$$\text{Not}[\text{ThermalPerformance}(x, \text{Good}) \wedge \text{SolarGain}(x, \text{Excessive})] \qquad (3)$$

$$\text{WindowConcentration}(\text{MyHouse}, \text{North}) \qquad (4)$$

The conclusion that interests us from statements (1) and (4) is that

$$\text{ThermalPerformance}(\text{MyHouse}, \text{Good}). \qquad (5)$$

We can also conclude from statement (3) that

$$\text{not}[\text{SolarGain}(\text{MyHouse}, \text{Excessive})]. \qquad (6)$$

If we now add the fact that

$$\text{WindowAwnings}(\text{MyHouse}, \text{None}), \qquad (7)$$

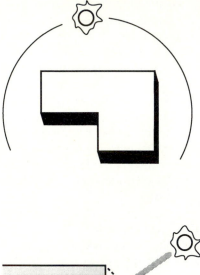

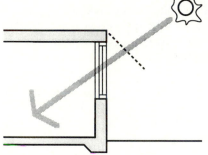

Figure 4.57   The desirability of solar orientation and window awnings.

we face a contradiction in that we can conclude from statement (2) that

$$\text{SolarGain(MyHouse, Excessive).} \tag{8}$$

This is in direct violation of statement (6). We could restore consistency by modifying any one of the statements (1), (2), (3), (4), or (7). The problems are knowing which statements to change and how to manage such modifications. If the modification is in rejecting statements, then possible strategies are to reject the smallest set of statements possible to restore consistency, to reject statements that represent the simplest explanation of the inconsistency, or to reject the statements that have least support, according to some *a priori* weighting.

In order to utilize such heuristics, it is necessary to keep track of derivations of proofs—that is, to keep a record of the *environment* in which statements are true, those sets of assumptions that support the proposition. Various methods have been proposed for doing this; an important one was developed by de Kleer (1986). Certain statements are identified as defaults; such assumptions that are believed true unless there is evidence to the contrary. When a new statement is derived a record is kept of the statements contributing to the derivation of that new statement. A contradiction indicates that a set of assumptions is inconsistent. From these records we know what statements are affected by contradictory assumptions.

Certain simplifications are made in de Kleer's method that do not appreciably affect the richness of the representation language. Statements are represented as Horn clauses. The proof procedure is basically goal directed, using linear resolution. Because there are no statements about negation, a clause cannot be a rule about how something is proven to be untrue, and a fact cannot be a negation. A statement such as (3) would be reformulated as

$$\perp \Leftarrow \text{ThermalPerformance}(x, \text{Good}) \wedge \text{SolarGain}(x, \text{Excessive}) \qquad (3.1)$$

where the $\perp$ symbol means "contradiction." If this proposition can be proven, then a contradiction exists.

It is also necessary to identify default assumptions; these are declared during the formulation of the system. Based on our intuitions, we might identify statements (1) and (2) as default assumptions. Placing most of the windows on the north side of a building is regarded as desirable in most circumstances—in the southern hemisphere, within a temperate zone, and provided there are no features in the design that negate that orientation. In the same way, awnings over windows are generally desirable, and if they are missing the building is prone to excessive solar gain. Again, we assume that there are no factors that override this statement—that the windows are protected by a reflective foil, for example. We may therefore decide that statements (1) and (2) are default assumptions and are subject to revision pending further information.

To make the problem more interesting we can add a further statement: that it is desirable to have good thermal performance and no window awnings. This reflects some aesthetic criterion for assessing buildings; we may decide that it is not a default assumption. The knowledge base and the facts now look like this:

$$\text{ThermalPerformance}(x, \text{Good}) \Leftarrow \text{WindowConcentration}(x, \text{North}) \qquad (1)$$

$$\text{SolarGain}(x, \text{Excessive}) \Leftarrow \text{WindowAwnings}(x, \text{None}) \qquad (2)$$

$$\perp \Leftarrow \text{ThermalPerformance}(x, \text{Good}) \wedge \text{SolarGain}(x, \text{Excessive}) \qquad (3.1)$$

$$\text{WindowConcentration}(\text{MyHouse}, \text{North}) \qquad (4)$$

$$\text{WindowAwnings}(\text{MyHouse}, \text{None}) \qquad (7)$$

$$\text{DesignQuality}(x, \text{High})$$
$$\Leftarrow \text{ThermalPerformance}(x, \text{Good}) \wedge \text{WindowAwnings}(x, \text{None}) \qquad (9)$$

The numbers identifying default assumptions are in italics. We can now attempt to prove the following theorem as a goal: ThermalPerformance(My House, Good). This can be proved according to statements (1) and (4). We represent this derived statement by attaching the labels of the default assumptions constituting its justification:

$$\text{ThermalPerformance(MyHouse, Good) } \{(1)\}. \qquad (5.1)$$

We can also prove the goal SolarGain(MyHouse, Excessive) by statements *(2)* and (7):

$$\text{SolarGain(MyHouse, Excessive) } \{(2)\}. \qquad (10)$$

In addition, we can prove the goal DesignQuality(MyHouse, High) by statements (8) and *(1)*:

$$\text{DesignQuality(MyHouse, High) } \{(1)\}. \qquad (11)$$

We can also prove the goal

$$\perp \ \{(1), (2)\}.$$

The members of any set of assumptions that allows $\perp$ to be true cannot coexist. At least one of these assumptions must be modified. The simplest course of action will be to retract one from the system. If that assumption constitutes part of a set of statements providing justification for a particular proposition, and if that set of justifications is the only one, then that proposition must also be retracted, as if the removed assumption never existed. Thus, in removing assumption *(1)* we must remove derived statements (5.1) and (11). In removing assumption *(2)* we need only delete derived statement (10). As a simple heuristic, if we wanted to affect only the smallest number of derived statements, then we would probably reject assumption *(2)*.

Several sets of assumptions may affect a proposition, and their interdependencies can be fairly complicated. In this example the link between the statement and contradictory assumptions is obvious, but in a knowledge base of any complexity the conjunction of contradictions may not be intuitively obvious.

In essence what we have is a system that enables us to reason consistently by keeping a record of statements that support a proposition. We have a means for bringing other knowledge to bear in resolving the matter. For example, if an inconsistency is detected and one of those supporting statements is removed, then we are able to modify any dependent propositions. The system therefore manages temporary inconsistencies during the reasoning process. A system for representing knowledge about resolving contradictions, updating justifications, and managing the process of retracting and modifying assumptions is often called a *truth-maintenance system* (TMS) or an *assumption-based truth-maintenance system* (ATMS). Due to combinatorial problems an assumption-based truth-maintenance system needs to carry out the process of retracting and modifying assumptions efficiently.

We return to non-monotonic reasoning in our discussion of abductive reasoning in Chapter 6, where assumptions become even more critical in reasoning from desired performances to design decisions.

4.8   Summary

We began this chapter by considering control in design from the point of view of the designer. We have addressed the issue of search in terms of backward chaining, forward chaining, goal-directed search, and the generation of facts by actions in a production system. The issue of goal satisfaction is addressed most simply by the idea of chaining backward through a set of interconnected rules. We ask which rule results in a goal being true. The preconditions of that rule are then treated as subgoals, and the process continues until there is some explicit statement in the facts base that matches a subgoal. There are refinements to this model. We can reason forward to goals by first determining whether certain facts are connected. We then follow that path, proving propositions along the way. Where knowledge is represented as actions, scheduling, focusing attention, and reasoning directly about the "contents" of actions may assist the generation of facts that match goals.

In our discussion of control, we assumed that failure is handled by backtracking. This was seen as a property of the theorem prover used to reason about meta-knowledge. We also considered the simple device of asking, in the event that there is no fact stating the truth of a particular proposition. This device is employed in some expert systems in a dialogue with the user of the system. Other approaches to failure handling are addressed in Chapter 7 in the context of the control of design generation.

Our approach to constraint manipulation is also based on search and backtracking. If a constraint on a variable is violated, then the procedure is to backtrack to another instantiation of the variable—provided the facts base can furnish alternatives. If there are no instantiations that satisfy the constraint, the goal fails. The issue of constraint manipulation is particularly interesting when we wish to produce a design description that satisfies constraints (see Chapters 6 and 7).

We also considered the issue of non-monotonic reasoning. In the event of an inconsistency in reasoning, there are other options to consider than simply rejecting the fact we are trying to prove. We considered the use of default assumptions and the maintenance of records of new facts that depend on those assumptions. Again the issue is interesting in the context of producing designs the knowledge for which is inherently contradictory (see Chapters 6 and 7).

We will also defer a discussion of the control of multiple knowledge sources, multiple abstractions, and multiple control levels to subsequent chapters. These issues assume increasing importance when we consider the production of designs rather than just their interpretation. A central theme to this chapter has been the idea that there can be a uniformity in the way we represent knowledge and the way we represent its control. This suggests that in capturing design knowledge we may also endeavor to capture the knowledge by which designers reason. Knowledge about control is a significant component of expert knowledge.

4.9   Further Reading

Some of the notions of reasoning are well described in a clear but non-technical manner in Harmon and King (1985). Reasoning plays a dominant role in many introductory texts on artificial intelligence, such as Winston (1984). A complete description, at a fairly sophisticated level, of some of the reasoning processes described in this chapter can be found in Nilsson (1982). Kowalski (1979) was influential in making the logic-based approach more accessible. Genesereth and Nilsson (1987) is a more advanced and more theoretically complete treatment than any of the preceding works, and is based on symbolic logic. It covers not only all the topics in this chapter but also most of the reasoning processes described throughout this book.

Exercises

4.1 Discuss the meaning of reasoning, control and meta-knowledge. What is their role in design systems?

4.2 Discuss the relative merits of backward and forward chaining as control strategies in design systems.

4.3 What are the advantages of production systems as a means of knowledge representation in design systems? What use are scheduling, focus of attention, and reasoning about actions in design systems?

4.4 Choose a particular design domain and discuss how an interpretive design system might make use of the concept of focus of attention.

4.5 How should an ideal expert system behave? Compare the description of an expert system in Section 4.5 with the ideal. What additional facilities would be needed to bring the systems described closer to the ideal? Distinguish between major conceptual problems and implementation issues.

4.6 What is the difference between an action-oriented system and an object-oriented system? What are the advantages of object-oriented systems for knowledge-based design?

4.7 What are the major drawbacks of assuming monotonicity in the way we treat design knowledge? What are the requirements of a system in order that it behave non-monotonically?

4.8 Find out what you can about truth-maintenance systems (TMS). What is the difference between a truth maintenance system and a logic programming language?

4.9 Experiment with an expert system shell or language (such as OPS5 or a similar system) and describe the control strategies used. How applicable are these to controlling design knowledge?

4.10 Implement a logic program to facilitate backward chaining (in PROLOG or some similar language) along the lines described in Section 4.3.1. What are the advantages and disadvantages of this way of making control knowledge explicit in such a system?

4.11 Enhance the program of Exercise 4.10 to handle forward chaining as explained in Section 4.3.2. In order to run as a logic program it may be necessary to delete a from the preconditions list d. Why?

4.12 Enhance the program of Exercise 4.11 to find a path connecting two propositions. Devise a logic program for proving the propositions along the path. What is the value of this method of goal-directed forward chaining?

4.13 Implement a logic program to demonstrate the workings of a production system (in PROLOG or some similar language) along the lines described in Section 4.4.1. Implement a simple version, in which actions are fired by retracting and asserting facts. Experiment with different ways of controlling the system.

4.14 Implement a simple production system in LISP. (Some of the basic control constructs are given in Appendix B.) Enhance the program to simulate backward and forward chaining.

4.15 Implement the kernel of an expert system shell in (in PROLOG or some similar language) along the lines described in Section 4.5.1. Experiment with forward chaining and goal-directed reasoning.

4.16 Implement the kernel of an expert system shell in LISP, based on Exercise 4.14, that operates as a production system. Devise an interface that facilitates a dialogue between the system and the user.

4.17 Implement a simple object-oriented reasoning system in (in PROLOG or some similar language) along the lines described in Section 4.6. How does this approach compare with rule-based approaches to representing knowledge described elsewhere in this chapter? Is there any fundamental difference in the way information is handled?

4.18 Experiment with an object-oriented programming language (SMALLTALK, LOOPS, CLOS, C^{++}, FLAVORS, or some similar language). How does the language facilitate or hinder design knowledge representation?

PART III

Knowledge-Based Design

In this part of the book we describe how design knowledge can be brought to bear in the interpretation and the production of designs. Interpretation involves determining the properties of a design not explicitly stated in its description, as well as evaluating and simulating. We discuss the formulation of knowledge-based design systems for interpretation. Chapter 5 describes and elaborates the interpretation of designs.

An understanding of interpretation is important before we address the issue of producing design descriptions through knowledge-based systems in Chapter 6. We discuss the use of interpretive knowledge—knowledge about how goals can be accomplished—and the use of syntactic, or generative, knowledge in producing designs. Chapter 6, the longest chapter, brings us to the center of what to expect from knowledge-based design systems. Chapter 7 expands on design processes implicit in much of the discussion in Chapter 6. Finally, Chapter 8 explores learning and the creative potential of knowledge-based design systems.

CHAPTER 5

The Interpretation of Designs

In this chapter we investigate the representation and control of knowledge pertaining to the interpretation of designs, focusing on design systems that interpret design descriptions. In this book we define the process of interpretation as one that maps the representation of a design in a computerized design system onto the *meaning* of that representation.

Building on our understanding of language and logic, we review what we understand of interpretation before discussing issues in the interpretation of designs. One of the most powerful tools at our disposal for modeling interpretation is the process of logical deduction, which can be used to construct simple expert systems. Thus we review the use of logical rules for interpretation, with only cursory reference to their control.

For our purposes interpretation is concerned with deriving certain properties of the design that are not explicitly stated in the design description. But we are generally interested in particular properties, such as those associated with the design's performance. We may make use of *derived* attributes to match against *desired* attributes. For example, if we wish to design an object that is sharp, and we have some ways to generate descriptions of objects with different characteristics, we can interpret each of these objects in turn until we arrive at one that matches our desired interpretation. This is clearer if we think of the object descriptions in terms of drawings. The interpretive knowledge is that which identifies characteristics of sharpness, such as acute angles and hardness. Thus the process of interpretation is instrumental in the satisfaction of goals.

We can also consider the role of interpretation in *evaluating* designs. We argue that evaluation is a process by which the performances of alternative designs are compared, or by which a single design is compared against a fixed requirement. An evaluation is a comparison between the meanings of a performance of two or more interpretations, generally resulting in statements about comparisons—for example, that the cost is too high, or the weight is too great.

One important area of interpretation relevant to design is the interpretation of geometrical and topological descriptions. This involves the determination of attributes implicit in those descriptions and also the graphical interpretation of those representations. Currently, computers do not readily facilitate the explicit representation of graphical images. Visual images constitute descriptions of designs in a language that must be translated into text and numbers. We will focus on some examples of expert systems that explore these issues.

An area of further interest is the use of design codes as repositories of interpretive knowledge. We discuss expert systems that attempt to exploit this knowledge in the context of interactive design systems.

5.1   The Role of Interpretation

In common with language and logic, design is concerned with interpretation. In natural language we are interested in determining that to which a set of sentences makes reference, and in logic we are interested in determining the truth of implicit statements, or theorems, from other statements. The term *semantics* is used frequently in various domains, including design and computing. It is worth clarifying what we mean by semantics in the context of our model, outlined in Chapter 2. Because this topic requires some amplification, we consider the relationship between language, logic, and design in some detail. We begin with a simple view before exploring the complexity of the issue.

In language, the study of meaning is concerned with the relationship between signs, symbols, and the entities to which they refer. In logic, meaning concerns the principles that determine the truth values of statements that can be derived from other statements.

5.1.1   Interpretation in Language

In *The Meaning of Meaning*, Ogden and Richards (1923) cite 16 favored definitions of meaning. The role of meaning that they proposed, still widely held today, serves as a reasonable working model of the role of language. There are

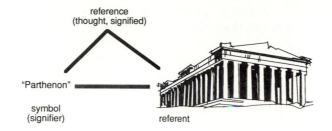

Figure 5.1 The "semiotic triangle" of Ogden and Richards (1923).

other views, but theirs is adopted here because it is so applicable to automated systems and to design.

Words and groupings of words constitute symbols. When used appropriately they become signs that refer to things—that is, ideas or objects. The hearer is involved in *interpretation*, which concerns establishing references from the incoming signs. The hearer has caught the speaker's meaning when there is a correspondence between that to which the speaker makes reference and that which the signs suggest to the hearer (Figure 5.1). Factors that impinge on this transaction constitute the "sign situation"—that is, those past and present "experiences" that determine psychological reactions to signs, experiences that we embrace under the general heading of knowledge.

Words have evocative as well as symbolic functions, but there is no inherent connection between words and the things to which they refer. The study of linguistics therefore steers away from a concern with the meaning of individual words to the study of the interpretation of signs. In our terminology the sign situation manifests itself as knowledge, which provides the mapping between signs, ideas, and objects.

5.1.2 Interpretation in Logic

There are two ways in which interpretation may be discussed in the context of symbolic logic. In the first, we consider the symbols of predicate calculus and their role as signs that refer to ideas and objects in the same sense as words in natural language. This is a valid but low-level view of the interpretation of a logic statement because it involves relatively simple mappings with natural language.

In the second sense, according to Kowalski (1979), any meaning that might be associated with a predicate symbol or logical statement is relative to the collection of statements constituting the *axiomatic system*. The interpretation of a set of logical statements is therefore that which can be logically inferred from them. Kowalski argues that all talk of meaning can be reexpressed in terms of logical implication. The meaning of a set of statements is therefore that which is not stated explicitly but which can be inferred from them.

The same view can be taken of a set of linguistic "statements." The process by which signs, made up of complexes of symbols, are decoded to give meaning can be characterized as deductive inference. The process of interpretation of designs entails the derivation of a new set of facts (those dealing with performance) from a given set of facts (those dealing with the description of the design). The interpretation of a set of logical statements is simply that which can be inferred from them. Deductive inference provides a mechanism for interpretation.

5.1.3   Interpretation in Design

We have argued that an artifact can be described by a set of statements (or axioms) in logic, specifically as facts. A description on its own is insufficient for the production of new facts. When a description is combined with other statements, generally regarded as knowledge, however, the derivation of meanings is possible. The interpretation of a design (or composition of elements) is similar to the problem of interpreting an utterance in natural language. The mapping is provided by knowledge.

Both understandings, from linguistics and from logic, lead to the view that design descriptions and interpretive knowledge combine to provide a logical system that enables the derivation of the meaning of the design. We can formalize this process by describing it as a mapping between an interpretation, knowledge, a design, and a transformation. In Section 2.6 the process of interpretation was described by the following:

$$I = \tau_1(K_i, D) \tag{5.1}$$

Interpretive knowledge, represented as K_i, consists of statements about the mapping between the design description, D, and the interpretation, I, a performance description. The interpretation of a design is then the result of the application of this interpretive knowledge to the design.

5.2   Issues in Design Interpretation

So far we have presented a simplified view of interpretation in logic, language, and design. We expand this view throughout the chapter as we consider the implementation of interpretive systems, but first it is worth considering some of the complexities of the issue. In the following subsections we review the difference between syntax and semantics in design, the issue of multiple design interpretations, the role of parsing in design, and the issues of incomplete knowledge and ambiguity.

5.2.1   Syntax and Semantics in Design

In Chapter 2 we discussed how the explicit description of artifacts can be characterized as an issue of syntax. The interpretation of design descriptions is an issue of semantics. In spatial terms, syntax is concerned with the configuration of elements, such as points, lines, labels, elements, and assemblies, and with explicit statements about properties of elements and configurations. Semantics is concerned with derived descriptions, generally performances. For example, given a syntactic description of a knife in terms of dimensions, material type, and weight, we may derive its meaning in terms of performance—that it cuts, that it cuts well, or that it cuts to a depth of 2.5 millimeters. The knowledge by which this performance is derived has to do with interpretation: if the material is hard and is of the right shape, then it cuts.

In this model properties of a design are not inherently syntactic or semantic, yet we see that it is possible to formulate hierarchies of meaning in terms of how properties can be derived from one another. What is semantics at one level is syntax at another (Figure 5.2).

We have characterized design as concerned with finding mappings between a space of interpretations (semantic space) and a space of design descriptions (syntactic space). Given a well-defined body of interpretive knowledge, a given

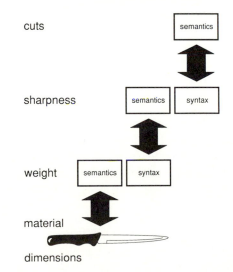

Figure 5.2   Syntactic and semantic levels.

syntactic description will correspond to a single point in the semantic space (this point may represent a multidimensional set of performances).

5.2.2   The Role of Parsing in Design Interpretation

In order to interpret a statement, it is generally considered necessary to understand its structure, or grammar. Hence parsing is considered an essential ingredient of interpretation. In discussing design we have assumed that the knowledge by which designs are generated—their grammars or sets of design actions—can be separated from that by which they are interpreted. Knowing the steps that have been traversed in the production of a design description, however, may turn out to be extremely valuable in interpreting that design. Knowing *how* may tell us *what* in certain cases. For example, knowing that a building plan is derived from a square that has been truncated in various ways may give us a clue as to its structural performance. Parsing may also be seen as a means of decomposing a design into its component parts.

The strategy of separating the role of interpretation from that of generation, as in this book, may be artificial, but, as we demonstrate, it has considerable utility. In making this distinction, however, we should bear in mind that the interpretation of a design may depend to some extent on knowledge of the processes by which the design comes about.

5.2.3   Incomplete Knowledge in Design Interpretation

Interpretation in language is also made difficult by the fact that usually the set of statements that we are interpreting is incomplete. A single sentence does not generally contain all the information we require in order to interpret it; there is a *context* that is not stated explicitly. The same applies to design descriptions.

Certain interpretations of a design will depend on information about context. For example, certain features of the thermal performance of a gear box in an automobile cannot be determined without reference to the location of the gear box itself. In order to determine its heat-dissipation characteristics, the heat dissipation of the adjoining and surrounding elements is needed. This information, although not normally considered part of the design description, must be incorporated into the reasoning process. Any uncertainty about the context in which the artifact is to be placed has a bearing on the interpretation process.

We may also encounter some incompleteness in interpretive knowledge. This is particularly important in the area of value judgments that may not be possible to externalize. Incompleteness becomes an important issue in evaluating design interpretations —that is, comparing the performances of alternative designs and making some kind of evaluation of their relative merits. Such evaluation requires knowledge about value judgments—that one performance is better than another in some respects.

5.2.4 Inconsistency, Ambiguity, and Uncertainty

In logic, of course, we assume the axioms to be well-defined and consistent. We do not necessarily have the same certainty in language. The knowledge by which interpretations are made is not necessarily well-defined, consistent, or complete. A similar problem arises in design. Two people may look at the same design description and arrive at two different interpretations. The problem may be due to ambiguity in the way the design is described or in differences in the respective knowledge bases applied to the interpretation task. We normally demand of a design system that it facilitate the production of consistent design descriptions. This stipulation need not apply, however, during the development of a design. Partial and intermediate stages of the design process may require that conflicting descriptions of the design be maintained concurrently. The interpretation of such partial and inconsistent descriptions would appear to be an important part of the design process. This issue was partly addressed in Chapter 4 in the context of non-monotonic reasoning.

Natural-language sentences are frequently ambiguous, which sometimes contributes to the richness of language, but can, of course, lead to confusion. The problem is magnified when the source is spoken rather than written. In this case the string being interpreted is not simply a statement, but rather a signal that is ambiguous and also subject to interference, or noise. It is worth referring here to the HEARSAY experiment (Erman et al., 1980). The HEARSAY programs are natural-language understanding systems that take spoken sentences within a limited domain and attempt to interpret them as written English sentences. One operational feature of the system formulates hypotheses that are likely assuming a given context. Various heuristics are brought into play in deciding among competing hypotheses. The process is quite different from simple logical deduction. In interpreting designs, we may have to deal with "noisy signals"—particularly if we are trying to interpret design sketches—and we may have to account for this in design systems (Figure 5.3).

The explicitness with which a design is described relative to any interpretations that are to be derived from it bears on the process of interpretation. For example, a line drawing of a three-dimensional object can be stored as a series of vectors in two dimensions—such as a perspective or isometric projection. Such a representation may be ambiguous and will not facilitate the derivation of geometrical properties (such as its volume) as readily as a three-dimensional model of the object. The knowledge required to interpret a two-dimensional line drawing involves complexities not encountered in the interpretation of a description containing explicit topological and dimensional information.

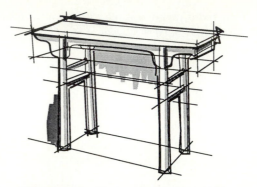

Figure 5.3   Design sketch for a table.

In the following sections we explore the role of interpretation in design systems by assuming that it can be usefully modeled as deduction.

5.2.5   Interpreting Designs by Deductive Inference

In Chapter 2 we discussed deduction and described how a new statement deduced from certain statements can be regarded as an interpretation. A statement derived from a set of logic statements generally provides an opportunity for the derivation of further statements. This suggests that there are hierarchies of meaning, in that certain meanings can be deduced from others. As demonstrated in Section 3.2, the hierarchical nature of a design may be represented explicitly in a facts base. This idea is new neither to computing (Dahl, Dijkstra, and Hoare, 1972) nor to computer-aided design. For example, the representation of numbers within a data structure may be interpreted as points and lines. Points and lines can be interpreted as primitive shapes. These shapes can be interpreted as design elements. Collections of elements can be interpreted as objects, in the sense of belonging to aggregation or part hierarchies, and certain properties of these objects and collections of objects can also be derived. (The hierarchical levels can usually be separated at the lower levels of description, though they may involve complex connections at the higher levels.) Elaborate programs have been devised for calculating the properties of designs described in this way. In knowledge-based systems we are interested in making this hierarchy explicit in the way information about designs is organized (the subject of Chapter 3) and also in the way knowledge is organized (the subject of Chapter 7).

Under our model of design, designs are interpreted by means of knowledge. This knowledge can be embedded within algorithms or represented in a form that makes the mappings between representation and meaning explicit. As discussed in previous chapters, a useful and intuitively appealing method of representing interpretive knowledge is as rules of the following form:

> **If** a_1 **and** ... **and** a_n
> **then** b_1 **and** ... **and** b_m.

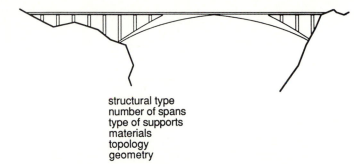

structural type
number of spans
type of supports
materials
topology
geometry

Figure 5.4 A bridge and some of its attributes.

This says that if a design possesses attributes $a_1, \ldots a_n$ (the antecedents), then infer that attributes $b_1, \ldots b_m$ (the consequents) are also true. The if-then construct is similar to the *implies* ($\Rightarrow$) of standard logic:

$$a_1 \wedge \ldots a_n \Rightarrow b_1 \wedge \ldots \wedge b_m$$

This statement can also be reformulated as a set of Horn clauses employing the *if* ($\Leftarrow$) connective:

$$b_1 \Leftarrow a_1 \wedge \ldots \wedge a_n$$

...

$$b_m \Leftarrow a_1 \wedge \ldots \wedge a_n$$

We can also incorporate disjunctions (**or**s, $\vee$) as well as conjunctions into such rules. The knowledge by which a design is interpreted may consist of many such rules.

Expert systems, introduced in Section 4.5, facilitate the representation of interpretive knowledge in this form and make it amenable to automation. Workable systems can be devised that operate on the basis of deductive reasoning; examples are given in subsequent sections.

5.3 Design Evaluation

A design, then, possesses attributes other than those by which it is explicitly described in a computer system (or in a set of working drawings, for that matter). For example, a bridge might be described in terms of its structural type (such as suspension, arch, or cantilever), the number of spans, the type of supports, the materials, and the topology and geometry of its members (Figure 5.4). A bridge possesses other attributes, however, such as stability, traffic-carrying capacity, impact on the environment, and aesthetic qualities. These attributes constitute performances implicit in the design description that can be derived from it. Given a design description, we should be able to derive a new set of descriptions in a performance space. This derived set of performances can then be compared with some other set of performances; such a comparison constitutes an evaluation.

If we think of performances in terms of attributes and values, then we may wish to compare the values of the same attributes for different designs. For example, the attribute "cost" of design1 may have a value of $50, while that of design2 may be $60. A typical evaluation of these two performances is that design1 is cheaper than design2. But we can also compare values against some "ideal" performance. For example, design2 may be evaluated as "too expensive" because it exceeds some threshold value of $55.

We can formulate a design system in which the attainment of an ideal performance constitutes an objective. In this case, the value of the ideal performance is most often zero or infinity, and is expressed in terms of minima or maxima. Thus the goal is to minimize the cost or to maximize the ratio of floor area to perimeter length.

Evaluation results in the production of statements such as the following: performance is satisfactory; performance is unsatisfactory; the performance of design1 is better than the performance of design2; this is the best performance possible given the constraints of the problem.

We can describe the design method by *simulation*, employed in operations research. A candidate design is generated, and its performance derived, by simulating the behavior of the design, using analogic or symbolic models. If the designer is satisfied with this performance, by evaluating it either against known performances or other criteria, that candidate design may be selected as the final design. If the designer is dissatisfied with the results of the evaluation, a new candidate design may be generated. With sufficient iterations this process may lead to a desired design, especially if the interpreted result gives a clue to the improvement of the candidate design.

It is quite usual to represent evaluative knowledge in terms of algorithmic functions. However, we can also represent certain evaluative knowledge in an intuitive way as rules (shown graphically in Figure 5.5).

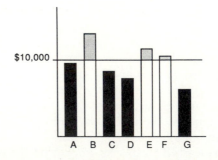

Figure 5.5   Those cases in which cost is too high.

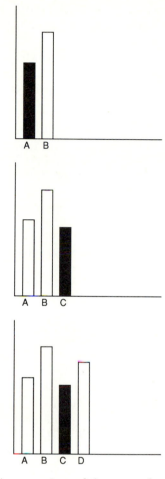

Figure 5.6   A comparison of the costs of successive designs.

If   cost **is greater than** $10,000
then   evaluation **of** cost **is** "too high."

We may also consider an evaluative heuristic such as the following to be useful during the design process (Figure 5.6).

If   cost **of** element under consideration
        is less than cost **of** previous element
then   prefer element under consideration.

In general terms we may have a rule that states the following:

If   performance **of** element under consideration
        is better than performance **of** previous element
then   prefer element under consideration.

where the interpretation **is better than** depends on the specific performance being evaluated. The knowledge for this may be stated as follows:

> **If**   performance **is** cost **and**
>   cost **of** element1 **is less than** cost **of** element2
> **then**   performance **of** element1 **is better than** performance **of** element2.

> **If**   performance **is** weight **and**
>   weight **of** element1 **is less than** weight **of** element2
> **then**   performance **of** element1 **is better than** performance **of** element2.

It may, however, be necessary to qualify this knowledge because a lower weight does not always mean a better performance. Thus the rule may read as follows:

> **If**   performance **is** weight **and**
>   low weight is desired **and**
>   weight **of** element1 **is less than** weight **of** element2
> **then**   performance **of** element1 **is better than** performance **of** element2.

There may be other rules to determine that low weight is desired, perhaps depending on the domain of the problem, such as aircraft design.

Evaluation is especially helpful when we use it to guide the search for a design. We can see that the process just described can assist in producing either a satisfactory solution or an optimal solution using repeated simulation. This process of producing designs is discussed further in Chapter 6, where it is compared with other methods. The process may occur at various stages of the design process, in configuring the whole design, or in selecting a suitable component for consideration for part of a design.

5.4   The Interpretation of Spatial Design Descriptions

Designers are interested in the spatial qualities of designs, often in terms of topology and geometry. They generally feel comfortable with drawings as a means of representing these properties. In computer-aided design systems, structured numbers and text constitute simple ways to describe spatial properties. It appears that we are not yet advanced enough in our understanding of computational and design processes to enjoy powerful automated "graphic reasoning" without going through this intermediate stage of text and numbers. If we are interested in visualizing spatial properties as drawings, information as text and numbers must be interpreted into graphical images.

In this section we first consider the interpretation of graphical descriptions to produce drawings. Second, we consider the interpretation of descriptions of design geometry and topology to determine other spatial properties. Third, we demonstrate how this can be achieved in rule-based systems in which graphical knowledge is represented as logical rules.

5.4.1 The Graphical Interpretation of Geometrical Descriptions

If we assume a particular convention in the way a theorem prover processes statements, images can be drawn by means of deductive reasoning. As demonstrated in Section 4.5.1, it is possible to define functions that permit information to be read to and written from display devices. Such functions can be given the same status as predicates, and return true if their operations can be carried out. Graphics primitives can also be implemented and are employed similarly to graphics commands in procedural languages. We can therefore define *Move* and *Draw* functions that can be incorporated into rules that draw objects. Thus a logical statement can be devised for drawing a rectangle:

$$
\text{DrawRectangle}(x_o, y_o, w, h) \Leftarrow Move(x_o, y_o) \land Draw(x_o, y_o + h) \\
\land \ Draw(x_o + w, y_o + h) \land Draw(x_o + w, y_o) \land Draw(x_o, y_o),
$$

where x_o and y_o are the coordinates of the origin of the rectangle, and w and h are the width and height respectively. Clearly, it is important we attempt to prove the propositions to the right of the "if" connective ($\Leftarrow$) in the order given. The logic is such that proving that a rectangle can be drawn results in the drawing of the rectangle.

As well as being described like this, objects can be described hierarchically, as with the simple plan in Figure 5.7 (discussed earlier in Section 3.2). The configuration is a stylized plan of a room, an office. Office is a composite object made up of various components, such as columns and walls. Each of these objects is made up of line segments, which are defined by end points, which have coordinate values. Each of the following predicates has two arguments. The first argument is the name of the object: a room, component name, or

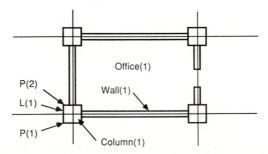

Figure 5.7 A composition described hierarchically.

graphic primitive in the form of a line or point. The second argument is a list in each case:

Composite(Office(1), Column(1) **and**
        Column(2) **and** ... **and** Wall(1), Wall(2) **and** ... Nil)

Object(Column(1), L(1) **and** L(2) **and** L(3) **and** L(4) **and** Nil)

...

Line(L(1), P(1) **and** P(2) **and** Nil)

...

Point(P(1), 0.00 **and** 0.00 **and** Nil)

Point(P(2), 0.00 **and** 0.50 **and** Nil)

The objects can also be drawn hierarchically by the following set of logic statements:

DrawComposites(Nil)

DrawComposites(a **and** b) $\Leftarrow$ Composite(a, c) $\wedge$
        DrawObjects(c) $\wedge$ DrawComposites(b)

DrawObjects(Nil)

DrawObjects(a **and** b) $\Leftarrow$ Object(a, c) $\wedge$
        DrawLines(c) $\wedge$ DrawObjects(b)

DrawLines(Nil)

DrawLines(a **and** b) $\Leftarrow$ Line(a, p_1 **and** p_2) $\wedge$
        Point(p_1, x_1 **and** y_1) $\wedge$ Point(p_2, x_2 **and** y_2) $\wedge$
        $Move(x_1, y_1) \wedge Draw(x_2, y_2) \wedge$ DrawLines(b)

Drawing a composite object is defined in terms of drawing the objects of which it is composed. Drawing objects is defined in terms of drawing lines. Drawing lines is defined in terms of points and the *Move* and *Draw* primitives. If we attempt to prove the theorem

DrawComposites(Office(1) **and** Nil)

then the theorem prover will draw Office(1) by constructing each of its components and subcomponents.

The formality of logic is somewhat compromised by the introduction of function calls to handle communications (Robinson, 1983), simply because so much depends on the method of resolution employed by the theorem prover. Attempts have been made by Pereira (1982) and Swinson (1983) to fit graphics into the logic of PROLOG. A line can be defined as a boundary of an infinitely extending half plane. Graphical objects can therefore be represented as the intersection of various half planes (Figure 5.8). Mapping the logic of set theory onto that of predicate calculus is intended to provide a "declarative graphics" medium.

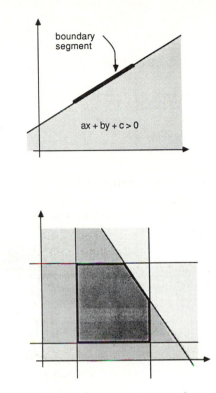

Figure 5.8 Half planes to describe geometry: a polygon defined in terms of half planes and boundary segments.

5.4.2 Deriving Spatial Properties from Geometrical Descriptions

As discussed in Section 3.2, geometrical information can be represented in predicate calculus notation. Predicate calculus has the advantage of being a relatively simple way to represent the idea of a relational database. CAD systems are making increasing use of relational databases, where the relational idea underlies the logic of database systems, if not their implementation. Geometrical information about a room, for example, can be described in the following way:

Room(Office(1), 3.0, 1.0, 4.0, 2.0)

This constitutes a fact or an assertion in a CAD database; the graphical interpretation is shown in Figure 5.9. The five arguments to the Room predicate include the name of the room and the common values of the vertices on the north, south, east, and west faces, respectively.

As indicated earlier, complex objects can be built up from a set of such descriptions. Here we are interested in how it is possible to formulate statements

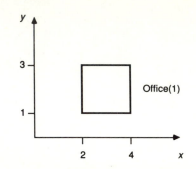

Figure 5.9   The geometry of a simple room.

from which we can deduce interpretations other than graphical ones. For example, we can infer topological relationships between spaces with statements such as the following:

$$\text{Direction}(a, \text{North}, b) \Leftarrow \text{Room}(a, n_a, s_a, e_a, w_a) \wedge$$
$$\text{Room}(b, n_b, s_b, e_b, w_b) \wedge (s_a > n_b)$$

(Note that we could also express this knowledge as an if-then rule.) We are stating that room a is north of room b if the south face of a is above the north face of b. Using similar rules for other directional relationships, if we attempt to prove the theorem

$$\exists x[\text{Direction}(x, \text{North}, \text{Office}(1))]$$

then we should be able to infer all the rooms that are north of Office(1)—that is, all the room names that can be instantiated to x in order for the theorem to be true.

Once certain implicit attributes are inferred in this way, higher levels of interpretation can be derived. More complex logical statements can be formulated by which it is possible to determine what rooms are adjacent to each other, and what rooms must be passed through to move from one space to another. It is therefore possible to derive circulation routes through a building from a database description:

$$\text{Path}(a, a, b, b)$$
$$\text{Path}(a, b, c, d) \Leftarrow (Link(a, e) \vee Link(e, a)) \wedge$$
$$Not(Member(e, c)) \wedge \text{Path}(e, b, e \textbf{ and } c, d)$$

This is a complicated, recursive definition (Path is defined in terms of itself), and it requires the use of a list structure. The first argument of Path is the room from which the journey is to commence; the second argument is the destination. The third argument is a list of rooms already visited, which must not be visited again. The final argument is the list of rooms that constitutes the path.

The first statement says that a journey from a room to itself is the same as the list of rooms already visited. The second statement says that there is a path from a to b if a is linked to another room that has not already been visited and that is connected to the destination by a path. A function is required for finding out whether two rooms are *Linked*; another function is required for finding out whether an element is a *Member* of a list. Also, of course, *Not* must be defined.

If the preceding rule constitutes part of an expert system knowledge base, we can tell the system to prove that there is a path between two particular rooms. In the process of proving this fact, the system will establish the path and return it as the value of a variable (x). When applied to the facts base for the building of Figure 3.50, an attempt to prove the theorem to find all paths from A and I

$$\exists x[\text{Path}(A, I, \text{Nil}, x)]$$

results in the following instantiations of x (depending on implementation) (Figure 5.10):

A **and** B **and** C **and** D **and** G **and** H **and** I **and** Nil

A **and** B **and** C **and** F **and** H **and** I **and** Nil

A **and** B **and** E **and** F **and** C **and** D **and** G **and** H **and** I **and** Nil

A **and** B **and** E **and** F **and** H **and** I **and** Nil

More complicated properties of design topologies can be produced similarly by deductive inference. Here we have considered only simple facts about the topology and geometry of a design in order to demonstrate how geometrical knowledge can be made explicit. The preceding example also suggests applications in graph theory.

5.5   Knowledge-Based Systems for Interpretation

In this section we apply some of the ideas discussed in the previous sections. The systems described are working systems that exploit the interpretation idea as part of some overall design strategy. Only those aspects of the systems pertinent to interpretation are discussed.

5.5.1   A System for the Interpretation of Graphical Descriptions

Various knowledge-based systems have been developed that couple computer graphics techniques with expert system methodologies. Hopgood and Duce

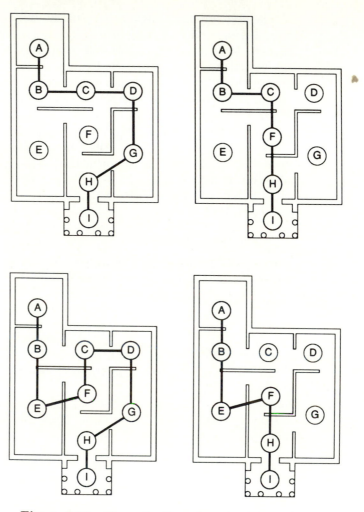

Figure 5.10   The paths through a plan joining A to I.

(1978) describe the potential of production systems for graphics. Arbab and Wing (1985) and Woodbury and Oppenheim (1988) discuss "geometric reasoning" with the aim of processing geometric information logically by reasoning from a knowledge-based understanding of shapes and their attributes. Szalapaj and Bijl (1985) outline the design and use of MOLE (Modeling Objects with Logic Expressions), a system that understands the sequence of graphical operations performed by the user and regenerates modified drawings. Le Texier et al. (1985) describe a graphical interface for a wind-loading code in structural engineering. As also discussed previously, Manago and Gero (1986) describe the issues of the communication between expert systems and graphics systems, while Oxman and Gero (1987) provide icon-oriented graphics in PREDIKT.

Facts allow us to make statements about the physical form of a design, but we must also be able to interpret those facts, establishing new information about the form of a design in a way that parallels a designer's ability to understand and interpret a drawing. Let us consider a simple example of the use of rules in interpreting the properties of a configuration of objects from a primitive description in terms of connected vertices.

We describe the formulation of a system for representing and reasoning about the spatial characteristics of physical objects (based on the system TOPOLOGY1 (Akiner, 1985; Gero, Akiner, and Radford, 1983)). For simplicity we restrict ourselves to the domain of orthonormal parallelopipeds, where the edges of objects are parallel to their faces. The objects are rectangular prisms. The aim of the system is to support a designer in the interpretation of configurations of designs. This requires knowledge about topological relationships between entities, about geometry, and about various attributes of objects and spaces. We can utilize first-order predicate logic to encode knowledge about spatial relations. Objects are defined in terms of facts about three-dimensional geometry. Points are given a label (V_1) and X, Y, and Z Cartesian coordinate values:

$$\text{Coord}(V_1, [X, Y, Z])$$

Vertices, enclosed in square brackets when listed, are attached to objects with facts such as

$$\text{Vertex}(OBJECT, V_1)$$

where V_1 is the label of a vertex associated with $OBJECT$. Knowledge is expressed as a collection of logical statements about relationship between entities. We can classify this knowledge as follows:

1. Knowledge about single-object topology—that is, knowledge concerned with the relationship of vertices, edges, and faces. For example,

   If   the two vertices belong to an object **and**
   the two vertices are not identical **and**
   two of their respective coordinates match
   then   an edge exists between two vertices of the object.

This can be represented more formally as follows:

   If $\text{Vertex}(object, v)$ **and**
   $\text{Member}([v_1, v_2], v)$ **and**
   $\text{Not}(v_1 = v_2)$ **and**
   $\text{Coord}(v_1, [x_1, y_1, z_1])$ **and**
   $\text{Coord}(v_2, [x_2, y_2, z_2])$ **and**
   $\text{TwoMatch}([x_1, y_1, z_1], [x_2, y_2, z_2])$
   then $\text{Edge}(v_1, v_2, object)$.

Other rules define predicates such as Member and TwoMatch.

2. Knowledge about single-object geometry—that is, knowledge concerned with such attributes as length, width, area and volume. For example,

> **If**   an edge exists between two vertices of an object
> **then**   the length of an edge is the distance between two vertices.

This can be represented more formally as follows:

> **If** $\text{Edge}(v_1, v_2, object)$ **and**
> $\text{Distance}(v_1, v_2, length)$
> **then** $\text{Length}(v_1, v_2, object, length)$,

where the Distance predicate determines the distance between two given vertices.

3. Knowledge about single-object attributes—that is, knowledge concerned with such attributes as geometric type, color, texture, price, and direction. An example described informally follows:

> An object is a cube if its length, width, and height are all equal.

4. Knowledge about topological relationships between objects. This knowledge is concerned with relationships such as adjacent to, below, above, and inside. These relationships can be organized hierarchically according to derivation. An example of a rule is as follows:

> **If**   the common coordinate of the bottom face of a is greater
>   than the common coordinate of the top face of b
> **then** a is above b.

More formally, this could be written as

> **If** $\text{Face}(a, \text{Bottom}, z_1)$ **and**
> $\text{Face}(b, \text{Top}, z_2)$ **and**
> $z_1 > z_2$
> **then** $\text{Above}(a, b)$.

5. Knowledge about geometrical relationships between objects—that is, knowledge concerned with such attributes as bigger than and longer than. For example,

> **If** a is greater than b in terms of volume
> **then** a is bigger than b,

or more generally,

> **If**   the value of c of a is greater than the value of c of b
> **then** a is greater than b in terms of c.

6. Knowledge about attributes of groups of objects—that is, knowledge concerned with comparing attributes such as more expensive than and cheaper than (this is a specialized case of the preceding general expression. For example,

> **If**   *a* is greater than *b* in terms of price
> **then**   *a* is more expensive than *b*.

7. Knowledge about circulation-network relationships—that is, knowledge about relationships such as connected to, linked to, and path to. For example, assuming that the spaces are rooms in a building,

> **If**   space *a* has a doorway to space *b* or
>   space *b* has a doorway to space *a*
> **then**   space *a* has access to space *b*.

We can see from these examples of categories of knowledge how a description of objects in terms of their coordinates can be interpreted to produce various meanings. We can also see the hierarchical nature of this process—for example, volume can be derived from area and height, and area can be derived from plan dimensions.

We should be able to ask an expert system armed with this knowledge certain questions about topologies, quantities, circulation networks, and the names of objects and spaces. Assuming that objects are organized in terms of compositional hierarchies, we should be able to ask such questions as the following:

What is above A in C?

What is between A and B in C?

How many As are there in C?

What is the path from A to B in C?

What are the names of the components in C?

where A, B, and C are the names of spaces and compositions of spaces.

We need some facility for translating these statements into system goals—theorems. This can be accomplished simply with templates that help match standard queries to system goals. We can consider these questions in the context of a house. Figure 5.11 shows the abstracted spaces of Frank Lloyd Wright's Carlson House. We should be able to pose the following queries about this building represented by the context "house." User queries are in bold type.

What is on top of living room in house?
Roof garden is on top of living room in house.

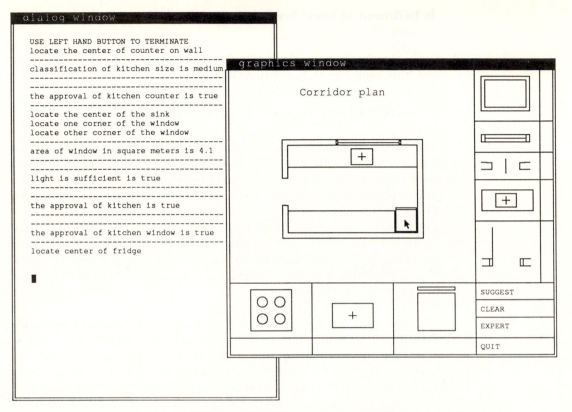

Figure 5.13   The evaluation of a kitchen design.

Figure 5.13 shows the computer screen during a typical session with a system for evaluating kitchen designs; the system is called PREDIKT (Oxman and Gero, 1987). The left window is for text input and output, while the right window is for graphic input and output. Although the system has several modes of operation, the one that interests us here is the *design evaluation mode.* The designer describes the design graphically using the graphic window shown in Figure 5.13. The graphical description is interpreted to produce facts about objects and their relationships, resulting in a facts base. The system then interprets and evaluates these facts using its knowledge base. (Comments in italics in the figure were added later.) The knowledge base is used with a standard expert system shell that communicates with the database of the graphics system.

5.5.3   Interpreting Graphical Descriptions of Engineering Objects

Here we describe a knowledge-based system for interpreting graphical descriptions of simple engineering structures (Balachandran, 1988). The system

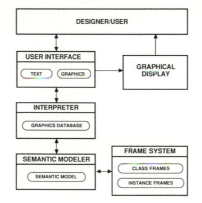

Figure 5.14 The organization of a system for interpreting and producing structures.

identifies certain features of objects and categorizes the objects into classes. It interprets information supplied graphically on a display screen and builds a descriptive model of the design as a frame representation. The architecture of the system is shown in Figure 5.14. The domain consists of a large number of object types, sub-types, and their components. The knowledge about individual objects is represented explicitly using a frame language that provides constructs for describing individuals and classes (see Chapter 3). An example of the hierarchy of the objects in the domain is shown in Figure 5.15.

Figure 5.16 is an example of a class declaration which defines the class truss. A class description is organized into several parts: a class name, superclasses, subclasses, attributes, and values. Procedural knowledge is used in tasks such as graphic display, graphic interpretations, and the evaluation of properties of graphic entities. In Figure 5.16 "truss-span," "truss-triangulation," "truss-redundants," and "truss stability" are the names of attached procedures to evaluate the properties of span, triangulation, redundancy, and stability.

The system also possesses classification knowledge, allowing an object or property to be identified from observations provided by the user. The *classifier* uses the class features contained in the type descriptions to successively work its way down a taxonomy, achieving increasingly specific classification. For example, the system could use a "beam" taxonomy to determine first that a given structure is a "beam" as opposed to a "frame" or a "truss," then that it is a "stable beam" as opposed to an "unstable beam," then that it is a "cantilever beam," and so on.

Figures 5.17 and 5.18 show screen displays of sample sessions with the system. The model inferred from the graphic description and the system's reasoning process are monitored in the text window. In the graphics window are the various iconic symbols about which the system is capable of reasoning,

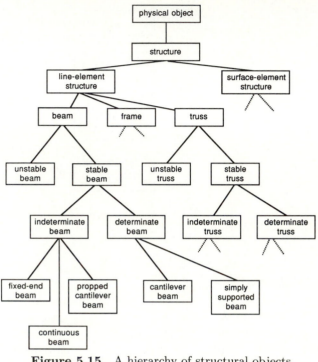

Figure 5.15   A hierarchy of structural objects.

```
TRUSS

name   : Truss
slots  : SuperClasses          : value       : line_element_structure
         SubClasses            : value       : [stable_truss
                                                unstable_truss]

         MembershipCriteria    : value       : [numberOfMembers
                                                numberOfJoints
                                                numberOfSupports
                                                typeOfJoints
                                                typeOfForces
                                                typeOfSupports
                                                locationOfSupports]

         NumberOfJoints        : valueClass  : [intersection integers
                                                greater_than (2)]

         NumberOfMembers       : valueClass  : [intersection integers
                                                greater_than (2)]

         NumberOfSupports      : valueClass  : [intersection integers
                                                greater_than (1)]

         LocationOfSupports    : value       : joint_locations
         Span                  : if_needed   : truss_span
         Triangulation         : if_needed   : truss_triangulation
         NumberOfRedundants    : if_needed   : truss_redundants
         Stability             : if_needed   : truss_stability
```

Figure 5.16   A description of the truss class.

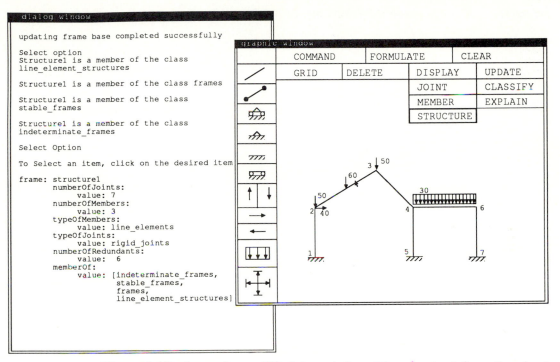

Figure 5.17   A sample run of the interpretation system showing information about a frame structure.

including members with fixed ends, members with pin joints at their ends, and various types of joints, loads, and dimensions.

Figure 5.17 shows that the system has interpreted the graphical description to determine that the structure is an indeterminate, stable frame with 7 joints (all rigid), 6 members, and 3 supports with a redundancy of 6. Figure 5.18 shows that the system has interpreted the graphics to determine that the structure is an indeterminate, stable truss with 18 pin-joints, 33 members, 3 supports, that the forces and supports are nodal at the joints, that the truss is triangulated without overlap, and that the redundancy is 1.

The interesting point about this kind of analysis is that the interpreted model of the design can then form part of the process by which other features of the design are generated. In the next chapter we demonstrate how this kind of information can be used to formulate a design problem for the calculation of member sizes by optimization techniques.

5.5.4   Simulation Knowledge

We consider here an expert system called WATERPEN (Thomson, 1987), that uses heuristic knowledge about water penetration through windows in buildings. The system interprets a description of a window assembly, which is de-

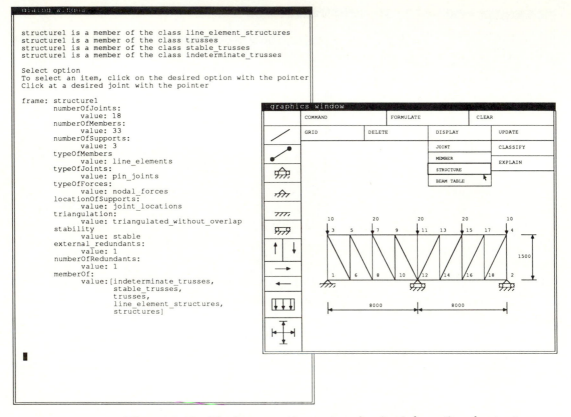

Figure 5.18   The interpretation system showing information about a truss.

scribed by the names of the components and lists of points representing their geometry. The interpretation determines the behavior of water through the assembly, and, by implication, the performance of the assembly. The system simulates water movement and uses rules to capture some of the qualitative knowledge about the complex behavior of water in a window section.

The system interprets a given design description to derive the performance of the design in terms of water penetration. The simulation is displayed graphically so it is evident to a designer whether water penetrates into the interior of the building or not. The designer adds the evaluation—that the performance is or is not satisfactory.

There is a graphical interface that allows for the display and manipulation of a design in the form of a diagrammatic cross section of a window. The knowledge base includes qualitative (commonsense) knowledge of physical processes, specifically about how water might penetrate through the interstices of the window assembly. The physics involved in calculating where water might pass through the frame is complex, especially if the effects of turbulent air

Figure 5.19 If a compartment is wet, then all its corners and sides are wet.

movement are taken into account. Therefore a simplified approach is taken. The knowledge base is largely composed of heuristics, comprising algorithms and rules. It is the rules that interest us most here.

The reasoning process is one of "proving" that certain elements are "wet." The design is interpreted in terms of compartments, surfaces, and corners. Two of the rules are represented and depicted in Figures 5.19 and 5.20:

> **If** a compartment is wet
> **then** all its corners and sides are wet.

> **If** a corner is adjacent to another corner on the
>   boundary of the same compartment **and**
>   the higher point is wet
> **then** the lower corner will also be wet.

The second rule accounts for some of the effects of gravity and surface tension along a surface. It also applies if the two corners are joined by a horizontal surface. Another rule depicted in Figure 5.21 is as follows:

> **If** one end of a common boundary (the line where two objects meet)
>   is wet
> **then** the other end and all the points and sides in between
>   will also be wet.

Figure 5.20 If the higher corner of adjacent corners is wet, then the lower corner is also wet.

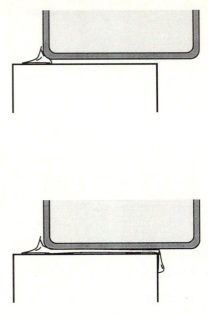

Figure 5.21   The transfer of moisture along a boundary.

This rule assumes the worst case, in which every seal between components of the window frame fails, and in which there is enough wind pressure to drive water through very small openings. Exceptions to this heuristic, of course, are accounted for in the knowledge base. Further knowledge accounts for the effects of drain holes, of water dripping from the corners at the top of a compartment, and of differences in air pressure between compartments.

The graphical representation of an example window-sill section is shown in Figure 5.22; the rain is shown as diagonal lines. It is assumed that the outer seal has failed. Whenever a rule is found to apply, a new fact is added to the collection of things that have been "proven wet," along with a source of the water. The knowledge is used to demonstrate in a graphical fashion the path that water may take through the various parts of the window section.

5.6   Design Codes and Standards as Interpretive Knowledge

In Section 4.5 we referred to design codes; here we consider them in greater detail. Design codes and standards represent a significant body of design knowledge. The knowledge is often causal and experiential, distilled from many sources. In most industries, codes are used to control the design and manufacture of products. Even though codes and standards generally have a formal

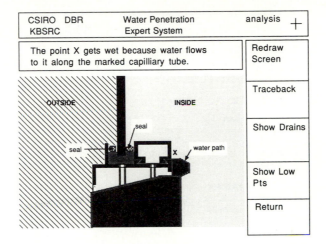

Figure 5.22   Simulating the behaviour of a window sill with WATERPEN (Thomson, 1987).

appearance, they rarely constitute logical, monotonic systems that can be immediately translated into the knowledge bases of deductive expert systems. Furthermore, the knowledge by which designers find their way through a particular code and gain access to that knowledge represents a considerable body of expertise. The representation of this meta-knowledge about design codes is not straightforward.

The functions of codes are discussed by Fenves and Liadis (1976) and Fenves and Wright (1977). In general, design codes are concerned with imposing standards to protect life and property, to prevent adverse conditions, and to protect the rights of the public and individuals. They are generally made up of *provisions* and *definitions*, often in the form of rules, which prescribe certain design requirements. In this section we confine our discussion to the knowledge contained within codes, much of which can be characterized as interpretive knowledge, and consider its exploitation in the design process.

Design codes can be used in two ways. First, before he or she begins designing, a designer may use the code in order to establish certain performance specifications for the design. Designers have some general concept of what they are designing, and they wish to find out what constraints are applicable. The second use of design codes is for compliance checking—that is, the knowledge of the code is employed in the interpretation of a proposed design. The interpretation of most interest is generally whether the design complies or does not comply with the code. We will demonstrate both of these uses with reference to some of the knowledge contained in a particular code. The first example shows a rule-based system deriving requirements, while the second shows a model-based system carrying out evaluation.

PART 49—ROOM SIZES AND HEIGHTS

SIZES OF HABIT-
ABLE ROOMS IN
RESIDENTIAL
BUILDINGS—
Kitchens excluded

49.1 * * * *

49.2 (1) The minimum floor areas prescribed by this Clause shall not apply to fully-enclosed kitchens.

Basic minimum
area

(2) A habitable room in a Class I, II, III or IV building shall have a minimum floor area of 7.5 m² and such area shall be increased, when necessary, to comply with subclauses (3), (4), and (5).

One-room
dwellings or flats

(3) In a dwelling-house or flat containing only one habitable room (not counting any fully-enclosed kitchen), such room shall have a minimum floor area of 18.5 m².

Dwellings and flats
containing more than
one habitable room

(4) In a dwelling-house or flat containing more than one habitable room (not counting any fully-enclosed kitchen)—
   (a) at least one habitable room shall have a minimum floor area of 14 m²; and
   (b) at least one other habitable room shall have a minimum floor area of 11 m².

Habitable rooms
incorporating
cooking facilities

(5) Where, in lieu of a fully-enclosed kitchen, an alcove or other space within a habitable room is provided for the preparation and cooking of food, the required minimum floor area of that room shall be increased by 3 m².

SIZES OF BATH-
ROOMS AND
SHOWER ROOMS
IN RESIDENTIAL
BUILDINGS—
Basic minimum
areas

49.3 (1) A bathroom or shower room in a Class I, II, III or IV building shall have a minimum floor area as follows:
   (a) Bathroom—2.2 m².
   (b) Bathroom provided with a bath and a shower that is not above the bath—2.8 m².
   (c) Shower room—1.1 m².

Additional
facilities
installed

(2) It shall not be necessary to increase the area of a bathroom or shower room to accommodate a wash-hand basin, but where a water closet pan is installed, or the bathroom or shower room is designed to accommodate clothes washing facilities, the minimum floor areas pre-scribed by subclause (1) shall be increased for each facility as follows:
   (a) Water closet pan—0.7 m².
   (b) Washing machine without washtub (where this is additional to the provision elsewhere of the clothes washing facilities required by part 46)—0.7 m².
   (c) Washing machine and washtub—1.1 m².
   (d) Copper and washtub—1.1 m².
   (e) Clothes-drying cabinet—0.5 m².

SIZES OF WATER
CLOSETS: ALL
BUILDINGS

49.4 A water closet in any class of building shall have minimum floor area of 1.1 m² and a minimum width of 810 mm.

HEIGHT OF
ROOMS—
Minimum height

49.5 (1) In a Class I, II, III or IV building, the following rooms shall have the minimum heights set out:
   (a) habitable rooms—2400 mm.
   (b) Bathrooms, shower rooms, water closets, and laundries—2100 mm.

Reduced height
permissible

(2) The Council may permit the height of a room referred to in sub-clause (1) to be less than 2400 mm or 2100 mm in height, as the case may be, over a portion of the room if the Council is satisfied that such reduced height will not unduly interfere with the intended functioning of the room.

Figure 5.23 Part of the Australian Model Uniform Building Code.

5.6.1 Establishing Performance Specifications from Design Codes

We will deal with a part of a typical building code dealing with the minimum requirements for room sizes and heights in buildings (Australian Model Uniform Building Code, 1982). Figure 5.23 shows the actual text of part of the design code, while Figure 5.24 shows the translation of this knowledge into if-then rules suitable for processing as a simple deductive expert system (Rosenman and Gero, 1985). Only some of the rules are shown. Figure 5.25 lists the dialogue between the human operator and the system, where the knowledge forms part of the knowledge base of an expert system shell with forward- and backward-chaining capabilities. The user input is in bold type.

In this example there is no design since the system is used to explore the requirements of the code before design begins. The user asks the system to find the minimum required floor area of a room, in this case a bathroom. The system prompts for information it needs to answer this question. If the human

R13	**If**	type **of** room **Is** bathroom **and**
		not requirement for increase in area of bathroom
	then	required minimum floor area **of** room **Is** 2.2.
R14	**If**	type **of** room **Is** bathroom **and**
		requirement for increase in area of bathroom **and**
		increase in area required **of** bathroom **Is** INCR
	then	required minimum floor area **of** room **Is** 2.2 + INCR.
R15	**If**	contents **of** bathroom **Include** wc **or**
		separate shower **or** wm
	then	requirement for increase in area of bathroom.
R16	**If**	increase in area for wc **of** bathroom **Is** AWC **and**
		increase in area for shower **of** bathroom **Is** AS **and**
		increase in area for wm **of** bathroom **Is** AWM
	then	increase in area required OF bathroom **Is**
		AWC + AS + AWM.
R17	**If**	**not** contents **of** bathroom **Include** wc
	then	increase in area for wc **of** bathroom **Is** 0.0.
R18	**If**	contents **of** bathroom **Include** wc
	then	increase in area for wc **of** bathroom **Is** 0.7.
R19	**If**	**not** contents **of** bathroom **Include** separate shower
	then	increase in area for shower **of** bathroom **Is** 0.0.
R20	**If**	contents **of** bathroom **Include** separate shower
	then	increase in area for shower **of** bathroom **Is** 0.6.
R21	**If**	**not** contents **of** bathroom **include** wm
	then	increase in area for wm **of** bathroom **Is** 0.0.
R22	**If**	contents **of** bathroom **Include** wm
	then	increase in area for wm **of** bathroom **Is** 0.7.

Figure 5.24 Part of the code of Figure 5.23 as expert system rules.

```
enter command
••••••••••••••••••••••••••••••••••••••••••••••••••••••••••••••••••••••••••••
? find required minimum floor area of room
••••••••••••••••••••••••••••••••••••••••••••••••••••••••••••••••••••••••••••

The type of room is ? - enter value (how/why/explain)

the options for type of room are:
1   bathroom
2   living room
3   other
4   unknown

? 1

requirement for increase in area of bathroom ? - y/n (how/why/explain)

? why

type of room is bathroom and
not requirement for increase in area of bathroom
   needed to prove
required minimum floor area of bathroom is 2.2

requirement for increase in area of bathroom ? - y/n (how/why/explain)

? how

The contents of bathroom include ? - enter value (how/why/explain)

the options for contents of bathroom are:

1   wc
2   separate shower
3   wm
4   other

? 1 3

-----------------------------------------------------------------------------
requirement for increase in area of bathroom is true
-----------------------------------------------------------------------------
increase in area for wc of bathroom is 0.7
-----------------------------------------------------------------------------
increase in area for shower of bathroom is 0.0
-----------------------------------------------------------------------------
increase in area for wm of bathroom is 0.7
-----------------------------------------------------------------------------
increase in area required of bathroom is 1.4
-----------------------------------------------------------------------------
••••••••••••••••••••••••••••••••••••••••••••••••••••••••••••••••••••••••••••
required minimum floor area of room is 3.6
••••••••••••••••••••••••••••••••••••••••••••••••••••••••••••••••••••••••••••
```

Figure 5.25   A dialogue between the expert system and the user.

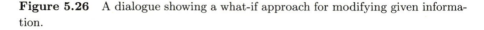

```
enter command
•••••••••••••••••••••••••••••••••••••••••••••••••••••••••••••••••••••••••••••••••••••
? what if contents of bathroom include separate shower
•••••••••••••••••••••••••••••••••••••••••••••••••••••••••••••••••••••••••••••••••••••

requirement for increase in area of bathroom is true

increase in area for shower of bathroom is 0.6

increase in area for wc of bathroom is 0.7

increase in area for wm of bathroom is 0.7

increase in area required of bathroom is 2.0

required minimum floor area of room is 4.2
```

Figure 5.26   A dialogue showing a what-if approach for modifying given information.

operator cannot provide this information, then he or she responds with "how," meaning "how do I answer this?" The system will attempt to find another rule in its knowledge base to see whether it can derive a proof for that proposition. If the user's responses indicate an interest in a bathroom containing a toilet and a washing machine, then the system finds that the minimum required floor area is 3.6 square meters.

The system can be asked to justify its findings, to explain what propositions have been proven true or untrue and why, and to explain why it did not arrive at some other value. A "what-if" query can also be employed to obtain alternative results for different answers to queries (Figure 5.26). The examples in Figures 5.25 and 5.26 show a goal-directed approach. The user asks the system to "find"—that is, to infer—a goal. The system chains backward from this goal until it is satisfied (or fails). Figure 5.27 shows an example of a forward-chaining approach, in which information is proffered and the system infers everything that it can from this information. In this case we see that not only requirements about floor area are obtained, but also requirements about height. The system uses forward chaining to infer as much as it can from the given information, but when it needs more information to infer some particular proposition it employs backward chaining.

5.6.2   Compliance Checking with Design Codes

As a design is checked for compliance with the code, certain design attributes emerge. In the case of building codes, for example, these attributes include the class of the building (whether it is residential, industrial, or public), the floor and window areas of the rooms, and the fire-resistance ratings of the walls. It is useful for an expert system for compliance checking to "anticipate" these

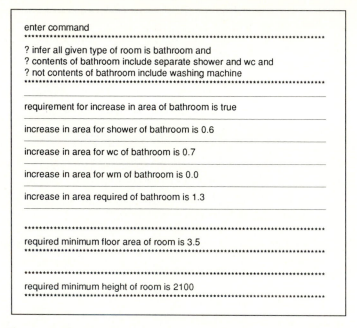

enter command

? infer all given type of room is bathroom and
? contents of bathroom include separate shower and wc and
? not contents of bathroom include washing machine

requirement for increase in area of bathroom is true

increase in area for shower of bathroom is 0.6

increase in area for wc of bathroom is 0.7

increase in area for wm of bathroom is 0.0

increase in area required of bathroom is 1.3

required minimum floor area of room is 3.5

required minimum height of room is 2100

Figure 5.27   A dialogue showing a forward-chaining approach.

attributes, perhaps in the form of a framework into which attribute values can be slotted as they are determined. In Section 3.5 we showed that it is possible and desirable to extract a model represented as a frame from a knowledge base made up of rules. Here we demonstrate the use of such a facility in an expert system. Among other advantages, this model facilitates communication between an expert system and a commercial computer-aided design system.

In this discussion we consider the rules of Figure 5.24 to be part of the knowledge base of an expert system. We need three components for such a system: (1) an expert system shell; (2) an interface that provides the communication between the rule base, the model, and the model of any external program (we can call this a knowledge-based interface, or KBI); and (3) a frame generator that helps extract and create the model implicit in a particular knowledge base made up of rules (although this could have been created *a priori* (Rosenman, Manago, and Gero, 1986)).

We can employ an expert system shell of the kind described in Section 4.5. Propositions about a design can be represented in the following forms:

<**predicate**>

where <**predicate**> takes the value of true or false;
for example, "The local planning board is satisfied."

<**object**> <**verb**> <**value**>

where <**verb**> can be any user-defined term and <**object**> can take the value <**value**>; for example, "The room **is a** bathroom."

> **<attribute>** of **<object>** **<verb>** **<value>**
>   where **<attribute>** of **<object>** can take the value **<value>**;
>   for example, "Length of room should be 3,600."
>
> **type of <object> is <class name>;**
>   for example, "Type of room is bathroom."

Rules are of the if-then form described previously. As discussed in Section 3.5.2, rules of this form can be readily translated into frame descriptions. As an intermediate step we need a facility for parsing rules to derive lists of options for objects. Consider, for example, the following rules,

> **If**   room **is a** bedroom **or** living room **or** sunroom
> **then**   type of room is habitable.

> **If**   bathroom **contains** shower **or** washing machine **or** washtub
> **then**   increase in area of bathroom is required.

The following options lists are created for these rules:

> room **is a**:   bedroom, living room, or sunroom
> **type of** room **is**: habitable
> bathroom **contains**:   shower, washing machine, or washtub

Of course, the options lists would be expanded by options found in other rules.

These options lists can be translated into frame representations. Thus an options list of the form

> **<object>** **<verb>**: $O_1, O_2, ... O_n$

is translated into a frame, where **<object>** is a frame name and $O_1, O_2, ...$ O_n are "kinds of" **<object>**. The options list

> room **is a**:   bedroom, living room or sunroom

is translated into the frame shown in Figure 5.28.

This kind of translation provides a model containing information in terms of objects, attributes, and values; there is no reference to an actual design. This kind of analysis of the knowledge base of the code can be carried out before subjecting any design proposal to the code.

It is likely that the rules in the knowledge base also carry useful structural information about the design domain that cannot be extracted by simple pattern matching as just shown. While it is possible to construct such a model in part from the rules of the knowledge base, it may be necessary to augment the model and refine it manually. Certain definitions may also have to be

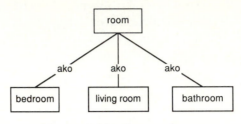

Figure 5.28   A room frame.

added. The model could, of course, have been created manually and "filled out" using the information in the rules.

The interface component (KBI) of this system is concerned with interpreting the database of the commercial CAD system and slotting information into the frames. Figure 5.29 shows three parts to a KBI: (1) the frame system, (2) an interpreter, and (3) a set of definitions. A database is included as an example of a description contained in an external system.

The frame system contains the model of the design or objects with which the system is concerned, including information about classes and superclasses of objects and object properties. A frame generator analyzes the rule base and constructs the view of the model that is implicit in the rule base.

The interface (KBI) links the expert system shell and the model by monitoring the requests made by the expert system. If information is already contained within the frame representation, then the system looks there rather than ask the user for the information. As new facts are determined in response to questions and as new facts are inferred, they are inserted into the appropriate slots in the frame representation.

The model constitutes a standard description of a class of designs to

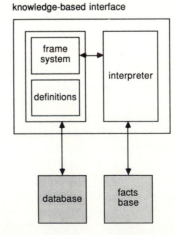

Figure 5.29   The organization of a knowledge-based interface.

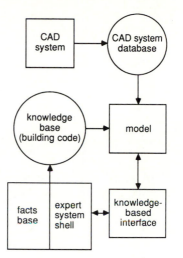

Figure 5.30   The organization of a model-based expert system

which different computer systems may have access. For example, if we take a description of a building in a CAD system, the CAD system's database would contain the following type of information: the number and type of rooms; the shape of the spaces; the dimensions (length, width, and height); and the elements contained in the rooms (doors, windows, fixtures, and furniture). How can this information be translated into a form that is useful to a design-code expert system? It is first necessary to extract the relevant data from the CAD system. In general, a specialized interface linking the data structure of the CAD system is required to translate it into the standard representation needed to communicate with the expert system (this is all that should be required in the way of tailor-made computer code). The interface could use the International Data Exchange Specification (IGES) or another interface between different CAD systems.

Figure 5.30 shows the components of a model-based expert system implemented and tested using EAGLE (Carbs Ltd., 1985). EAGLE, a typical general-purpose three-dimensional CAD modeling system, includes database management facilities. It comprises a series of computer programs that enable an operator to create a three-dimensional graphical model of an artifact. The graphical model may be built up of graphical sub-models that are used repeatedly in different designs. The graphical sub-models, such as components or component assemblies, may be defined from drawing office catalogs applicable to all projects using specific dimensional data. Alternatively, the graphical sub-models may be defined parametrically.

Figure 5.31 shows a two-bedroom apartment (called *flat1*). A graphical model was constructed using EAGLE, and the database information was interpreted to form part of the model. Figure 5.32 shows a typical design session in

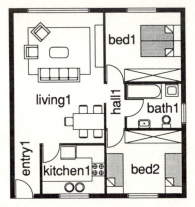

Figure 5.31   A two-bedroom apartment.

which the expert system, comprised of the expert system shell with a knowledge base made up from the building code, interprets the description of the two-bedroom apartment and determines the performance of the bathroom according to the code. The system makes the necessary interpretations about the actual area of the bathroom and the required minimum floor area, using the information within the model about the elements contained in the bathroom. The system then evaluates the area of the bathroom. Evaluative knowledge is built into the knowledge base. By comparing the actual area with the required area, the system makes the evaluation that the area of the room is "satisfactory."

The result of the interaction with the user is the minimum area required for a particular room. This information is stored in the frame representation so that the system can direct its attention to other spaces in the building. The same questions concerning the building have to be answered again, but for different spaces. The frame representation facilitates the management of multiple iterations through the elements of a design using the same knowledge base.

A model-based system should also be able to handle a wider range of user queries than suggested by the simple interfaces described in Section 4.5. The system should be able to "anticipate" certain queries and to accept various forms of queries, restating them to the knowledge base in a suitable form. Such a query might be, "Do I have to increase the area of the bathroom?" It should also be possible to find out things about the knowledge base, as well as about a particular design. For example, we should be able to pose the query, "Is a sunroom a habitable room?"

HI-RISE (Maher and Fenves, 1985) is a well-known system that utilizes a design model as just discussed. Because this system is concerned with configuring and then evaluating alternative structural designs, rather than interpreting a design description, it is discussed in Chapter 6. Another expert system containing knowledge about design codes is FIRECODE (Buis et al., 1986; Hay, 1984), commissioned by the Building Research Association of New Zealand

```
*******************************************************************************
find diagnostic on area of room
*******************************************************************************

the current context is bedroom1 do you want to change it? (y/n)
?y
place the pointer on the appropriate room and click the mouse

The area of room is ?
area of bathroom1 is 6.48
The type of room is bathroom?
yes bathroom1 is a bathroom
The classification of building is ?
classification of flat1 is Class II
The contents of bathroom include bath?
yes contents of bathroom1 include bath 1

The contents of bathroom include shower?
shower is a kind of element of bathroom
as there are some objects: element of bathroom in the input model
_____
I assume that contents of bathroom1 include shower is false
_____
The contents of bathroom include wc?
yes contents of bathroom1 contains wc1
_____
requirement for increase in area of bathroom is true
_____
increase in area for wc of bathroom is 0.7
_____
The contents of bathroom include clothes washing facilities?
yes contents of bathroom1 include wm1
washtub is a kind of element of bathroom
as there are some objects: element of bathroom in the input model
_____
I assume that contents of bathroom1 include washtub is false
_____
increase in area for clothes washing of bathroom is 0.7
_____
The contents of bathroom include clothes drying cabinet?
clothes drying cabinet is a kind of clothes washing facilities
as there are some objects: clothes washing facilities in the input model
_____
I assume that contents of bathroom1 include clothes drying cabinet is false
_____
increase in area for clothes drying of bathroom is 0.0
_____
increase in area required of bathroom is 1.4
_____
required minimum floor area of room is 3.6
_____
as area of bathroom1 is 6.48
*******************************************************************************
diagnostic on area of room is area of the room is ok!
*******************************************************************************
```

Figure 5.32 A typical design session with the system of Figure 5.30.

to help a designer check whether a building meets the requirements of a draft standard for fire safety. One section of the code dealing with means of escape was implemented as an advisory expert system that determines whether fire escape routes in a particular building design are adequate. FIRECODE is implemented using a frame-based expert system shell and three control mechanisms: (1) backward chaining, using inference rules; (2) the use of procedures as actions; and (3) a reasoning mechanism for the classification of instances. The latter uses explicit classification rules to place objects within a classification hierarchy (more accurately, a "class specialization lattice"). The system is interactive; a sample dialogue is shown in Figure 5.33, in which the user checks whether the entire story of a building meets the code requirements. The summary shows the interpretations that the system derived from the description of the building provided by user responses to system queries.

Welcome to the FireCode system.

Copyright © 1986, Building Research Association of New Zealand &
                    Department of Computer Science,
                    University of Auckland, NZ.

This system is intended to check a building to ensure that it meets the
requirements of Part 2 and Part 6 of the draft standard DZ4226
CODE OF PRACTICE FOR DESIGN FOR FIRE SAFETY

Within this session
Do you want the clause numbers to be displayed as the regulations are examined by
the system? **y**

You may examine one of the following:
   1 – Check the useclass and grading adjustments of an area
   2 – Check all the features of an area
   3 – Check all the areas of a storey
   4 – Check a whole fire compartment
Which of these do you want to choose > **3**

CHECKING THE STOREY

Within the storey
Does the storey have any intermediate floors or part storeys? **n**

How many areas are there within the storey > **2**

Within the intended use of the first area of the storey
The following are valid use classes:
  Assembly areas   : AS, AM, AL, AO
  Sleeping areas   : SC, SD, SA, SS, SH
  Commercial areas   : CO, CS
  Manufacturing areas : MXL, ML, MM, MH MXH
  General Use areas   : GL, GM, GH, GP
Which of these is the use class of the area (e.g. ML)
– If you are unsure of the useclass, or do not know it, type UNKNOWN > **unknown**

Figure 5.33   A sample dialogue with FIRECODE.

The following are descriptions of area uses:
1 – space used for assembly purposes
2 – space used for sleeping
3 – space used for commercial purposes
4 – space used for manufacturing or storage purposes
5 – space used for functions which are auxiliary to main use
Type the number that corresponds to the area you are interested in > **2**

Is the area
1 – enclosed room used by institutional care because of disability
2 – enclosed room used for confinement or detention
3 – a unit or accommodation in a building that shares fire exits with another unit
4 – accommodation in a building not sharing exits
5 – composed of fully detached accommodation or units
Type the number of the most appropriate description > **4**

Within the first area of the storey
The sleeping area can be one of the following:
1 – dormitory or hostel
2 – detention quarters
3 – ward containing more than 2 beds
4 – bunkroom
5 – bedroom
Type the number that corresponds to the description that best fits > **1**

Do you know the occupant load of this area? **n**

How many people was the area designed for > **30**

What is the floor area (in square metres) > **120**

What is the length of the longest path leading from the area
– measured from its origin in the area to the nearest final
exit or safe path (in metres) > **what**

The length of a path is important as it must be possible for all the occupants to leave a
building within 2.5 minutes. The length of the path in this case is regarded as the length
of the longest of any of the paths leading from within any point of the area under
consideration to the point at which it reaches a safe path or a final exit.

Within the first area of the storey
What is the length of the longest path leading from the area
– measured from its origin in the area to the nearest final
exit or safe path (in metres) > **14**

Is the area a dead end, or does it contain a dead end? **n**

Area checked.

Within the intended use of the second area of the storey

The system asks the user questions about the second area, as above. And then ...

Does the area have sprinklers? **why**

I am attempting to determine the longest path allowable in the area. An area with
sprinklers is assumed to spread fire more slowly. This allows a greater path length
within a sprinklered area. (See regulation 6.4.3.1)

Figure 5.33   A sample dialogue with FIRECODE (continued).

Within the second area of the storey
Does the area have sprinklers? **y**

...

Area checked.

Within the storey
How many exitways are there from the storey > **2**

...

Check complete.

Summary of session:

Checking Area
The fire severity rating assumed is low. The occupant response assumed is poor
The occupant mobility assumed is good. The access to safety assumed is good
If any of the above assumptions are doubtful or incorrect, you should examine these
gradings separately. The area has a useclass of ss.
The area has no special features to check.
The maximum allowable length of any path leading from the area is 100 m. The
minimum allowable width of a path within this area, or leading from it is 800 mm.

Checking Area
The fire severity rating assumed is low. The occupant response assumed is poor.
The occupant mobility assumed is poor. The access to safety assumed is limited.
If any of the above assumptions are doubtful or incorrect, you should examine these
gradings separately. The area has a useclass of sc.
For all floors above the ground floor, where the occupants are confined to bed and there
is not an alternative path of travel leading to another storey at the same level but in a
separate fire compartment then at least one alternative path of travel should be through a
fire separation. The first separation should have the appropriate FRR for a storey
separation in the useclass, with a fire control door equipped with a hold open device. See
regulation 6.4.7.5. All changes in floor levels must have ramps. See regulation 6.4.7.6.
The maximum allowable length of any path leading from the area is 18 m. The minimum
allowable width of a path within this area, or leading from it is 1200 mm.

Figure 5.33   A sample dialogue with FIRECODE (continued).

5.7   Summary

We have defined the process of the interpretation of a design as that by
which information is extracted that is implicit in a design description and
a knowledge base. This information often relates to the performance of the
design in some way.

We began by looking at the role of interpretation in language, logic, and
design, demonstrating how interpretation can be modeled simply in design systems by deduction. There appear to be hierarchies of interpretation, in that
certain derived properties of a design lead to the derivation of other properties.
This represents a simplified view of interpretation; we discussed some of the
complexities of the issue. Many features of a design can be inferred from its
explicit description; because not all interpretations are of interest, there must
be some strategy to guide the interpretation process. Generally this can be

achieved by setting up interpretation goals—statements whose truth we wish to test against the design description. Therefore interpretation can exploit proof procedures. As simple as this process appears, we need to be aware of three factors. First, just as parsing has a role to play in interpretation in natural language, we cannot always divorce knowledge of the generative process from effective design interpretation. Second, the knowledge by which we make interpretations is not always complete. Finally, our interpretive knowledge is not always consistent, unambiguous, and certain.

We considered how evaluation fits into our model of interpretation. Evaluation involves making comparisons between interpretations; we demonstrated how it can be modeled quite simply with deductive reasoning. We also focused on interpretation in the context of spatial and geometrical descriptions of designs, demonstrating how graphical images can be seen as a kind of interpretation of a spatial description and how spatial descriptions can be interpreted to infer other properties of a design. Finally, we looked at the role of design codes as repositories of interpretive knowledge.

Interpretation may serve various purposes in design systems. We may simply want to know about the qualities of a design not explicitly stated in design descriptions. We may also wish, however, to incorporate interpretive knowledge in the production of design descriptions, by using that knowledge to suggest the form of our artifact, if at all possible, or to guide the generative process. These topics are discussed in the next chapter.

5.8   Further Reading

While no books specifically on knowledge-based design interpretation have been published, a number contain examples of such systems, in addition to the examples scattered throughout journals.

Numerous examples also appear in the proceedings of the International Conference on the Applications of Artificial Intelligence to Engineering (Sriram and Adey, 1986a, 1986b, 1987a, 1987b, 1987c; Gero, 1988a, 1988b, 1988c) and of the International Federation of Information Processing (Latombe, 1978, Bø and Lillehagen, 1983; Gero, 1985a, 1985b, 1987a; Yoshikawa and Warman, 1987). Additional books of interest are Dym (1985) and Kostem and Maher (1986). Both are collections of conference papers that begin with tutorial material and contain papers on interpretations of designs.

Maher (1987) is a commissioned monograph, the first part of which describes expert systems technology; the second part considers applications, including many that fall into the category of interpretation of designs.

Collections of contributions relating expert systems to various aspects of design can be found in Pham (1988) and Rychener (1988).

Exercises

5.1 What are the similarities and differences in the use of "interpretation" in language, logic, and design. Why is interpretation a useful design term?

5.2 What mechanisms are required in knowledge-based design systems to overcome some of the difficulties in interpreting design descriptions? Distinguish between those mechanisms that can be implemented easily and those that pose conceptual difficulties.

5.3 Define "evaluation." What is its role in design systems?

5.4 Why is the interpretation of geometrical descriptions so important in design? Compare the advantages and disadvantages of procedural and knowledge-based approaches to interpreting geometrical descriptions. How could both approaches be combined effectively in a knowledge-based design system?

5.5 How could user interaction in the systems described in Sections 5.6.1 and 5.6.3 be enhanced? Distinguish between those enhancements that can be implemented easily and those that pose conceptual difficulties.

5.6 Discuss how simulation can be implemented in knowledge-based systems. How could procedural and knowledge-based approaches to simulation be combined effectively in a design system?

5.7 Construct a knowledge base for interpreting a design description in a particular domain.

CHAPTER 6

Producing Designs

So far in this book we have discussed the representation of knowledge, the control of knowledge, and how designs can be described and interpreted. These issues are important if computers are to aid the design process. Also, however, we would expect knowledge-based design systems to aid in the area of design decision making—in other words, in the *production* of design descriptions. In this chapter we look specifically at how we utilize design knowledge in the production of design descriptions.

We refer to two approaches to the production of designs—goal directed and generative. Toward the end of the chapter we discuss the combination of these approaches. However, we can characterize the knowledge concerned with each quite differently. The first approach can be characterized as *goal-directed* or *semantic*. Given a set of design goals—intended performances of the design—an attempt is made to derive a description of a design consistent with these performances. This is achieved by the application of knowledge about how certain aspects of design descriptions map onto their performances—interpretive knowledge. Let us begin, for example, with the idea that our design is to be easy to hold, manually operated, and able to cut wood. By inspecting each of these goals, we bring knowledge to bear about materials and their properties to arrive at a description of a carpenter's saw (or perhaps another implement not previously invented). This approach generally involves successive refinement through a decompositional hierarchy of goals, and the resolution of the inevitable conflicts between subgoals encountered along the way.

The second approach, which can be characterized as *generative* or *syntactic*, involves the exploration of a space of designs defined by some means, perhaps in terms of design actions, grammar rules, or prototype descriptions. One characterization of the generative approach arises from the discussion in the previous chapter. In Chapter 5 we discussed the interpretation of a design description to derive its implicit properties, generally performance attributes. We showed how these derived descriptions of the design can be compared with desired properties in order to test the adequacy of our design. This involves evaluation—comparing performances. We can characterize the generative approach as one of proposing a design description, interpreting it, evaluating its performance, and then proposing a new design description in response to that evaluation. This is one way of producing designs that match our intentions, but as we also pointed out, the approach entails some problems.

In general terms, as outlined in Chapter 2, the production of design descriptions can be characterized by the symbolic mapping

$$D = \tau_4(K_i, K_s, I, V), \tag{2.4}$$

where D represents a design description, V a vocabulary of elements, I an interpretation, K_i and K_s are knowledge used in the production of designs, and τ_4 is a transformation by which a design is produced given the vocabulary, the desired interpretation (goals), and appropriate knowledge. Even when K_i and K_s are both operative in the design process, they can be discussed separately.

We begin this chapter by concentrating on the place of interpretive knowledge (K_i) in producing design descriptions. We pursue six approaches to this:

1. Interpret design performance specifications, or requirements, to produce a design.

2. Interpret design specifications to produce a problem formulation that can be handled by an optimization algorithm.

3. Use plausible reasoning.

4. Use fuzzy reasoning.

5. Use constraint propagation.

6. Use assumption-based reasoning.

In each case the reasoning process may be characterized as *abductive*, as opposed to *deductive*. Through examples we show that, in some cases, the same knowledge and system architecture can be used for both abduction and deduction, depending on how we are able to exploit the mapping between design descriptions and design performances in the particular domain.

We then consider generative, or syntactic, knowledge for producing design descriptions. We develop the model of generation as a stage/state process within a space of design possibilities. For this kind of design system to be useful, it should contain an appropriate representation of the vocabulary of its domain. The system also requires knowledge about how this vocabulary can be manipulated to produce design descriptions. This knowledge "guarantees" that the designs it produces conform to the syntax of the system, much as phrase structure rules ensure syntactically correct sentences in natural language. It is usual, however, that a design must conform to a set of performances—much as a syntactically correct sentence should also map onto to some meaning. In design, generation must be combined with interpretation, evaluation, and "regeneration."

We discuss how the process of generation might be integrated with the interpretive process to produce a system that can generate design descriptions that meet performance requirements. The final part of this chapter, which addresses the control of these operations in order to produce designs, is illustrated with examples of some design systems.

6.1   Issues in Goal Satisfaction

Here we discuss how knowledge about the interpretation of design descriptions can be used to produce descriptions of designs from statements about desired interpretations. Desired interpretations constitute goals: we have some idea about what we want to be true of the design when its description is interpreted. In goal satisfaction we start with that intended interpretation and produce a design description. This has been expressed formally as

$$D = \tau_2(K_i, I) \tag{2.2}$$

where K_i is interpretative knowledge, I is the intended interpretation, and D is the design description that results from the transformation τ_2.

It is important to understand the problems inherent in attempts to satisfy design goals. With knowledge-based systems, particularly those in which the knowledge is based on rules, these problems can be made explicit as issues in logic. Such problems, however, are inherent in any attempt to formalize design based on goal satisfaction, including algorithmic methods.

In Section 2.2 we referred to the rapport human beings enjoy in relation to deductive reasoning. Given the statement

if A **then** B

and given that we know A, then it follows that B is also true (Figure 6.1). We appear to carry this mode of operation into various areas of reasoning endeavor. In a court of law, or while investigating a misdemeanor, we are generally prepared to accept statements arising from arguments that follow the line

if <event> **then** <evidence>

Figure 6.1 Deductive reasoning: if A then B.

In car maintenance or in the area of medical diagnosis, we countenance arguments based on the form

> **if** <condition> **then** <symptom>

with little controversy. In the area of design too, most people will accept facts that emerge from statements based on these forms:

> **if** <description> **then** <performance>
> **if** <form> **then** <function>
> **if** <explicit description> **then** <implicit description>
> **if** <description> **then** <meaning>

The following three statements are rules that follow some of these forms:

> **If** the suspect fires the revolver **and**
>   is not wearing gloves **and**
>   the gun is not touched again
> **then** the gun bears the suspect's fingerprints.

> **If** there is a simple fracture in the arm
> **then** the arm exhibits swelling.

> **If** a beam is subject to excessive bending moment
> **then** it will fail.

We refer to statements that follow this form as deductive rules (Figure 6.2). There may be shortcomings in our knowledge, but usually we can scrutinize a knowledge base of rules following these forms and be confident that the deductive reasoning process will produce valid results. The human species is content not merely to interpret, but seeks to create, to fabricate, and to synthesize. This requires the reversal of the preceding statements (Figure 6.3):

> **if** <evidence> **then** <event>
> **if** <symptom> **then** <condition>
> **if** <performance> **then** <description>
> **if** <implicit description> **then** <explicit description>
> **if** <meaning> **then** <description>

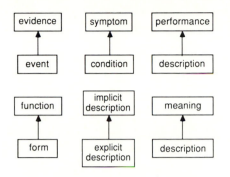

Figure 6.2 Deductive rules.

Some examples of rules following these forms are

> **If** the gun bears the suspect's fingerprints
> **then** the suspect has fired the revolver **and**
> was not wearing gloves **and**
> the gun was not touched again.

> **If** the arm exhibits swelling
> **then** there is a simple fracture in the arm.

> **If** a beam fails
> **then** it has been subjected to excessive bending moment.

It should be apparent that there are difficulties in reasoning with such statements. Just because a gun bears the suspect's fingerprints, it does not follow inevitably that the suspect fired the gun; there are other causes of swelling than simple fractures; there are other ways in which a beam can fail other than because of excessive bending moment. When we reason in this way we generally regard the consequent of a rule as a hypothesis that is supported by

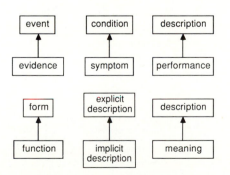

Figure 6.3 Abductive rules.

the evidence of the preconditions. The general form is as follows:

if <evidence> **then** <hypothesis>

We refer to statements that follow these forms as *abductive rules*. If we intend to use these forms for knowledge representation, then we must address three major technical issues: the closed-world assumption, conflict, and one-to-many mappings.

6.1.1   The Closed-World Assumption

The following statements are examples of deductive rules for interpreting descriptions of designs (Figure 6.4):

If   the motor is connected to the wheels by a drive shaft
then   the energy of the motor will be transferred to the wheels.

If   a room is air-conditioned
then   it will be cool in summer.

If   a garden design is based on the geometry of Platonic solids
then   it will appeal to the intellect.

When we turn this knowledge around and use it as a basis for making design decisions, or as generators of design descriptions, the statements take on the abductive form:

If   the energy of the motor is to be transferred to the wheels
then   the motor should be connected to the wheels by a drive shaft.

The problem that arises immediately from this statement is that there may be other ways of transferring energy from motors to wheels, such as gears,

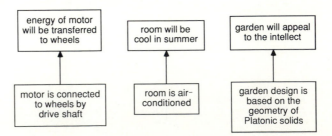

Figure 6.4   Examples of deductive rules.

Figure 6.5 What does knowing that A implies B and B is true say about A?

belts, and chains. They are precluded by this statement, effectively limiting the number of design options. This also applies to the following statement:

> **If** a room is to be cool in summer
> **then** it should be air-conditioned.

There are other ways in which a room can be made cool in summer, such as by adequate passive solar cooling. Let us also consider the following statement:

> **If** a garden design is to appeal to the intellect
> **then** it should be based on the geometry of Platonic solids.

There are other ways in which a garden design may appeal to the intellect, perhaps in terms of its colors, the nature of the design or construction process, or the symbolic references it makes. We encounter particular problems in this area of value judgments. We determine that we like something, come up with reasons for liking it, then use those reasons as a prescription for producing other things we like.

Problems arise if we wish to reason deductively with abductive rules. The key to the problem is that we must assume that the deductive knowledge from which the abductive rules are derived represents a *closed world*—that is, a world in which no knowledge exists other than that stated. This can be made clear by considering a more abstract world.

$$A \Rightarrow B$$
$$B$$

we can infer A if we assume that the system is a closed world, and that B cannot be derived except from A (see Figure 6.5). We sometimes represent this assumption explicitly with the "if and only if" symbol ($\Leftrightarrow$) (Figure 6.6):

$$A \Leftrightarrow B$$

Figure 6.6 A is true if and only if B is true.

Figure 6.7   A implies B, and C implies B.

We cannot adopt this convention, however, if there is more than one rule that produces the same conclusion (Figure 6.7):

$$A \Rightarrow B$$
$$C \Rightarrow B$$

If we are prepared to accept the closed-world assumption, we may simply reverse the rules (Figure 6.8):

$$B \Rightarrow A$$
$$B \Rightarrow C$$

Given B, we can infer A and C or A or C. There is some validity in making the closed-world assumption if we formulate our knowledge around this assumption. This is the case in the knowledge of design codes as discussed in the previous chapters. Design codes are not only descriptive, telling us whether or not a design is satisfactory, but also prescriptive, requiring that the constraints of the code be met.

Even if we can formulate our knowledge such that the closed-world assumption applies, we must also take account of the possibility of conflicting propositions. It may not be sensible, necessary, or possible for both A and B to be true. For example, we may not accept both air-conditioning and passive cooling because they represent alternatives.

Thus, given

$$A \Rightarrow B$$
$$C \Rightarrow B$$
$$B$$

we can say, in a closed world, that at least A or C must be true. If we then pose

Figure 6.8   B implies A, and B implies C.

the question "Give me the requirements for X (to be true)" in a system with its knowledge formulated as deductive rules, all the alternative requirements for X could be true—a useful feature in a design context.

6.1.2 Conflict Resolution

Sets of statements may appear perfectly consistent as deductive rules, but when translated into abductive form they may lead to contradictions. This can be demonstrated with a simple set of statements. First consider the following deductive rules (Figure 6.9):

> **If** an object is made of steel **and**
> it has appropriately arranged sharp teeth
> **then** it will cut wood.

> **If** it has no sharp edges
> **then** it is safe to use.

> **If** it cuts wood **and**
> it is safe to use
> **then** it is suitable as a carpenter's tool.

Steel objects with teeth running down one side make good wood-cutting tools (even though we could offer other descriptions of good wood-cutting tools); objects that do not have sharp edges are generally safe to use; and safe wood-cutting tools are suitable for carpenters. Armed with these rules, we may attempt to assess whether a handsaw is a suitable tool for a carpenter. It will certainly cut wood, by the first rule. When we apply the second rule we find

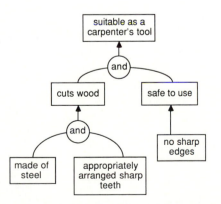

Figure 6.9 An inference net for a carpenter's tool.

that a handsaw does have sharp edges. We have not, however, proven that it is unsafe; we have established only that we cannot prove its safety by this rule. When we attempt to apply the third rule, we find that nothing can be said in terms of the handsaw's suitability as a carpenter's tool. We could make the knowledge base more useful by adding more rules about other ways that an object can be deemed safe. The knowledge is too incomplete to be useful, but is not inconsistent.

If the statements are translated into abductive rules, however, we do encounter inconsistencies. In converting to abductive rules we assume that the closed-world assumption applies and that no other rules impinge on the domain. We wish to use this knowledge in order to produce a description of a carpenter's tool. It is as if we wish to design a tool that we can supply to carpenters (Figure 6.10).

> **If**   it is to cut wood
> **then**   it should be made of steel **and**
>   have appropriately arranged sharp teeth.

> **If**   it is to be safe to use
> **then**   it should have no sharp edges.

> **If**   it is to be suitable as a carpenter's tool
> **then**   it should cut wood **and**
>   be safe to use.

If we study the logic of these statements we see that there is a contradiction. The problem is that the object is to have sharp teeth and yet no sharp edges. We can represent the problem more abstractly. The deductive knowledge base

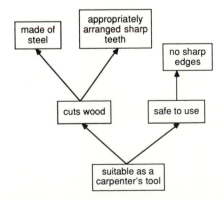

Figure 6.10   The abductive version of Figure 6.9.

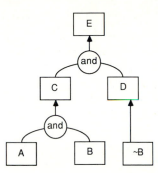

Figure 6.11 An abstract version of Figure 6.9.

can be simplified as follows (Figure 6.11):

$$A \wedge B \Rightarrow C$$
$$\sim B \Rightarrow D$$
$$C \wedge D \Rightarrow E$$

If we translate these statements into abductive rules, assuming that they represent a closed world, we produce the following (Figure 6.12):

$$C \Rightarrow B \wedge A$$
$$D \Rightarrow \sim B$$
$$E \Rightarrow C \wedge D$$

Given that E is true, we can infer C and D. However, C produces B and D produces $\sim$B. That both B and $\sim$B are true is a contradiction, and thus the system is not logical.

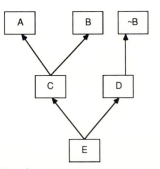

Figure 6.12 An abstract version of Figure 6.10.

In this example the source of the contradiction is fairly obvious, and after a little thought we can see how to reformulate the statements so that this kind of contradiction does not occur. With large knowledge bases of abductive statements, however, such contradictions may not be as easy to detect and rectify. As we demonstrate in Section 6.2.6, it may not even be desirable to remove such contradictions; we may need them to produce design alternatives. Clearly, some mechanisms other than straight logical deduction are required to handle this kind of reasoning.

6.1.3   One-to-Many Mappings

We often must establish a mapping between names of entities and the name of the class to which those entities belong:

    If x is a bicycle **or** a car **or** a bus
then x is a vehicle.

Deductive rules such as this one provide us with a many-to-one mapping: all of the objects satisfying the requirements on the left side of the rule fall within the category specified on the right side of the rule. This is also the case when we describe objects in terms of geometrical parameters (Figures 6.13–6.15).

    If   there are two rooms a and b **and**
          the southern extremity of a is north of the northern extremity of b
then a is north of b.

    If   the object on the drawing is a square between
          70 and 120 centimeters **and**
          the square is hatched with pattern No.2
then   the object on the drawing is a concrete column.

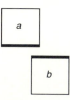

Figure 6.13 a is north of b.

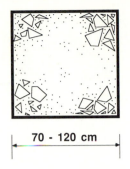

70 - 120 cm

Figure 6.14   A concrete column.

> **If**   the saw has at least 5 teeth per centimeter **and**
>   is no more than 2 millimeters in thickness
> **then**   the saw is good for joinery work.

The sense of these rules will generally be accepted subject to some refinement. If these statements are re-expressed as abductive rules, then the mappings become one-to-many mappings.

We may use the first of these three rules to produce a description of the locations of two rooms a and b, given that we want a to be to the north of b.

> **If** a is to be north of b
> **then** the southern extremity of a is to be north
>   of the northern extremity of b.

The difficulty here is that there will be many ways in which the southern extremity of a is north of the northern extremity of b (Figure 6.16). This problem also arises when the other rules are converted to abductive rules

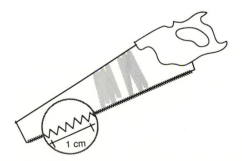

1 cm

Figure 6.15   A saw suitable for joinery work.

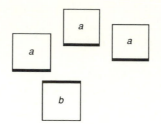

Figure 6.16   Some of the many ways that *a* can be north of *b*.

(Figures 6.17 and 6.18):

> **If**   the object on the drawing is to be a concrete column
> **then**   the object on the drawing must be a square between
>   70 and 120 centimeters **and**
>   the square must be hatched with pattern No. 2.

> **If**   the saw is good for joinery work
> **then**   the saw must have at least 5 teeth per centimeter **and**
>   must be no more than 2 millimeters in thickness.

Rules such as these define *spaces of designs* in terms of constraints. This approach is interesting and useful if there are several such rules, each imposing different constraints on the space of designs. For example, in the case of the simple configuration of two rooms, we may require that the kitchen is north of the dining room, that the kitchen is half the size of the dining room, and that their west walls are aligned. Armed with abductive rules telling us how each of these performances can be achieved in terms of geometrical coordinates, we effectively constrain the space of possible designs. With more constraints a design may emerge eventually, but problems arise where these

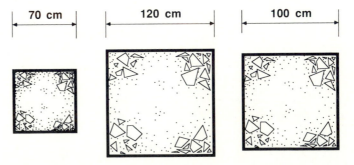

Figure 6.17   Some of the columns that satisfy the rule.

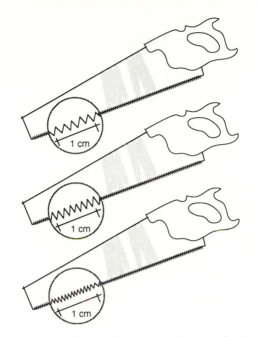

Figure 6.18   Some of the saws that satisfy the rule.

constraints conflict. We discuss the handling and propagation of constraints in Section 6.2.5.

6.2   Producing Designs through Interpretive Knowledge

Having considered some of the issues in abductive reasoning, we can discuss some approaches to reasoning from goals, as well as what makes goals interesting in the context of knowledge-based design systems. First, for certain domains and for certain formulations of the design task, we can assume that there is a direct mapping between the design specification (the requirements) and the design description resulting from those specifications. This certainly provides the simplest approach to reasoning from goals. In some other cases we may be able to formulate our design knowledge such that there is a direct mapping between some statements of a set of design requirements and a problem formulation that can be handled algorithmically. These approaches use deterministic knowledge.

A further approach is to express our knowledge in terms of fuzzy variables and to apply the calculus of fuzzy logic to determine descriptions that satisfy goals. Because of the inherently tentative nature of the mapping between performance and description, we may prefer to state design knowledge in probabilistic terms in order to decide among competing design descriptions. These approaches would appear most suitable when there are small numbers of design decisions to be made.

Third, we may also reason from the constraints within our design knowledge base to produce new constraints. These new constraints may be more effective in defining the space of design possibilities. Finally, using "truth maintenance", we can tackle head-on the issue of conflicts among competing decisions and reason.

Most design systems based on goal satisfaction will incorporate a combination of approaches. The chapter includes discussions of expert systems for design that employ some of these approaches.

6.2.1   Interpretation of Design Specifications

In this approach we make several assumptions. First, we assume that the deductive form (**if** <description> **then** <performance>) of our knowledge constitutes a closed world. We also assume that we are dealing with one-to-one mappings between performances and design descriptions—that is, one set of performances matches one design. Finally, we assume that there are no conflicts in the abductive form (**if** <performance> **then** <description>) of the knowledge. The knowledge is formulated to account for these assumptions. In the example presented here they are carried through to the detailed description of the design.

In this approach we can treat abductive rules as if they are deductive rules, and we can reason with them using a deductive expert system shell. In this sense the process can be seen simply as the interpretation of specifications to produce descriptions of designs (Figure 6.19).

The use of abductive inference in the interpretation of performance specifications is demonstrated by RETWALL (Hutchinson, 1985; Hutchinson, Rosenman, and Gero, 1987), an expert system for the selection and design of earth-retaining structures. The procedure followed by RETWALL to produce a design description is first to select a set of general design descriptors based on an interpretation of the desired performance, given primarily in terms of site conditions. Then the system selects a final choice based on an implicit cost criterion, and finally refines the properties of the structure by defining its

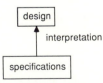

Figure 6.19   Interpreting specifications to produce a description of a design.

parameters (geometry, dimensions, and reinforcement). A typical rule is as follows:

> **If** height of earth-retaining structure (in millimeters)
>   is greater than 1,800 **and**
>   height of earth-retaining structure (in millimeters)
>   is less than 10,000 **and**
>   reinforced concrete wall is aesthetically acceptable **and**
>   labor and materials are available for reinforced concrete
> **then** possible type of earth-retaining structure is concrete cantilever **and**
>   reinforced concrete wall is suitable for this application.

RETWALL then employs a deductive-reasoning mechanism (using either backward or forward chaining) to arrive at a suitable description of a structure that meets the given requirements. The system reduces the range of options available by judicious inquiry about soil and topographical conditions as well as the designer's preferences. When it has made a decision about the type of retaining wall, it proceeds to determine dimensions and other parametric conditions. Thus, the prototypical design is refined incrementally, depending on the design specifications. Some further examples of rules are given next. They are divided into categories, but the categories make no difference in the way the knowledge is processed.

1. Rules to determine whether an earth-retaining structure is required. For example,

> **If** type of application for wall is APP **and**
>   type of application for wall is marine **and**
>   site case most applicable (as shown in diagram) is 1 **and**
>   horizontal distance to level area (d1) (in millimeters) is D1 **and**
>   horizontal distance to level area (d2) (in millimeters) is D2 **and**
>   height difference shown (h) (in millimeters) is H **and**
>   tidal range (h1) (in millimeters) is TR **and**
>   height of low-water level (h2) (in millimeters) is HLWL **and**
>   X is D1/H **and**
>   X is less than 3
> **then** earth-retaining structure is required **and**
>   height of earth-retaining structure (in millimeters) is H.

2. Rules to determine whether an embankment/cut should be used. For example,

> **If**   not(type of application for wall is marine) **and**
>   not("are you aware of any technical or other reason
>   for a retaining structure") **and**
>   geometry of the site is sufficient to allow the construction
>   of an embankment/cut on site **and**
>   consequence of failure of earth-retaining structure
>   is moderate **and**
>   groundwater flow through proposed alignment is low
> **then**   earth-retaining structure is earth embankment/cut.

3. Rules to determine blockwork wall suitability and to refine the design. For example.

> **If**   height of earth retaining structure (in millimeters)
>   is greater than 2,600 **and**
>   backfill type is 2 or 3 **and**
>   base type is 4
> **then**   blockwork wall type is not standard.

4. Rules to determine soil classification. For example,

> **If**   material type is gravel **and**
>   sieve analysis result is 2 **and**
>   Atterburg result is 3 **and**
>   grading is well graded
> **then**   soil classification of backfill is GW-GC.

5. Rules for selecting type of earth-retaining structure. For example,

> **If**   type of application for wall is A **and**
>   not(type of application for wall is heavy vertical load/abutment
>   **or** temporary **or** marine **or** emergency) **and**
>   height of earth-retaining structure (in millimeters)
>   is less or equal to 3,200 **and**
>   blockwork wall is aesthetically acceptable **and**
>   labor and materials are available for blockwork wall
> **then**   possible type of earth-retaining structure is blockwork wall **and**
>   blockwork wall is suitable for this application.

Various options are presented by the system as diagrams from a stored set of images associated with various propositions. The final design is also presented graphically. Figure 6.20 shows a screen display of part of the dialogue in which the designer provides answers based on the interpretation of

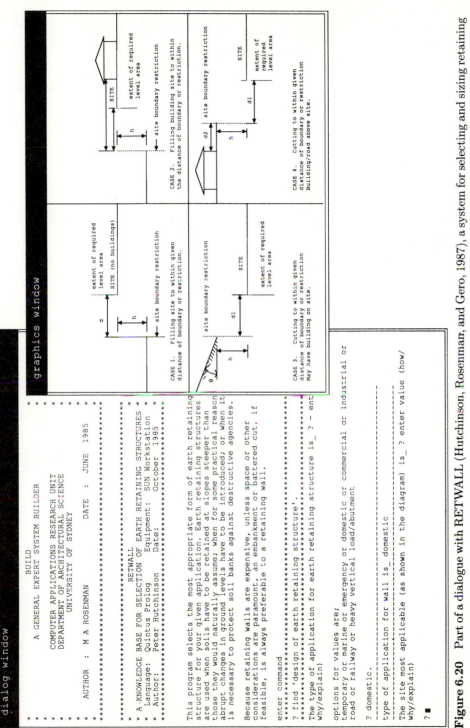

Figure 6.20 Part of a dialogue with RETWALL (Hutchinson, Rosenmar, and Gero, 1987), a system for selecting and sizing retaining walls.

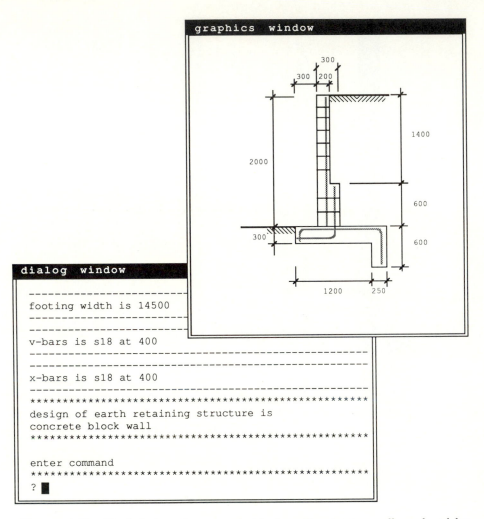

Figure 6.21 Final design of a concrete block earth-retaining wall produced by RETWALL.

the diagrams. Figure 6.21 shows a screen display of the final design; its cross section has been selected and sized for the imposed loads.

This approach to design—one in which design requirements are interpreted to produce a design description—has been exploited to advantage in many design systems. Another example is provided by some of the rules in HI-RISE (Maher, 1984), a system for the preliminary design of the structural frame of multistory buildings, which we introduced in Section 5.7.2. The following rule shows one of the heuristics used to estimate the depth of the floor for the design of a lateral load-resisting system:

> **If**   occupancy is commercial **and**
>   material is steel
> **then** assume 4 inches of concrete on a steel deck **and**
>   beams in both directions using a span/depth ratio of 24.

The knowledge in these examples was formulated as if the closed-world assumption applied to the deductive form of the knowledge, as if we were dealing with one-to-one mappings between performance and design description, and assuming that there were no conflicts in the abductive form of the knowledge. This approach is particularly suitable when we are interested in knowledge about selecting design components or selecting solutions or partial solutions to a task, or when we wish to select a prototype that is to be refined. Prototype refinement is discussed in Section 6.3.2.

6.2.2 Interpreting Design Specifications for Problem Formulations

In the example presented here, we maintain the assumptions given in the previous section in formulating a design specification in terms of an optimization problem in which conflicts are handled algorithmically. We can adopt optimization as the decision-making approach when the values of the variables are numeric and there are many conflicting decisions, each dependent on the other, and when the knowledge for their resolution is represented algorithmically.

What constitutes a design description and what is a statement about the performance of a design depends on one's point of view. We can consider the process of design by interpreting specifications as one in which statements about the design are translated into other statements. To illustrate this point, consider rule 4 from the example in the preceding section:

> **If**   material type is gravel **and**
>   sieve analysis result is 2 **and**
>   Atterburg result is 3 **and**
>   grading is well graded
> **then** soil classification of backfill is GW-GC.

(GW-GC is an acronym for a particular class of soil.)

At first glance, a soil classification may appear far removed from a design description. One of the design requirements, however, is that the retaining wall accommodate the peculiarities of the site. One of the site characteristics is its soil. In order to help determine something about the retaining wall, site characteristics must be interpreted in terms of soil classification. This classification is then of a form that can be used by other knowledge. Effectively we

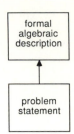

Figure 6.22   Interpreting a problem statement as a formal, algebraic description.

are converting parameters from one form into another so that they can be useful to the next step in the decision-making process. The required performance
is restated in terms that facilitates the application of further knowledge to
produce descriptions of the design. We can apply the same principle to the
translation of informal descriptions of a problem into a formal, algebraic description (Figure 6.22).

In the systems considered so far we have demonstrated one uniform representation of knowledge—that is, as rules. As we have discussed, there are
other methods of knowledge representation, including procedures encoded as
algorithms. One important body of algorithmic knowledge for translating performances into design descriptions is represented by different techniques of
optimization. Although optimization techniques do not fall within the scope
of this book, there is some value in discussing how a description of a design
specification can be brought to the stage at which it can be interpreted by
an optimization algorithm. Sometimes this description of a problem is said to
be in *canonical* form—in optimization, a set of algebraic and inequality expressions. We discuss how we can take statements about desired performance,
generally in terms of constraints, and translate them into a canonical form
suitable for processing as an optimization problem.

Because a designer rarely describes problems directly in optimization
terms, there is some advantage in devising design systems that can achieve
the necessary interpretation. We must translate the desired performance expressed as a pseudo-English description into canonical algebraic expressions.
We choose to represent the canonical description as frames. Subsequently, this
information can be used by a body of knowledge that "recognizes" appropriate features of the formulation, then selects and runs appropriate optimization
algorithms. Knowledge about problem recognition is discussed in Section 6.7.

The canonical form consists of a collection of facts about objects and
their properties in terms of variables, as well as relationships between objects
in terms of constraints and objectives. It is presumed that the designer can at
least establish the constraints and design goals that govern the design. Some
examples of necessary translations appear next.

Variables

Variables, which form the basis of any optimization problem description, can be treated in terms of object-attribute-value triplets. The following statement is an example of an informal description of some knowledge about a design domain:

living room length **equals 2 times** kitchen width

This can be partially interpreted as follows:

Object	Attribute	Value
living_room	length	2L
kitchen	width	L

The statement

bathroom length **is** 2.7 meters

can be interpreted as

Object	Attribute	Value
bathroom	length	2.7

Constraints

The design constraints reflect the requirements that must be satisfied by the design. These constraints can be stated by declaring a maximum or minimum value for an attribute of an object or by specifying equality or inequality relationships among object attributes. For example,

maximum living room area **is** 30.0 square meters

sum of living room width **and** kitchen width **equals** house width

Objectives

The design goals or objectives can be declared by the designer and are optimized. The design objectives are stated as to maximize or minimize an object attribute or a function of object attributes. For example,

minimize house cost

maximize house area

Frames can be used to represent the attributes and values of objects. Procedural knowledge, which is required to carry out certain tasks in the modeling process, can be attached to a particular attribute of an object, using such specifications as if needed, if added, and if removed (see Sections 3.1.4 and 3.1.5). The class inheritance mechanism means that when an instance of an object is created, it inherits all the attributes of the class frame. Frames are used to represent the constraints and objectives as well as the design variables. The knowledge for the problem-formulation processes can be represented as production rules (or actions).

As an aside, an interesting aspect of this operation is that by which algebraic expressions are simplified. Optimization techniques generally demand that expressions are in a particular form—usually, such that there are no brackets. We can capture basic algebraic knowledge in terms of rewrite rules:

$$a \times (b + c) \rightarrow a \times b + a \times c$$
$$(a + b) \times (c + d) \rightarrow a \times (c + d) + b \times (c + d)$$

We can represent algebraic expansion rules in this way and treat them as actions (see Section 4.4.1):

$$\text{Action}(\text{DistributiveRule1}, a \times (b + c), a \times b + c \times d))$$
$$\text{Action}(\text{DistributiveRule2}, (a + b) \times (c + d), a \times (c + d) + b \times (c + d))$$

Given a particular expression, such as,

$$X_1 + (X_2 + X_3) \times (Y_1 + Y_2) - Y_3$$

we can devise a control mechanism such that the firing of actions produces new forms of the expression. The successive firing of actions, such as those just given, results in the expanded expression

$$X_1 + X_2 \times Y_1 + X_2 \times Y_2 + X_3 \times Y_1 + X_3 \times Y_2 - Y_3$$

This expression may constitute part of a constraint, or an expression to be optimized. Because of the complexity involved in expanding certain expressions, we also require control knowledge to select among ambiguous expressions or alternative expansions. We need not go into these control issues here. An implementation of algebraic manipulation in LISP is described by Balachandran (1988) and Balachandran and Gero (1987b) based on the MACSYMA program (Mathlab Group, 1977).

Let us now consider two examples. The first deals with the optimum dimensions of building floorplans, while the second is concerned with the optimum design of beams. For the sake of clarity, the examples show the designer describing the problem in a textual form. We would expect, however, that our system would use a "graphical interpreter" to create this information (see Section 5.5.3).

The Floorplan Dimensioning Problem

This problem is concerned with finding the optimal dimensions of a small rectangular floor plan for which the topology and dimensionless geometry

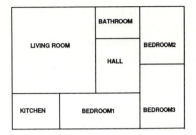

Figure 6.23   Dimensionless representation of a house layout.

have been given (Mitchell, Steadman, and Liggett, 1976). The problem is described in Figure 6.23. The following description is given to a system for interpretation as an optimization problem. The reserved words and symbols used by the system are shown in bold type.

Parameters
house **and** living room **and** kitchen **and** bathroom **and** hall **and**
   bedroom 1 **and** bedroom 2 **and** bedroom 3 **are a kind of** rectangle
bathroom width **is** 2.45 meters
hall width **is** 1.83 meters
kitchen width **is** 3.05 meters
bedroom 2 width **is** 3.35 meters
living room unit_cost **is** 30
kitchen unit_cost **is** 20
hall unit_cost **is** 10
bedroom 1 unit_cost **is** 10
bedroom 2 unit_cost **is** 10
bedroom 3 unit_cost **is** 10
bathroom unit_cost **is** 25

Topological Relationships (Topological Constraints)
house length **equals sum of** living room length **and**
   bathroom length **and** bedroom 2 length
house width **equals sum of** living room width **and** kitchen width
living room width **equals sum of** bathroom width **and** hall width
bathroom length **equals** hall length
bedroom 1 width **equals** kitchen width
bedroom 2 length **equals** bedroom 3 length
sum of living room length and hall length **equals sum of**
   kitchen length **and** bedroom 1 length
sum of living room width **and** kitchen width **equals sum of**
   bedroom 2 width **and** bedroom 3 width

Cost Relationships
house cost **equals sum of** living room cost **and** kitchen cost **and**
bathroom cost **and** hall cost **and** bedroom 1 cost **and**
bedroom 2 cost **and** bedroom 3 cost

Area Constraints
maximum living room area **is** 30 square meters
minimum living room area **is** 15 square meters
maximum kitchen area **is** 12 square meters
minimum kitchen area **is** 5 square meters
maximum bathroom area **is** 6.5 square meters
minimum bathroom area **is** 4.5 square meters
maximum hall area **is** 7.2 square meters
maximum bedroom 1 area **is** 18 square meters
minimum bedroom 1 area **is** 10 square meters
maximum bedroom 2 area **is** 18 square meters
minimum bedroom 2 area **is** 10 square meters
maximum bedroom 3 area **is** 18 square meters
minimum bedroom 3 area **is** 10 square meters

Objectives
maximum house area
minimize house cost

```
 name   :  bathroom

 slots  :  a-kind-of
                      value        :  rectangle

           length
                      expression   :  (hall length)

           width
                      value        :  2.45
                      unit         :  m

           unit-cost
                      value        :  25
```

```
 name   :  living-room

 slots  :  a-kind-of
                      value        :  rectangle

           width
                      expression   :  (bathroom width)
                                      +
                                      (kitchen width)

           unit-cost
                      value        :  30
```

Figure 6.24   The objects of a floorplan-dimensioning problem, described using frames.

```
name  :  objective-1

slots  :  objective-function

                    expression  :  (house area)

          optimality-criteria

                    value      :  maximize
```

```
name  :  objective-2

slots  :  objective-function

                    expression  :  (house cost)

          optimality-criteria

                    value      :  minimize
```

Figure 6.25   The objectives of a floorplan-dimensioning problem, described using frames.

Figures 6.24, 6.25, and 6.26 show the frames representing the objects, objectives, and constraints of this problem. There is an economy in this form of representation. Note that there is no need to specify that the house area as the sum of the other areas because house is declared as a rectangle and its length and width are specified in terms of the room lengths and widths. Figures 6.27 and 6.28 show screen displays of a session with the system. Figure 6.27 shows the canonical optimization model constructed by the system,

```
name  :  constraint-1

slots  :  lhs
                    expression  :  (living_room area)

          rhs
                    value  :  30.0
                    unit   :  sqm

          predicate

                    value      :  less_than_or_equal_to
```

```
name  :  constraint-5

slots  :  lhs
                    expression :  (living_room length)
                                  +
                                  (hall length)

          rhs
                    expression  :  (kitchen length)
                                   +
                                   (bedroom1 length)
          predicate

                    value      :  equal_to
```

Figure 6.26   The constraints of a floorplan-dimensioning problem, described using frames.

```
dialog window

MATHEMATICAL MODEL FOR THE FLOOR PLAN PROBLEM


maximize 7.33 * X1 + 7.33 * X2 + 7.33 * X3

minimize 84.00 * X1 + 80.96 * X4 + 30.48 * X5 +
         73.15 * X3

subject to


4.27 * X1 < = 30
4.27 * X1 => 15
3.05 * X4 <= 12
3.05 * X4 => 5
1.83 * X2 <= 7.2
2.44 * X2 <= 6.5
2.44 * X2 => 4.5
3.05 * X5 <= 18
3.05 * X5 => 10
3.35 * X3 <= 18
3.35 * X3 => 10
3.96 * X3 <= 18
3.96 * X3 => 10
X1 + X2 - X4 - X5 = 0.0

continuous variables X1 to X5

X1 = living_room_length
X2 = hall_length
X3 = bedroom3_length
X4 = kitchen_length
X5 = bedroom1_length
enter command
==> ■
```

Figure 6.27   The canonical optimization model constructed by the system for the floorplan problem.

including the representation of the design variables. In Figure 6.28 we can see the interpretation that the system derives from the original description.

A Beam Design Problem

This problem is concerned with the optimal design of a simply supported, single-span, wide-flange beam as shown in Figure 6.29. The goal is to select the dimensions of the beam given the objective of minimizing its weight. The description of the beam can be entered graphically as described in Section 5.6.3, augmented as necessary by textual information. Here we present the description as text only:

Parameters
beam span **is** 7.5 meters
beam uniformly_distributed_load **is** 60.0 kiloNewton per meter
beam **is_a_kind_of** I_section
beam deflection **equals** 5 * w * L ^ 4/384 * E * I
L **equals** beam span
w **equals** beam uniformly_distributed_load
E **equals** beam elastic_modulus
I **equals** beam second_moment_of_area
d **equals** beam web_depth

```
┌─────────────────────────────┬─────────────────────────────┐
│ dialog window 1             │    dialog window 2          │
├─────────────────────────────┼─────────────────────────────┤
│ PROBLEM RECOGNIZER          │  PROBLEM SOLVER             │
│                             │                             │
│ The following information has│ NISE algorithm is being executed│
│ been identified.            │                             │
│                             │  Allowable error percentage is│
│ VARIABLES                   │  assumed as zero            │
│                             │                             │
│ continuous variables = 5    │  Solution process in progress│
│ discrete variables = 0      │                             │
│ all the variables are       │  Sol.No  Area (sqm)  Cost (units)│
│ continuous                  │                             │
│                             │    1     104.15    1429.90  │
│ OBJECTIVES                  │    2      51.03     819.70  │
│                             │    3      88.43    1140.31  │
│ number of objectives = 2    │    4      72.45     933.02  │
│ the objective functions are │    5      98.16    1305.50  │
│ linear                      │                             │
│                             │                             │
│ CONSTRAINTS                 │  Decision Variables         │
│                             │                             │
│ Number of '<=' constraints = 7│ Sol.  X1    X2    X3    X4    X5│
│ Number of '=>' constraints = 6│      (m)   (m)   (m)   (m)   (m)│
│ Number of '=' constraints = 1│                             │
│ All the constraints are linear│  1   3.26  1.71  2.77  1.52  3.46│
│                             │  2   5.30  1.71  4.22  1.52  5.49│
│ MODEL                       │  3   3.26  1.71  4.22  1.52  3.46│
│                             │  4   8.53  1.71  4.22  2.76  5.49│
│ Number of constraints = 14  │  5   5.30  1.71  4.22  1.52  5.49│
│ Objective functions are linear│                           │
│ Constraints are linear      │                             │
│                             │  All segments have been explored│
│ The model is multiobjective │  There is no other solution outside│
│ linear programming problem  │  the boundary               │
│                             │                             │
│ ?- ■                        │  --> ▢                      │
└─────────────────────────────┴─────────────────────────────┘
```

Figure 6.28   The problem interpretation and solution processes. The left-hand window shows the information generated by the system when it attempts to recognize the algebraic model of the floorplan problem. The right-hand window shows the set of optimal solutions generated for the problem.

D **equals** beam overall_depth
t **equals** beam web_thickness
beam bending_stress **equals** w * L ^ 2 * D/{16 * I}
beam shear_stress **equals** w * L/{2 * d * t}
Fy **equals** beam yield_stress

Constraints
maximum beam bending_stress **equals** 0.660 **times** beam yield_stress
maximum beam shear_stress **equals** 0.370 **times** beam yield_stress
maximum beam deflection **equals** 1/360 **times** beam span
sum_of beam web_depth **and** 2 **times** beam flange_thickness **equals** beam overall_depth
quotient_of beam web_depth **and** beam web_thickness **is less than** 180
quotient_of beam web_depth **and** beam web_thickness **is less than** 816/{0.37 * Fy} ^ 0.5
1/15 times beam span **is less than** beam overall_depth **is less than** 1/8 times beam span
1/5 times beam overall_depth **is less than** beam flange_thickness **is less than** 1/3 times beam overall_depth

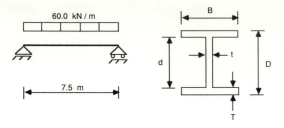

Figure 6.29   A simply supported beam with uniformly distributed load.

0.003 meters **is less than** beam flange_thickness **is less than** 0.100 meters
0.003 meters **is less than** beam web_thickness **is less than** 0.100 meters

Objective
minimize beam weight

Figure 6.30 shows the frame representation of the beam. Figures 6.31 and
6.32 show screen displays of a session during which the preceding problem
was solved. Figure 6.31 shows the canonical model constructed by the system,
while Figure 6.32 shows the interpretation derived from the description.

In these examples, designs are produced by bringing knowledge to bear in
constructing a problem formulation suitable for solution by some algorithmic
method that takes account of conflicting goals and defines a design space.
Later, we address the issue of conflict more directly in Section 7.1.3.

```
name   :  beam

slots  :  a-kind-of
                    value       :  steel l_section

          span
                    value    :  7.5
                    unit     :  m

          udl
                    value    :  60
                    unit     :  kN / m

          bending_stress

                    expression :   w * L ^ 2 * D / {16 * I}

          shear_stress

                    expression :   w * L / {2 * d * t}

          deflection

                    expression :   5 * w * L ^ 4 / {384 * EI}
```

Figure 6.30   A beam description as a frame.

```
dialog window

MATHEMATICAL MODEL FOR THE BEAM DESIGN PROBLEM

minimize 109.99 * X1 * X2 + 549.50 * X3 * X4

subject to

0.184 * X5 - 13.75 * X1 * X5 ^ 3 + 13.75 *
X1 * X4 ^ 3 - 13.750 * X3 * X4 ^ 3 <= 0.0

172.76 * X1 * X5 ^ 3 - 172.76 * X1 * X4 ^ 3 +
172.76 * X3 * X4 ^ 3 => 1.0


X5 => 0.457
X5 <= 0.875
X1 - 0.20 * X5 => 0.0
X1 - 0.33 * X5 <= 0.0
X2 => 0.003
X2 <= 0.100
X3 => 0.003
X3 <= 0.100
X4 - 100 * X3 <= 0.0
X4 - 84.844 * X3 <= 0.0
X4 + 2 * X2 - X5 = 0

continuous variables X1 to X5

X1 == beam_flange_width
X2 == beam_flange_thickness
X3 == beam_web_thickness
X4 == beam_web_depth
X5 == beam_overall_depth

enter command
==> █
```

Figure 6.31   A canonical model of the beam problem.

6.2.3   Plausible Inference from Design Specifications

In certain design tasks we are confronted with a small number of independent design decisions and must choose among them. These decisions may represent alternative descriptions of a design or of parts of a design. One simple example is the decision whether to select a raised timber floor or a concrete slab on the ground as the basic method of subfloor construction in a small building (Figure 6.33). Each of these possible descriptions could be regarded as a hypothesis, and the various performance requirements could be regarded as contributing evidence to either hypothesis. Knowledge may take the general form

if <evidence> **then** <hypothesis> (p)

where p is a measure of the support the hypothesis has from the evidence. Let us call p a "plausibility factor."

We may formulate a design system in which knowledge that maps performance requirements to descriptions is represented in the form of several of these rules. We then require a reasoning mechanism that will propagate measures of support for the hypothesis through a network of such rules. We adopt the hypothesis that has the highest plausibility given the accumulated evidence. We will call the plausibility measure attached to any proposition,

```
dialog window 1                 dialog window 2

PROBLEM RECOGNIZER              PROBLEM SOLVER

The following information has   NLP algorithm is being executed
been identified.
                               This program solves nonlinear
VARIABLES                       programming problems using the
                                sequential linear programming
continuous variables = 5        method.
discrete variables = 0
all the variables are
continuous                      Initial feasible solution found

OBJECTIVES
                                Solution process in progress
number of objectives = 1
the objective function is
nonlinear and quadratic         The solution process completed

CONSTRAINTS
Number of '<=' constraints = 7
Number of '=>' constraints = 5  SUMMARY OF RESULTS
Number of '=' constraints = 1
Constraints are linear and
nonlinear types                 minimum weight of beam = 187.62

MODEL
                                Decision Variables
Number of constraints = 13
Objective function is           beam_flange_width = 0.271
nonlinear and quadratic         beam_flange_thickness = 0.03
Constraints are linear and      beam_web_thickness = 0.01
nonlinear                       beam_web_depth = 0.764
                                beam_overall_depth = 0.824
The model is nonlinear
programming problem             --> ■
?-
```

Figure 6.32   An interpretation of the original beam-problem description. The left window shows the interpretation derived from the canonical model constructed by the system. The right window shows the final solution obtained by using the sequential linear programming method.

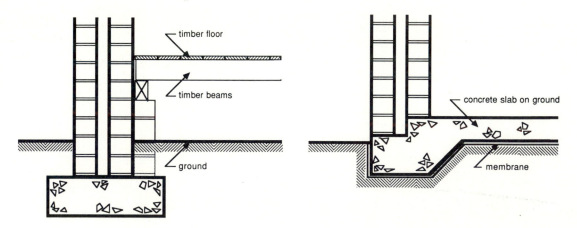

Figure 6.33   Raised timber floor (left) and slab on ground (right) sub-floor construction.

including a hypothesis, a measure of belief (MB). As an example of this approach, consider the following set of "evidential rules" to support the hypothesis that a raised timber floor should be used:

> **If** no termite activity
> **then** possible to use raised timber floor (0.2).

> **If** timber is readily available
> **then** possible to use raised timber floor (0.6).

> **If** a crawl space is required
> **then** possible to use raised timber floor (0.8).

We can also chain rules of this kind together. Suppose that the proposition that "a crawl space is required" can also be determined by some evidential rules:

> **If** flexible servicing is required
> **then** a crawl space is required (0.7).

> **If** there are sloping site conditions
> **then** a crawl space is required (0.3).

The evidence for the hypothesis that we should use a "concrete slab on the ground" can be stated as follows:

> **If** concrete is readily available
> **then** possible to use concrete slab on ground (0.3).

> **If** no crawl space is required **or**
>   a tiled-floor finish is required throughout
> **then** possible to use concrete slab on ground (0.9).

The following rules determine that no crawl space is required:

> **If** fire hazard zone
> **then** no crawl space is required (0.4).

> **If** flat site conditions
> **then** no crawl space is required (0.5).

The network depicting the linking of the propositions in the rules is shown in Figure 6.34. The numbers against the arcs indicate the plausibility factors (p). It is necessary to supply some indication of the measures of belief for

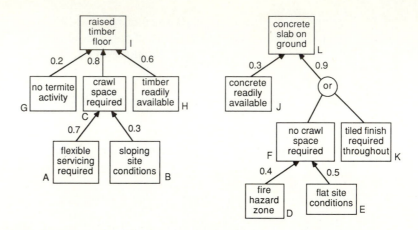

Figure 6.34   The network linking evidential propositions and hypotheses.

each of the nodes at the branch ends of the trees. We may assign the value 1.0 to those propositions that we are sure of; we assign a value of 0.0 to those that are not true. In some cases, however, it may be useful to say that we are only partially certain that a particular proposition holds, and assign it a value between 0.0 and 1.0. To make the example interesting, let us give the following measures of belief for each of the evidential propositions:

no termite activity	0.5
flexible servicing is required	0.6
sloping site conditions	0.1
timber is readily available	0.6
concrete is readily available	0.5
fire hazard zone	0.8
flat site conditions	0.8
tiled-floor finish required throughout	0.8

The task is to find the measure of belief given to the two hypotheses "raised timber floor" and "concrete slab on ground." In the process we must find the measure of belief given to "crawl space required" and "no crawl space required." Let us tackle the first of these, "crawl space required." The measure of belief in this hypothesis contributed by "flexible servicing is required" can be calculated by multiplying the measure of belief in the evidential proposition by the plausibility factor (p) of the appropriate rule. This produces a measure of belief for the hypothesis ("crawl space required") of 0.42 (0.6 multiplied by 0.7). This is demonstrated in Figure 6.35. The evidential proposition "sloping site conditions" also contributes to the hypothesis. We revise the measure of belief in the hypothesis by the following formula:

$$MB_3 = MB_1 + MB_2(1 - MB_1)$$

Figure 6.35 Calculating the contribution of "flexible servicing required" to "subfloor cavity required."

where MB_3 is the new measure of belief in the hypothesis, MB_1 is the existing measure of belief, and MB_2 is the weight of the new evidence. The effect of combining evidence in this way is shown in Figure 6.36.

The proposition "crawl space required" now has a measure of support of 0.44. We can continue with this kind of calculation, as shown in Figure 6.37. If there is a disjunction of evidential propositions contributing to a hypothesis, we simply take the evidence with the largest measure of support and ignore the rest. (In the event of a conjunction of evidential propositions, we take the one with the least support into account.)

After propagating all the measures of belief, we produce Figure 6.38. On the basis of these calculations we would adopt the "concrete slab on ground" hypothesis, which has the greatest measure of belief when all evidence is taken into account.

The approach used here is based on that developed in the MYCIN expert system for medical diagnosis (Shortliffe and Buchanan, 1975; Buchanan and Shortliffe, 1984). MYCIN, however, also takes account of measures of disbelief in a proposition, making it possible to decide on the extent to which evidence points to a hypothesis being untrue.

The preceding formula is based on simple probability theory. It assumes, however, that propositions providing support for any hypothesis are conditionally independent. For example, it would be unreasonable to add the following rule to the system just described:

> **If** ducted heating is to be installed after building is completed
> **then** crawl space required (0.6).

This requirement for ducted heating is related to the requirement for flexible servicing, already accounted for by one of the rules. The preceding rule would distort the distribution of measures of belief. In some cases the effects of

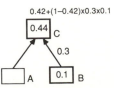

Figure 6.36 Combining evidence from two propositions.

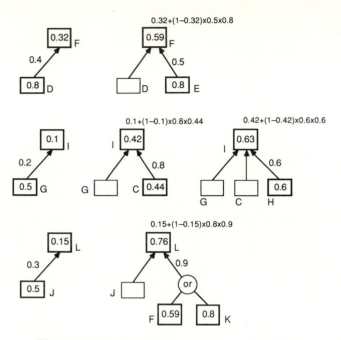

Figure 6.37   Propagating measures of belief.

interdependence of evidential propositions may be more subtle. Reasoning systems have been proposed that take account of non-additive combinations of evidence (Schafer, 1976; Garvey, Lowrance, and Fischler, 1981), though the explicit representation of plausibilities relating to combinations of evidence can be prohibitive.

There are several other approaches to the propagation of plausibility through inference networks, notably in the PROSPECTOR expert system, which computes the likelihood of discovering minerals at particular geological sites (Duda, Hart, and Nilsson, 1976; Gaschnig, 1982). This system takes

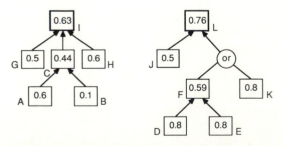

Figure 6.38   The final measures of belief in "raised timber floor" (0.67) and "slab on ground" (0.76).

the Bayesian notion of prior probabilities into account. Examples of alternative approaches to uncertain inference are provided by Garvey, Lowrance, and Fischler (1981), O'Neill (1987), and Quinlan (1982, 1983).

Much of the effectiveness of plausible reasoning systems depends on the judicious selection of plausibility factors. In certain domains, such as medical diagnosis, some appeal can be made to statistical data in linking symptoms (evidence) to causes (hypotheses). Such data rarely exists for design domains. This raises questions about the intuitive appeal of representing design knowledge in terms of plausibility factors, and hence the appropriateness of this method for representing design knowledge.

6.2.4   Fuzzy Reasoning from Design Specifications

Although designers may not be used to putting numbers on propositions to express uncertainty, they may be more comfortable with the idea of *imprecision*. Imprecise reasoning is that in which we are able to make statements such as the following:

> **If**   termite protection is adequate **and**
>   the slope is moderate
> **then**   a crawl space is highly suitable.

We want to be able to reason with such statements mathematically and objectively. A calculus for manipulating these kinds of statements has been formulated by Zadeh (1975, 1983) in the theory of *fuzzy sets* or *fuzzy logic*. Terms such as "adequate," "moderate" and "highly suitable" are called *linguistic variables*, or *fuzzy predicates*.

Essentially, the theory enables us to reason with partial set membership. According to classical set theory, an element is either a member of a particular set or it is not. Thus we can define a set of objects called "kitchen utensils" and decide that forks, spoons, and wire whisks are members, but that jackhammers, lawn mowers, and chain saws are not. But class membership is not always simple to determine. Some objects, such as screwdrivers, felt-tip pens, and thermometers, are somewhere on the borderline of the kitchen-utensils set (Figure 6.39). The problem is more severe when we attempt to describe sets with qualifiers. For example, in the context of a steel structural building frame, we may wish to decide whether the cross-sectional area of a structural element is "large" or "small." In terms of classical set theory we assume that an element is a member of either of these two sets, the set of steel structural elements that are "large" or the set of steel structural elements that are "small." We require a cutoff value between the two sets. Thus something less that 100 square centimeters is small; anything above 100 square centimeters is large. An element must lie in either set, and an element cannot be both large and small.

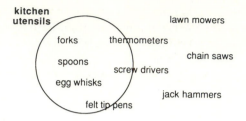

Figure 6.39   The problem with sharp boundaries between classes.

We may feel unhappy about this crisp boundary between large and small. The language of fuzzy sets enables us to specify membership functions that account for partial set membership. Thus a steel member with a cross-sectional area of less than 4 square centimeters may be regarded as small with a set membership of 1.0, whereas a steel member with an area of 100 square centimeters may be regarded as small with a set membership of 0.9. The set membership of 550 square centimeters in "small" is perhaps very low, say 0.01. We can plot the distribution of set memberships of "small" and "large" against a continuum of steel sizes. Typically, these distributions tend to be bell-like (or half a bell), as illustrated in Figure 6.40. Thus the *degree of set membership* can lie somewhere between 0.0 and 1.0, and the membership functions maps these values against some numeric value—area in this example.

In order to use set membership values, we need a few simple rules about how they can be combined. Let us consider three rules here. The first is that if an element e is a member of two sets S_1 and S_2, then the membership value of e for the conjunction of the two sets $(S_1 \cap S_2)$ is the minimum of the membership values for each of the sets.

$$\mu(e \in S_1 \cap S_2) = \min[\mu(e \in S_1), \mu(e \in S_2)]$$

where μ denotes a membership function. We can translate this into logic: the set membership of the conjunction of two fuzzy propositions is the minimum of the membership of the two propositions.

The second rule is that if an element e is a member of either of two sets

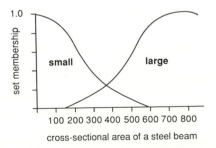

Figure 6.40   A distribution of set membership.

S_1 or S_2, then the membership value of e for the disjunction of the two sets $(S_1 \cup S_2)$ is the maximum of the membership values for each of the sets. (We used this rule for combining evidence in Section 6.2.3.)

$$\mu(e \in S1 \cup S_2) = \max[\mu(e \in S_1), \mu(e \in S_2)]$$

We can translate this into logic: the set membership of the disjunction of two fuzzy propositions is the maximum of the membership of the two propositions.

The third rule is simply that the membership value of e of the complement of a set S is the difference between 1.0 and the membership value of e in S:

$$\mu(e \sim S) = 1.0 - \mu(e \in S)$$

In logic this can be translated as: the membership value of the negation of a fuzzy proposition is 1.0 minus the membership value of the proposition.

We will demonstrate the application of this calculus with the rule given at the beginning of this subsection, plus a new rule by which we can determine whether a concrete slab is *highly desirable*:

> **If**   not(crawl space is *highly suitable*) **or**
>   owner preference for a tiled floor is *medium*
> **then**   concrete slab is *highly desirable*.

These two rules form an inference network (see Figure 6.41). We must assume that the linguistic variables for each of the propositions at the leaf nodes can be given numerical values. Let us suppose that termite protection is measurable on a scale of 0 to 100. The first rule describes termite protection in terms of adequacy. Thus we need a membership function to map the percentage of termite protection against the degree of set membership for each of the linguistic variables. This distribution is illustrated in Figure 6.42. We may

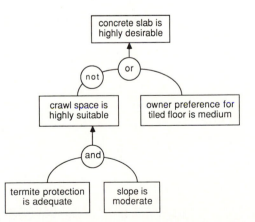

Figure 6.41   An inference network of rules containing fuzzy predicates.

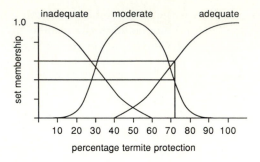

Figure 6.42 The set membership distributions for termite protection.

also define similar distributions for slope and tiled floor preference (Figures 6.43 and 6.44).

Suppose we now decide that the amount of termite protection is 72%, the slope is 3.6%, and the preference for tiled floors is 66%. From the graphs of Figures 6.42 to 6.44, we derive the following table of set memberships:

termite protection	72%	inadequate	0.0
		moderate	0.4
		adequate	**0.6**
slope	3.6%	flat	0.4
		moderate	**0.8**
		steep	0.0
tiled-floor preference	66%	low	0.0
		medium	**0.8**
		high	0.4

Those set membership values that apply to the design rules are indicated in bold type. Because of the gross nature of the computations involved in this kind of reasoning, we show values to only one decimal place.

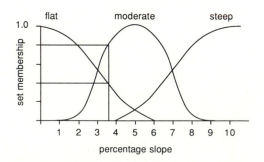

Figure 6.43 The set membership distributions for slope.

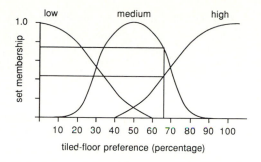

Figure 6.44   The set membership distributions for tiled-floor preference.

Let us consider the first design rule. In the event of a conjunction of terms, we take the maximum of the membership values to represent the degree of membership of the consequent, "crawl space is highly suitable."

$$\min[0.6, 0.8] = 0.6$$

In the second rule we take the negation of "crawl space is highly suitable":

$$1.0 - 0.6 = 0.4$$

Because the terms of the second rule are in disjunction, we take their maximum:

$$\max[0.4, 0.8] = 0.8$$

The membership value for "concrete floor slab" in the set "highly desirable" is therefore 0.8. This can be regarded as a "support set value," and is meaningful if we can compare it with some rival proposition. Thus we may have other rules relating to the desirability of a raised timber floor. Assuming both decisions required high desirability, we would probably select the material that had the highest membership of the "highly desirable" set. Thus if the support set value for "concrete floor slab is highly desirable" is 0.8 and that for "raised timber floor is highly desirable" is 0.2, then we would select a concrete floor slab as the design decision.

It is interesting to test the sensitivity of this kind of reasoning by supplying different values to the leaf nodes of Figure 6.41, especially where one of the linguistic variables has a 0.0 membership value:

termite protection	85%	inadequate	0.0
		moderate	0.0
		adequate	**0.8**
slope	9%	flat	0.0
		moderate	**0.0**
		steep	0.9
tiled-floor preference	5%	low	1.0
		medium	**0.0**
		high	0.0

With some thought we would expect the fact that the slope is not moderate to effectively discount any effect of the termite protection (irrespective of the membership value of "termite protection is adequate"). This would render the proposition, "crawl space is highly suitable" as not true. According to the second design rule, and irrespective of the membership value of "tiled-floor preference is desirable" we would expect to conclude that "concrete floor slab is highly desirable" with a maximum support set value. This intuition is borne out by the calculations.

Considering the first design rule, the combination of membership values produces the following:

$$\min[0.8, 0.0] = 0.0$$

The result of the second rule is

$$1.0 - 0.0 = 1.0$$
$$\max[1.0, 0.7] = 1.0$$

The proposition "concrete floor slab is highly desirable" has a support set value of 1.0.

There are similarities between this method of reasoning and the method of plausible inference discussed in the preceding section. Here we are not employing weights of evidence, but rather numerical values that can then be mapped onto linguistic variables. To make this simple example illustrate the general idea, we have supplied numerical values for termite protection, slope, and tiled-floor preference. In each case we could have provided linguistic values and assumed a set membership value of 1.0. Thus we could have said that the termite protection is "moderate" (set membership of 1.0). We would then need to map this onto the set membership value for "adequate" required by the rule. According to Figure 6.42, the set membership of "adequate" would be 0.0. This should then be taken into account in the reasoning process in the usual way. (We could imagine sets of distributions in which there is more overlap between sets than indicated in this example.) Provided we can supply the set membership functions, this type of reasoning is quite powerful as the basis of an expert system, and the numerical part of the process can be concealed effectively from the user of the expert system.

The advantage of evidential and fuzzy reasoning as tools for abduction is the way in which they skirt around the problem of contradiction. Given that we require something that is safe and cuts wood, statements such as the following can be rendered innocuous:

> **If** x is to be safe
> **then** x should *not* have sharp edges.

> **If** x is to cut wood
> **then** x should have sharp edges.

With plausible reasoning we could say,

> **If** x is to be safe
> **then** x should *probably not* have sharp edges
>   (*with some probability value*).

> **If** x is to cut wood
> **then** x should *probably* have sharp edges
>   (*with some probability value*).

With fuzzy reasoning we could say,

> **If** x is to be *moderately* safe
> **then** x should not have *very* sharp edges.

> **If** x is to cut wood *adequately*
> **then** x should have *reasonably* sharp edges.

We may also consider combining these approaches. The resolution to the problem of conflicting decisions in both case is to bring knowledge to bear in making a choice. Other approaches include making some kind of compromise, so that both requirements can be partially accommodated.

6.2.5   Producing Designs through Constraint Propagation

Constraint propagation is applicable to both abductive and deductive forms of knowledge—that is, whether we are reasoning from design descriptions to performances, or from performances to design descriptions. We discuss constraint propagation here, however, because of the way that it enables us to address the issue of one-to-many mappings in our knowledge.

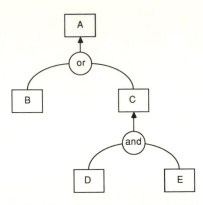

Figure 6.45   B or C implies A; D and E imply C.

As discussed earlier, we use logical statements to prove whether a particular proposition is true or not. For example, suppose we are given the following set of statements (two Horn clauses and a fact):

$$A \Leftarrow B \lor C$$
$$C \Leftarrow D \land E$$
$$B$$

We can determine that A is true (Figure 6.45). If we had only the two Horn clauses, however, we might wish to know what must be true in order for A to be proven. In order for A to be proven we need the following (Figure 6.46):

$$B \lor (D \land E)$$

(Even if, according to our facts base, this statement is not true, A may still be true. It just happens that it is not provably true by the Horn clauses given.)

In this simple logic system, every statement we make that bears on the truth of A is a *constraint*. It says something about what must be true in

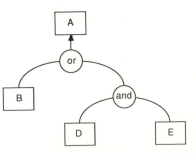

Figure 6.46   In order to prove A, B must be true or both D and E must be true.

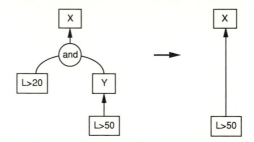

Figure 6.47 Reducing a set of numerical constraints.

order for A to be true. Transforming constraints into a form that makes their influence on A more explicit and succinct can be called constraint propagation. Eliminating C in the preceding statements is an example. In this example the process can be treated mechanically, and could be carried out by an automatic theorem prover.

Thus we see that constraints need not be statements limiting the values of variables (a common use of the term "constraint" in optimization). The following statement, however, provides an example of constraints in which this is the case (Figure 6.47):

$$X \Leftarrow (L > 20) \wedge Y$$
$$Y \Leftarrow (L > 50)$$

In order for X to be true,

$$L > 50$$

The reduction of the two Horn clauses into a single inequality is a further example of constraint propagation. The constraint of $L > 20$—rendered redundant by the alternative, less-restrictive constraint that $L > 50$—has therefore been eliminated. The type of reasoning carried out here goes beyond that of normal theorem proving, though it is not difficult to devise control methods for this kind of algebraic reasoning.

If we have some knowledge about the domain under consideration then we may be able to exploit this in the propagation of constraints. Consider, for example, the following two statements.

$$\text{SuitableColor}(x) \Leftarrow \text{RedLike}(x) \wedge \text{Cheerful}(x)$$
$$\text{Cheerful}(x) \Leftarrow \text{YellowLike}(x)$$

We should be able to deduce that we have a suitable color if we have something that is

$$\text{RedLike}(x) \wedge \text{YellowLike}(x)$$

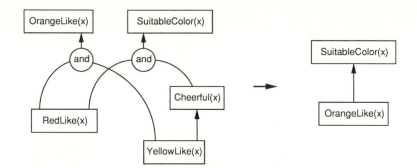

Figure 6.48   An inference network for determining a suitable color.

With the aid of some knowledge about colors,

$$\text{OrangeLike}(x) \Leftarrow \text{RedLike}(x) \wedge \text{YellowLike}(x)$$

we should also be able to deduce the useful constraint (Figure 6.48):

$$\text{OrangeLike}(x)$$

Two constraints have been reduced to one. This very simple example demonstrates the basic ideas of constraint propagation. There are several advantages of constraint propagation as a reasoning strategy in design. Constraint propagation can make the process more efficient. In some cases it can come up with useful constraints that can lead us to a design description. Also, new constraints can often tell us more about the space of design possibilities than the original form of our knowledge. Let us consider these issues in turn.

The approach to reasoning discussed in Chapter 4 is essentially concerned with constraint *checking* rather than constraint propagation, which can lead to greater efficiency in producing design descriptions from interpretive knowledge. We can demonstrate this by considering the goal-checking process described in Chapter 4.

We begin with the following statements:

$$\text{SuitableColor}(x) \Leftarrow \text{RedLike}(x) \wedge \text{Cheerful}(x)$$
$$\text{Cheerful}(x) \Leftarrow \text{YellowLike}(x)$$

In a goal-checking approach, we can effectively ignore the rule

$$\text{OrangeLike}(x) \Leftarrow \text{RedLike}(x) \wedge \text{YellowLike}(x).$$

The process of goal satisfaction, employing either forward or backward chaining, will not produce anything useful from this rule. Furthermore, with no facts base the goal, $\exists x \, [\text{SuitableColor}(x)]$, will tell us nothing about the design.

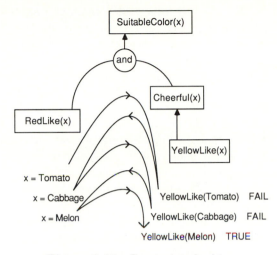

Figure 6.49 Constraint checking.

In the constraint-checking approach, we require some candidate facts to enable us to make instantiations of x. Consider for a moment that our knowledge about suitable colors relates to "designing" a salad. The candidate ingredients are as follows:

RedLike(Tomato)
RedLike(Cabbage)
GreenLike(Cabbage)
RedLike(Melon)
OrangeLike(Melon)
YellowLike(Melon)

Using linear resolution, we would attempt to prove our goal by finding something that is RedLike (x = Tomato), then check that it is also YellowLike (Figure 6.49). With the failure of this subgoal we would backtrack to find something else that is RedLike, and so on. The goal is eventually satisfied when we find an object that matches both of the constraints. This is a very simple example, but with something more complicated we could be involved in a lot of checking and a great deal of abortive search.

If we use constraint propagation to derive the constraint that the object is to be OrangeLike, however, we considerably reduce the amount of search and backtracking necessary. We have reduced two constraints to one, which in this case can be satisfied by a simple lookup of the facts base. With constraint propagation we are effectively deferring the selection process until the last moment. This becomes very important in domains where there are many possible design states.

Another advantage of constraint propagation is that we do not need a description of the design in order to find out something about it. As demonstrated by the preceding example, we do not need a list of candidate vegetables before knowing that the one we want is orange in color. Knowing about this constraint tells us something about the design.

In some domains the propagation of constraints can even lead directly to a solution. For example, if we wish to know what part of a building site we can build on, the successive consideration of site constraints can lead to the definition of the building outline (see Section 1.2.4).

Even if constraint propagation does not lead directly to a solution, it can sometimes tell us something valuable: the size and character of the design space. Whereas goal-directed theorem proving may produce individual designs, constraint propagation may also describe the design space. For example, by understanding the constraints of the design codes, we may better appreciate the scope available for meeting other design requirements. We say "may" because certain knowledge may lend itself to constraint checking; other knowledge, to constraint propagation. We may also expect the best results to be achieved by a combination of both approaches.

An effective demonstration of constraint propagation is provided by the MOLGEN system for designing chemical experiments (Stefik, 1981a). One interesting feature of this system is the three-stage process of constraint formulation, constraint propagation, and constraint satisfaction. In constraint *formulation*, constraints are selected and ordered for consideration. In constraint *propagation*, new constraints are created from old ones. Constraint *satisfaction* involves finding values for variables so that any set of constraints on the variables is satisfied. Because the domain of chemical experiments is a little foreign to most designers, let us consider these ideas in the context of a spatial design task.

The task we consider is the location of three elements—a house, a garage, and a swimming pool—on a building site (Figure 6.50). To help make the example manageable, we limit the world to rectangular gridded building sites and elements that each occupy only a single grid unit. The constraints on the location of elements are (1) the house must be adjacent to the garage, (2) the pool cannot be adjacent to the garage, (3) the pool is north of the garage and north of the house, (4) the site is 4 grid units by 4 grid units in area, and (5) part of the site cannot be built upon (Figure 6.51). We can represent these constraints in the form of a rule:

A location plan is suitable **if** there is a list of cells s and the cell position of the house h, the garage g and the pool p are members of s, and are not members of the list of cells in the exclusion zone z. Certain other constraints must also be met: the cell occupied by the house must be next to the cell occupied by the garage; the pool cell cannot be next to the garage cell; the pool cell must be north of the house cell; and the pool cell must be north of the garage cell.

Figure 6.50   A house, a garage, and a swimming pool.

We can represent this rule more formally:

SuitablePlan(h, g, p) $\Leftarrow$ SiteCells(s) $\wedge$
Member(h, s) $\wedge$
Member(g, s) $\wedge$ Member(p, s) $\wedge$
ExclusionZone(z) $\wedge$ Not[Member(h, z)] $\wedge$
Not[Member(g, z)] $\wedge$ Not[Member(p, z)] $\wedge$
NextTo(h, g) $\wedge$ Not[NextTo(p, g)] $\wedge$ NorthOf(p, h) $\wedge$ NorthOf(p, g)

We require a fact about site cells—simply a list of the coordinates of each of the grid cells:

SiteCells([1, 1] **and** [1, 2] **and** [1, 3] **and** ... **and** [4, 4])

The exclusion zone is indicated by the list of the cells that cannot be occupied by elements:

ExclusionZone([2, 1] **and** [2, 2] **and** [2, 3])

We also require some rules (not shown here) for determining whether an element is north of another element and for defining when an element is a member of a list.

As this knowledge is formulated, there is sufficient information to decide whether a plan is suitable. For example, the goal

SuitablePlan([3, 3], [4, 3], [1, 4])

is a theorem of the system. There is also sufficient information for a theorem prover to actually produce suitable layouts. The goal

$\exists h \exists g \exists p$SuitablePlan($h, g, p$)

will exhaustively return values for $h, g,$ and p that conform to all the constraints. Assuming a theorem prover that uses linear resolution and proceeds

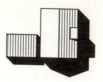

house is adjacent to garage

pool cannot be adjacent
to garage

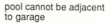

pool is north of garage and
north of house

site is 4x4 units; structures
must not occupy gray area

Figure 6.51   The constraints of the house, garage, and swimming-pool problem.

from left to right in proving subgoals, we would expect the system to select
a location for each of the elements and then check the locations against the
adjacency and directional constraints. This represents a constraint-checking
approach to producing a design description. The disadvantages of this approach is that it gives us no immediate indication of the size and character
of the design space and that it involves a large amount of search. Let us now
consider the constraint propagation approach.

The objective is to carry our reasoning as far as possible before making a
commitment about the locations of elements. The constraint for this domain
can be represented graphically. The first constraint is that the house and the
garage are adjacent. As indicated in Figure 6.52, we can show this in two
ways. We can also represent the constraint that the pool is not to be next

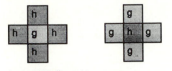

house is adjacent to garage

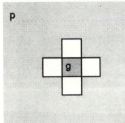

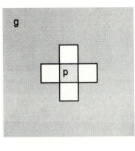

pool cannot be adjacent to garage

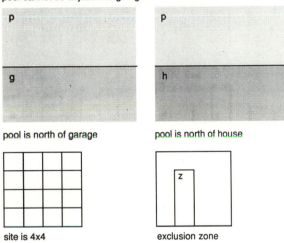

pool is north of garage   pool is north of house

site is 4x4   exclusion zone

Figure 6.52   Abstract representation of the constraints of Figure 6.51.

to the garage in two ways. These are examples of constraints that cannot be represented spatially except in terms of a partial commitment to the location of one of the objects. Note, however that there is no commitment to an actual grid location. The constraint that the pool is to be north of the garage, and the constraint that the pool is to be north of the house, can both be represented as a single line separating two zones. The site constraint can be represented by the site boundary; the exclusion zone is represented by the boundary to that zone.

The interesting part of the exercise comes when we attempt to combine constraints. In some cases we must select which form of each of the constraint representations we wish to adopt. First, we find that the "pool not next to

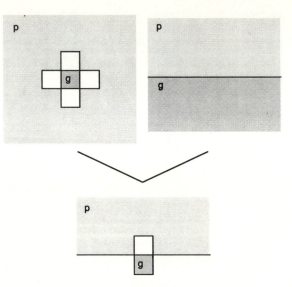

Figure 6.53   Combining the constraints "pool not next to garage" and "pool north of garage."

garage" and "pool north of garage" constraints combine to produce the constraint diagram of Figure 6.53. This constraint diagram can be combined with the "house next to garage" constraint to produce Figure 6.54. When the "site" constraint is added, this produces an absolute boundary to the area in which the pool may be located (Figure 6.55). Without committing ourselves to any location, we have reduced four constraints down to one spatial diagram. Unfortunately we cannot carry the process much further without making some tentative commitments. If we split the last diagram into two representations, then we can add the "pool north of house" constraint to each of these (Figure 6.56). If not for the exclusion zone, we would be able to overlay either diagram on the site in order to gain an appreciation of candidate locations of elements. If we commit ourselves to possible locations of the garage, we can apply the "exclusion zone" constraint to produce 13 diagrams showing candidate positions of elements (Figure 6.57). These diagrams represent the solution to our problem.

These 13 diagrams are much more succinct than the 200 or so solutions that would be produced by constraint checking. We have some idea of the size and character of the design space at a glance. One thing we notice is that, for each house and garage position, there are many possible locations for the pool. There are also certain cells on the site that cannot be occupied by the pool, and certain cells that cannot be occupied by the garage or the house. The amount of searching that must be carried out to arrive at these

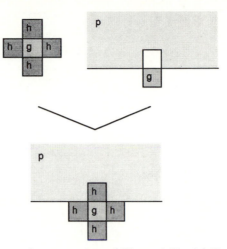

Figure 6.54 Combining the constraints of Figure 6.53 with "house next to garage."

constraint diagrams is generally considerably less than that required for constraint checking. Also note that this approach makes it easier to control the addition of extra constraints and the relaxation of constraints.

Of course, this approach depends on the availability of knowledge for propagating constraints in a useful manner. If we engaged in an exhaustive exploration of every pairwise combination of constraints and of every possible splitting of constraints, then the process would be prohibitive. Knowledge to assist in efficient constraint propagation will depend on the domain. Here are a few rules and strategies that may be pertinent to this task.

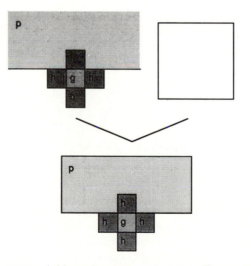

Figure 6.55   Adding the site constraint to Figure 6.54.

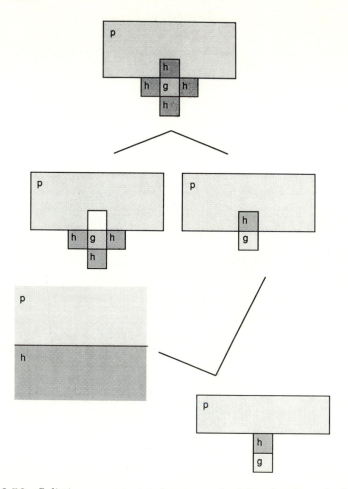

Figure 6.56 Splitting a constraint diagram and adding "pool north of house."

Identify the element that is most constrained (the garage) and consider its constraints first.

If it is necessary to make a locational commitment, then locate only the most constrained element.

If two constraints cannot be combined readily, then split one of them into two.

We have not considered here possible language for describing the constraints depicted graphically in Figures 6.52 to 6.57. We require some system of representing constraints that lends itself to this kind of manipulation.

Certain sets of constraints may not lend themselves readily to constraint propagation—for example, when we have to satisfy a large set of similar constraints, such as a large set of adjacencies. Constraint propagation is useful

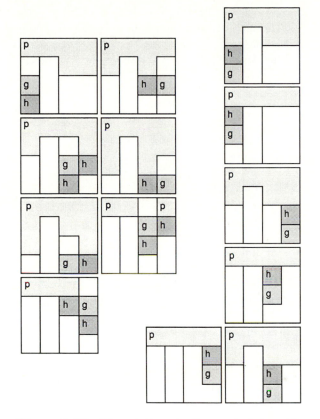

Figure 6.57 The solutions to the constraint problem.

only when the constraints serve to reduce the design space appreciably. In some cases the constraints may fail to define a design space—for example, if the pool had to be north of the garage and south of the house. The task is "over-constrained," and we might bring knowledge about constraint relaxation into play. To do this we must keep track of assumptions and dependencies, the subject of the next subsection.

6.2.6 Producing Designs through Assumption-Based Reasoning

We have referred to the problem of conflict in abductive forms of design knowledge. One solution is to view rival design descriptions, or decisions, as competing, and to apply knowledge to see which one has the greatest support. We have also discussed the problem as one of manipulating constraints that define design spaces. In the truth-maintenance approach, we reason from performances to candidate design descriptions, treating these descriptions as assumptions that are on hold until something causes us to change our mind about them. This approach is a useful way to propagate constraints while withholding commitments to design decisions as long as possible.

The issue can be demonstrated with reference to the subfloor example. Consider the following abductive design rules:

If   subsoil is sand
then   use slab on ground.

If   flexible servicing is required
then   use raised timber floor.

Consider also the fact that

subsoil is sand

The conclusion is that we are to

use slab on ground

If we then add the fact that "flexible servicing is required," we produce a new fact that we are to "use raised timber floor." Assuming that the two types of sub-floor construction are mutually exclusive, we face a contradiction. The truth-maintenance idea, introduced in Section 4.7, enables us to detect the source of the contradiction, to maintain the truth status of facts, and to reason about the resolution of contradictions. Basically, we separate beliefs from facts. Thus beliefs are those statements that we are prepared to assume are true until we have some evidence to the contrary. When a new fact is derived, we keep a record of those beliefs on which that fact depends. When a contradiction is detected, we delete the beliefs contributing to that contradiction, along with the facts that depend on the beliefs.

To illustrate the idea in relation to design, let us consider a simple example that was originally presented by Dietterich and Ullman (1987) in the context of their reasoning system, FORLOG. FORLOG is a truth-maintenance system

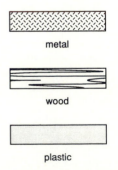

metal

wood

plastic

Figure 6.58   Three materials for the plate problem.

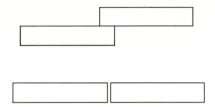

Figure 6.59 Overlapping and abutting configurations of two plates.

with a syntax similar to that of PROLOG. Here we focus on the explanatory power of the example rather than on the system.

We consider a design domain in which one of the tasks is to join two plates together; perhaps the material of one of the plates is known. In other words, we have the performance (or requirement) that two plates are joined, and we need a description that tells us how they are joined—what kind of joint is required. There are likely to be many ways in which the plates can be joined. If there is more than one solution, then we would like to gain some understanding of the nature of the design space.

The materials of these two plates can be the same or different; they can be metal, wood, or plastic (Figure 6.58) in any combination—that is, there are nine possible combinations. There are two possible configurations of these elements: overlapping or abutting (Figure 6.59). There can also be three types of bonds—weld, adhesive, or bolt (Figure 6.60). Even with this small number of individual decisions there are 54 possible configurations of materials and bonding methods (many more in a realistic problem). Not every configuration is suitable for any given application.

The specifications for this problem can be given in terms of partial descriptions of a design—for example, that two elements are to be joined together, say the edge of a bench top and the base support for some item of machinery (Figure 6.61). We know the material for one of these elements (the bench is of wood), and we wish to know what kind of joint should connect them and whether they should be bolted, glued, or welded. We also want to know the material of the other element. In this case the problem is under-constrained;

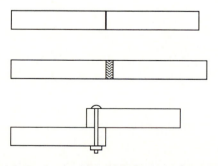

Figure 6.60 Adhesive, weld, and bolt joints between two plates.

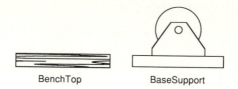

Figure 6.61   A bench top and a machine base.

we would like to know under what conditions different solutions are possible. We want to gain some appreciation of the space of design solutions and to be able to produce some designs from that space.

First, we represent the knowledge about possible configurations of plates. If two materials are required to be joined together, to form a joint, j, then they can form an overlapping or abutting configuration. The bond between them can be "weld," "adhesive," or "bolt." Let us represent this knowledge by the following set of statements:

$$\text{Joint}(j, m_1, m_2) \Rightarrow \text{Configuration}(j, \text{Overlap}) \qquad (A)$$

$$\text{Joint}(j, m_1, m_2) \Rightarrow \text{Configuration}(j, \text{Abut}) \qquad (B)$$

$$\text{Joint}(j, m_1, m_2) \Rightarrow \text{Bond}(j, \text{Weld}) \qquad (C)$$

$$\text{Joint}(j, m_1, m_2) \Rightarrow \text{Bond}(j, \text{Adhesive}) \qquad (D)$$

$$\text{Joint}(j, m_1, m_2) \Rightarrow \text{Bond}(j, \text{Bolt}) \qquad (E)$$

These rules can be regarded as the bases for *beliefs*, or simply beliefs. (The letters identifying those statements that are beliefs appear in italics.) We should note two points about these rules. First, like those that follow, the rules are essentially of the abductive form. The deductive rule "**If** two plates overlap, **then** we have a joint," has been changed to "**If** we have a joint, **then** we have two overlapping plates" (or "**If** we require a joint, **then** provide two overlapping plates"). Second, if we assert that "we have a joint," the result of rule A would contradict the result of rule B. All of the consequents of the preceding rules, A and B and C to E, cannot be simultaneously true. Thus each of the statements is represented as a belief, subject to subsequent revision if necessary.

We can also formalize the statement that if two materials are to be joined together then they must be of known material. This is not treated as a belief, but is assumed to be always true.

$$\text{Joint}(j, m_1, m_2) \Rightarrow \text{Material}(m_1) \wedge \text{Material}(m_2) \qquad (F)$$

We can also state that if there is a material then it is metal, wood, or plastic. These are also beliefs, for reasons similar to the ones we gave about statements A to E.

$$\text{Material}(m) \Rightarrow \text{Type}(m, \text{Metal}) \qquad (G)$$

$$\text{Material}(m) \Rightarrow \text{Type}(m, \text{Wood}) \tag{H}$$

$$\text{Material}(m) \Rightarrow \text{Type}(m, \text{Plastic}) \tag{I}$$

We need some rules to determine whether the materials on the two plates to be connected are the same. We can say that if two materials are to be joined together, then the materials must be the same type. Also, the Type of joint will be Same and the Type of the joint will be same as the Type of material. Thus if both materials involved in the joint are wood, then it will be a wooden joint. Alternatively, if two materials are to be joined together, the materials can be different, in which case the joint will be labeled Different. These rules can be treated as assumptions.

$$\text{Joint}(j, m_1, m_2) \Rightarrow [\text{Type}(m_1, u) \wedge \text{Type}(m_2, u)$$
$$\wedge \text{Same}(j) \wedge \text{Type}(j, v) \wedge \text{Type}(m_1, v)] \tag{J}$$

$$\text{Joint}(j, m_1, m_2) \Rightarrow [\text{Type}(m_1, u) \wedge \text{Type}(m_2, v)$$
$$\wedge \text{Not}(u = v) \wedge \text{Different}(j)] \tag{K}$$

We must also represent expertise about joints. It is impossible to bolt materials that abut. If two materials are in an abutting configuration, then they cannot be bonded with a bolt.

$$\text{Configuration}(j, \text{Abut}) \Rightarrow \text{Not}[\text{Bond}(j, \text{Bolt})] \tag{L}$$

We must also state that it is impossible to weld two different materials. If a configuration is Different, then the bond cannot be Weld.

$$\text{Different}(j) \Rightarrow \text{Not}[\text{Bond}(j, \text{Weld})] \tag{M}$$

One can also only weld metal to metal. If the configuration is Different and not of type Metal then the bond cannot be Weld.

$$\text{Same}(j) \wedge \text{Not}[\text{Type}(j, \text{Metal})] \Rightarrow \text{Not}[\text{Bond}(j, \text{Weld})] \tag{N}$$

Adhesive is not used to join two metal plates together. If the configuration is Same and not of type Metal then the bond cannot be Adhesive.

$$\text{Same}(j) \wedge \text{Type}(j, \text{Metal}) \Rightarrow \text{Not}[\text{Bond}(j, \text{Adhesive})] \tag{O}$$

Statements A to O represent our knowledge base. Let us now consider the reasoning process. We start by asserting that there is to be a joint (with the name Joint1) between two objects. Let us designate one object a bench top (BenchTop); the other is the base support of some machine that is to be supported by the bench (BaseSupport).

$$\text{Joint}(\text{Joint1}, \text{BenchTop}, \text{BaseSupport}) \tag{1}$$

We know that the BenchTop is made of wood (Figure 6.62).

$$\text{Type}(\text{BenchTop}, \text{Wood}) \tag{2}$$

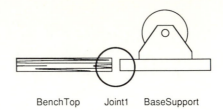

BenchTop   Joint1   BaseSupport

Figure 6.62   A wooden bench top and a machine base of unknown material.

Statement 1 results in the following new facts being asserted from assumptions A and B:

$$\text{Configuration(Joint1, Overlap) } \{A\} \qquad\qquad (3)$$
$$\text{Configuration(Joint1, Abut) } \{B\} \qquad\qquad (4)$$

The beliefs on which these statements depend are indicated in brackets {}. Statement 1 also results in the following new facts by statements C, D and E:

$$\text{Bond(Joint1, Weld) } \{C\} \qquad\qquad (5)$$
$$\text{Bond(Joint1, Adhesive) } \{D\} \qquad\qquad (6)$$
$$\text{Bond(Joint1, Bolt) } \{E\} \qquad\qquad (7)$$

Fact 1 also produces the following facts under rule F:

$$\text{Material(BenchTop) } \{\} \qquad\qquad (8)$$
$$\text{Material(BaseSupport) } \{\} \qquad\qquad (9)$$

These facts are not dependent on any beliefs. Statement 9 enables rules G, H, and I to produce the following facts:

$$\text{Type(BaseSupport, Metal) } \{G\} \qquad\qquad (10)$$
$$\text{Type(BaseSupport, Wood) } \{H\} \qquad\qquad (11)$$
$$\text{Type(BaseSupport, Plastic) } \{I\} \qquad\qquad (12)$$

Rules J and K produce the following sets of facts:

$$\text{Same(Joint1) } \{J, H\} \qquad\qquad (13)$$
$$\text{Type(Joint1, Wood) } \{J, H\} \qquad\qquad (14)$$
$$\text{Different(Joint1) } \{K, G\} \ \{K, I\} \qquad\qquad (15)$$

Statement 13 says that if we assume H, that the base support is wood, and also use the logic of rule J, then the materials are the same. Statement 14 says that the joint is wood based on the same assumptions. According to statement

15, if we assume that the base support is metal (G) or plastic (I), and also use the logic of rule K in both cases, then the materials are different.

We appear to have gone as far as we can with this line if reasoning. Now we must consider some of the constraints imposed by statements L to O. By constraint propagation we can determine that certain decisions result in contradiction (indicated by $\perp$). We need some statement that certain combinations of assumptions produce contradictions. The effect of rule L is to define a contradiction: bolts cannot be used where two materials abut. This means that assumptions B and E cannot coexist, as indicated by the following statement:

$$\perp \ \{B, E\} \tag{16}$$

Rule M tells us that we cannot have different materials and a weld. This means that assumptions K and C cannot coexist:

$$\perp \ \{K, C\} \tag{17}$$

By means of simple constraint manipulation from rule N, we discover that we cannot have a weld (C) where the materials are different—that is, where we have metal (G) or plastic (I):

$$\perp \ \{J, H, C\} \tag{18}$$
$$\perp \ \{J, I, C\} \tag{19}$$

Rule O tells us that we cannot have the same materials (J), and that the materials cannot be metal (G) and that we cannot use adhesive (D), as coexistent facts.

$$\perp \ \{J, G, D\} \tag{20}$$

By reasoning with statements 17, 18, and 19 we can conclude that welding (C) cannot be used in this problem:

$$\perp \ \{C\} \tag{21}$$

In order for there to be a weld, the materials must be the same and they must not be wood or plastic. By this reasoning we have been able to deter-

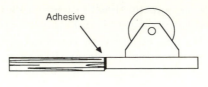

BenchTop   Joint1   BaseSupport

Figure 6.63   The final state of the bench and machine-base problem.

mine something about the design space. Without committing ourselves to the materials and the method of joining them, we have concluded that the design will not incorporate welding. The designs could be enumerated quite simply by considering all combinations of choices whose combined assumptions do not match a contradictory set. Of course, we may wish to defer this operation in the expectation that further considerations during the design process might further constrain the design space.

We may decide to assert the statement that the plates will abut. This statement already exists as statement 4:

$$\text{Configuration(Joint1, Abut) } \{B\} \tag{4}$$

The effect of this assertion is to negate assumption A and to include rule B. This implies

$$\bot \ \{A\} \tag{21}$$

The joint cannot be an overlap. As a consequence of statement 16 we can conclude that

$$\bot \ \{E\} \tag{22}$$

The connection cannot be a bolt. Only D, the use of adhesive, is left as a bonding technique (Figure 6.63).

We should note that the system has not determined the material of the base support. The option can be kept open until further information is available. The advantages of this approach to design reasoning is that different lines of reasoning are effectively pursued in parallel. These lines of reasoning can be combined when necessary. The system avoids commitment to any particular description of the design until a decision is demanded by other parts of the design problem.

In the example we have been following, any contradictions can be resolved from within the system. In other cases, we must bring other knowledge to bear in the resolution of the conflict. Assumption-based reasoning provides a good starting point for reasoning about these kinds of contradictions brought about by a surfeit of constraints.

6.3   Producing Designs through Knowledge about Syntax

So far in this chapter we have referred to interpretative knowledge concerned with the mapping between semantics and syntax—mapping performance onto a description of a design—and how this can be used in producing descriptions of designs. In the remaining sections we will focus on the use of syntactic knowledge to produce designs. We described this formally in Chapter 2 as

$$D = t_3(K_s, V), \tag{2.3}$$

where K_s is syntactic or generative knowledge and V is a vocabulary. D is produced through some transformation, t_3, of these.

There are two ways of looking at the issue of syntactic knowledge in design systems: in terms of information processing and in terms of vocabularies of design components. With regard to the first point of view, in a design system we need some means of describing designs once they have been produced. We can consider syntactic knowledge as that which defines the description language to which statements must conform, much as we have rules of syntax in natural language or computer languages. Rules of grammar in natural language can be used to generate sentences. We may also use syntactic knowledge in design systems to generate design descriptions. Thus syntactic knowledge facilitates the configuring of "words" about designs.

The second point of view is that syntactic knowledge is concerned with how design elements are configured. We assume that words in a design description relate to elements and their properties. Syntactic knowledge is concerned with the configuration of elements—their topology and geometry, or form. If we consider that design elements are represented in terms of words and their configuration, then both viewpoints map onto each other. In this book we generally refer to syntax as concerned with the configuration of elements; this maps onto the way information about designs is described in a computer.

We can represent syntactic knowledge in several ways. The first and simplest is as a set of *standard* or *pre-existing designs*; alternative configurations of design elements are already determined. It is as if the design space is defined by a catalog: we require a house design, so we look through the catalog of standard plans and select the one that maps onto a performance that closely matches some required performance.

The second way of representing syntactic knowledge is through *generic descriptions of designs*—that is, descriptions of classes of designs, including the degree of variation permitted within the classes. This method includes the description of prototypes or sets of prototypes. Thus we may have some ideas about a "courtyard house," and a "U-shaped house," an "oil bearing" or a "pneumatic bearing." Certain geometries are associated with these "types" that we can exploit in producing a new design.

The third way of representing syntactic knowledge is through *design generators*: the space of designs is defined by means of some algorithm that produces all the designs—all the possible configurations of elements—in that space. (We use the term "algorithm" in the most general sense, to include rules, actions, and procedures.) Represented in this way, syntactic knowledge is analogous to rules of grammar in natural language and rewrite rules in computational theory generally. Thus we can have rules that transform the design description from one state into another. These rules can also be seen as actions that operate on design elements (see Section 4.4). An example of an action is "place element A next to element B." We can also represent generative knowledge by means of procedures. Although we do not discuss procedures in detail here, we refer to how knowledge can be brought to bear in selecting the appropriate procedure to solve a particular problem.

In the following subsections we consider some of the important issues in syntactic knowledge and then look in more detail at its control. We then consider some examples of how syntactic knowledge can be brought to bear in producing designs.

6.3.1   Issues in Knowledge about Syntax

Here we consider several issues connected with syntactic knowledge, including the specificity of the generator, how we control generation, the importance of decomposition, and the use of control abstractions.

Specificity of the Generator

It is not difficult to devise generative knowledge for producing compositions of design elements. If we are interested in a "syntax of shapes," we could define general shape operators, such as union, intersection, and difference. We could begin the generative process with a few shapes and produce "designs" by considering every possible combination of shapes according to these operations. The space generated, however, is likely to be larger than useful. We would not generally say that the operators represent a syntax, but rather a calculus of shapes (March and Stiny, 1985). We could say the same thing about the manipulation of a "vocabulary" of toy building blocks. If we are interested in building toy houses and bridges, then a set of operators that permitted every possible combination of blocks is not likely to be useful (Figure 6.64). In design we are generally interested in a somewhat restricted universe of discourse, rules that allow only particular combinations of elements.

There are only certain ways that elements may come together to constitute a legitimate design within a particular domain. Thus in designing a concrete beam, the reinforcing rods go inside, not outside the beam. In roof detailing,

Figure 6.64 Some arrangements of blocks within a space defined by a calculus of configurations.

the gutter is attached to the fascia, not the chimney (Figure 6.65). We are interested in capturing this type of knowledge when dealing with syntax.

Control

A key issue in syntactic systems is how to produce "meaningful" designs. A grammar system in natural language says little about the production of sentences that convey meaning. In a design system we may be able to produce a set of designs that embodies a design syntax, but we are interested in those designs that exhibit particular performances. In the case of generic design descriptions, design grammars, or sets of actions we are not interested in every configuration of design elements that is possible, but only those that map onto some particular performance.

The production of meaningful designs is essentially one of control. If the design space is defined by a set of standard designs, then the control process will involve selection. In the case of prototypes or sets of prototypes we may have to select the appropriate prototype, then instantiate the generic description—find specific values for parameters that ensure that the design

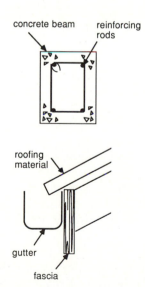

Figure 6.65 Knowledge about configuring elements.

stage 1

A	→	aU
A	→	bV

stage 2

U	→	uX
U	→	nY
V	→	vZ

stage 3

X	→	q
Y	→	r
Z	→	s
Z	→	t

Figure 6.66 A system of rules that demonstrates serial independence.

meets specific requirements. In the case of generation, we require some way to ensure that the generative steps in the process lead toward designs that exhibit intended performances.

In this section we focus most of our attention on the issue of control in generative systems. At any state in the development of the design, several actions may apply; in another state there is probably a different set of applicable actions. Actions, or rules, appear to be competing. The question is, how do we select the action at each state that is going to produce a design that meets our intended performances? Different approaches include generate and test; mapping performance to actions; generate, test, and modify the design in the light of some evaluation; prototype selection and refinement; and hierarchical control abstractions. Each of these is discussed in Section 6.3.2.

Decomposition

There are computational advantages if design goals can be satisfied independently, such that the satisfaction of one goal does not negate the satisfaction of another. Generative systems may also exhibit this decomposable property. There are three ways in which generative systems may be decomposable.

First, it may be possible to divide the actions into sub-sets that can be treated independently. For example, as we see in Section 6.6.3, in designing a roof detail in a building, we may take actions concerned with fixing rafters to the tops of walls, with fixing fascias and gutters, and with fixing the roofing material. The design process may be broken down into a progression through each of these "stages." Each stage can be considered independently; the stages

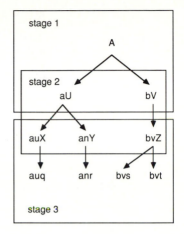

Figure 6.67   The search space resulting from the rules of Figure 6.66.

may follow a set order. A system based on actions of this kind exhibits a kind of "serial independence," illustrated symbolically with the generative rules of Figure 6.66 and the resultant search space of Figure 6.67.

The second way in which a generative system may exhibit the property of decomposability is when each action can be considered independently. That is, the system is "commutative"—the order in which actions are considered does not affect the outcome. This is easily demonstrated with abstract symbols (Figure 6.68). A non-commutative system is shown in Figure 6.69. Of course, we would not expect the actions in a design system to be entirely commutative

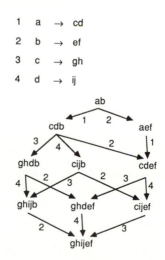

Figure 6.68   A commutative system.

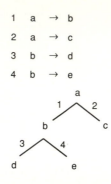

Figure 6.69 A non-commutative system.

or non-commutative in this manner, but rather to exhibit a mixture of properties. Clearly there are advantages in the control of commutative systems because we need be less concerned about conflict resolution strategies, such as backtracking. The order in which actions are fired, however, will probably affect the efficiency with which a solution is achieved.

A third sense in which a generative system may be regarded as decomposable is if the intermediate states of the design exhibit partial properties of the final design—that is, if the intended performance of the final design serves as a useful evaluator of intermediate states. In a design world of colored toy blocks, if we wish to produce a design made up entirely of yellow blocks, and that is the only performance we are interested in, then any states that contain configurations of anything other than yellow blocks will be excluded from further consideration. Thus we implement only actions that operate on yellow blocks. The control of this kind of design system is relatively simple because we have some criteria for evaluating partial states that will lead us toward the desired end state. Design tasks are generally not this obliging, though we may be able to formulate systems to exploit this property in certain ways.

Control Abstractions

We have discussed generative knowledge as concerned with the configuration of design elements. Generative knowledge as design actions readily captures this idea; such actions operate on design elements. But we can also think of other design "operators," such as "determine the location of the gear train" or "decide on the type of casing material," which represent actions at a different control abstraction than abstractions concerned with configuring elements. We could regard these operators as stages, steps, or tasks. There may also be operators that operate above these that might come under the heading of "strategies." In the next section we demonstrate with an example how this hierarchical view can be used to advantage in the control of generative systems. A thorough discussion of the hierarchical nature of design processes is deferred until Chapter 7.

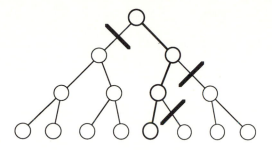

Figure 6.70   Pruning a search tree.

6.3.2   Control of Generative Systems

In this section we consider several major approaches to controlling generative systems: generate and test; mapping performance to actions; generate, test, and modify; prototype selection and refinement; and hierarchical control abstractions. We look at more complex control structures in Chapter 7.

Generate and Test

Generate and test is perhaps the simplest approach to controlling generative systems in order to produce designs that exhibit desired properties. Actions are brought to bear in producing a design that is then interpreted for compliance with the required performance. If it does not comply, further designs can be generated and tested against the performance. There are various strategies by which this kind of operation can be implemented (analogous to the control strategies discussed in Chapter 4), including exhaustive search of the design space and backtracking on failure.

There are two major variations of the generate-and-test approach. The testing part of the process may occur only at the final state or during intermediate design states. There are obvious advantages in being able to test intermediate, partial designs because this effectively results in the pruning of the tree of states constituting the design space (Figure 6.70). With this approach, it is possible to be considerably more economical in the production of a design to match the required performance. For this to work, there must be useful evaluators of partial design states. This is where the idea of decomposition, outlined in the preceding section, becomes important.

Such systems may use diagnosis. Beginning with the MYCIN experiment (Shortliffe and Buchanan, 1975), diagnosis has become one of the major areas in which expert systems have made a substantial contribution. So far such programs are directed toward diagnosing existing systems or artifacts. The application of this technology to the design process is at an early stage.

Performance/Action Mappings

In a second approach to the control of generative systems, knowledge maps desired performance directly onto an action that achieves that performance. We may be able to structure the knowledge after the following form:

if <performance> **then** <action>

Examples of this kind of knowledge include the following:

If   energy is to be conveyed from the motor to the wheels
then connect the motor to the wheels with a drive shaft.

If   tool is to cut wood **and**
      is made of metal
then arrange sharp teeth down one side.

If   A is to be above B
then locate A on top of B.

The advantage of such rules is that if we require particular performances, it is a simple matter to select actions that achieve them. This provides a valuable mechanism for selecting among competing actions at any state in the design process. It should be apparent that these are abductive rules, similar to those described in Section 6.1. In order for these rules to be useful in controlling generation, certain assumptions about a closed world must apply, and the knowledge must be organized to handle conflicts between actions. In Section 7.3 we demonstrate how this knowledge can be brought to bear in a useful way in a generative system.

Generate, Test, and Modify

We may also consider a variation of the generate-and-test model. If a design that has been generated does not match the intended performance we can bring other knowledge to bear in modifying the design (Figure 6.71). This method may be appropriate where we have "weak" criteria for evaluating partial design states—that is, our knowledge about interpreting the performance of intermediate states relative to the final state is uncertain. The knowledge that facilitates modification can detect deficiencies in the design, discover their causes, and implement remedial action, a combination of deductive and abductive knowledge.

Prototype Selection and Refinement

The syntactic knowledge that defines a design space need not be only in the form of rules and procedures, but may also exist in the form of an explicit gen-

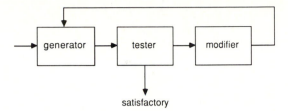

Figure 6.71   A generate, test, and modify system.

eralized design knowledge. The idea of a prototype (P) was described formally in Chapter 2 as consisting of the design description to be produced, D_p, the vocabulary to be used, V, interpretive knowledge, K_i, syntactic knowledge, K_s, and the interpretations or requirements, I:

$$P = (D_p, V, K_i, K_s, I). \tag{2.8}$$

We bring interpretive knowledge to bear in selecting an appropriate protoype; this can be seen as an abductive process. We may formulate our knowledge about selecting the appropriate prototype along the following lines:

if <requirements or performance> **then** <prototype>

Alternatively, prototypes may be selected by any of their other parameters $(D_p$ and $V)$, by any combination of these parameters, or by their label.

Once a prototype is selected, it is instantiated and its knowledge used to attempt to find values for the variables inherited during the instantiation. If either insufficient knowledge or insufficient facts are available, a new set of requirements is generated. The new requirements are used to search for and select more prototypes that are likely to help determine the values of the design variables. The process is represented graphically in Figure 6.72.

Thus prototype refinement consists of formulating requirements; selecting a prototype; producing an instance that expands the requirements, the vocabulary (and hence design variables), and the relevant knowledge; and using the knowledge in the instance to find values for the design variables. If values cannot be produced using the knowledge in the prototype, they are reformulated as requirements; further prototypes capable of producing these values are searched for and the process repeated. This is the basis of routine design.

The structural design system described by Gero, Maher, and Zhang (1988a) uses a set of prototypes that are combined as necessary to answer the design problem at hand. Oxman and Gero (1988) describe a system for designing by prototype refinement in architecture.

Hierarchical Control Abstractions

Where our generative knowledge can be decomposed in terms of discrete problem-solving tasks, or stages, we may be able to use control knowledge

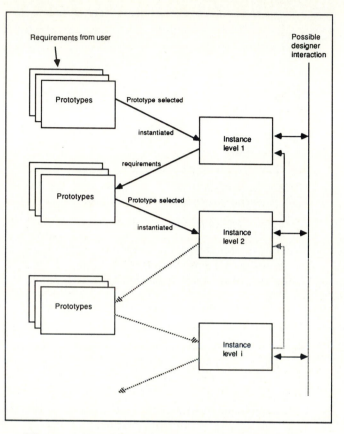

Figure 6.72   The design process using prototypes.

of the following form:

> **if** <stage> **and** <performance> **then** <action>

One example of a rule of this form is

    If   the stage under consideration is fixing of roof to walls **and**
          the wall is brick
then   fix top plate to wall.

Control knowledge in this form provides a certain economy in that only actions appropriate to a particular stage are considered. We demonstrate this approach in detail with an example in Section 6.6. The exploitation of hierarchical control abstractions is discussed further in Chapter 7.

6.4 Producing Designs through Generation

Even if we have no control knowledge at our disposal for producing meaningful designs, there is some value in considering the straightforward application of generative knowledge as a useful means of defining design spaces and producing designs. In this section we discuss the application of syntactic knowledge without reference to knowledge about performance, or semantics.

The idea of purely generative systems brought to the service of design has a substantial history. Interested readers are directed to summaries by Mitchell (1977) and March and Stiny (1985). One of the most direct applications of generation to the production of designs is embodied in the idea of a shape grammar. Just as we can devise grammars in the form of rewrite rules for producing strings of characters (phrase structure grammars, for example), we can also devise rules for transforming geometrical shapes. It is possible to define graphical languages similarly to the way we might define textual languages.

6.4.1 Shape and Set Grammars

It has been demonstrated that grammars expressed as production rules can facilitate the generation of designs (Spillers, 1972; Stiny and Gips, 1978). The technique has been employed to produce, among other things, line drawings resembling building plans, notably those of Palladio (Stiny and Mitchell, 1978) and Frank Lloyd Wright (Koning and Eizenberg, 1981). Applications in engineering include the setting out of structural systems (Fenves and Baker, 1987). The term *shape grammar* is employed to describe a particular type of algorithm for performing arithmetic operations on geometric entities called "shapes". These algorithms are formalized as production rules. The general form of a rule is

$$A \rightarrow B$$

Such a rule can be applied to a shape when there is a transformation that makes A a sub-shape. When the rule is implemented, A is replaced by the sub-shape B.

As described by Stiny and Gips (1978), shape grammars are defined for shapes made up of lines and points. Because these are insufficient to facilitate the mappings suggested by the preceding algorithm in the most general case, labels are incorporated into the formalism. Labels are given a geometrical status by being attached to points, allowing them to be treated as components of shapes. Labels can also serve as descriptors of shapes. "Parametric" grammars, a variation of shape grammars, make it possible to perform operations on classes of shapes.

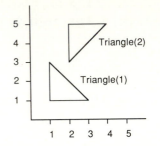

Figure 6.73   Two triangles.

The convention adopted for describing shapes (suitable for computer manipulation) is to represent them as sets of "maximal lines". A shape is described with the smallest number of lines possible (Krishnamurti, 1980 and 1981; Earl, 1985). As a new subshape is added to a composition, the representation is reformulated to retain the maximal line representation.

A shape can be decomposed into subshapes in whatever way required. Spatial ambiguities are allowed, and subshapes can be recombined and decomposed in different ways. Consequently, the implementation of shape grammars on a computer is a matter of some complexity. A shape editor for simple shapes has been implemented by Krishnamurti and Giraud (1984). We present the logic of a much simpler and more limited system in the next subsection.

The shape-grammar formalism appears to accord with the way a designer exploits the ambiguity of line drawings. Lines such as those that represent axes of symmetry, circulation, and construction lines typically change their meaning to become design elements as the design progresses. New shapes emerge from existing ones.

Set grammars, a conceptually simpler form of shape grammars, manipulate compositions made up of discrete elements. The distinction of set from shape grammars was made by Stiny (1980, 1982). In a set grammar, designs are made up of discrete elements that retain their integrity as they are operated on by a grammar. The disadvantage is that the composition of elements cannot be as readily decomposed and recombined to create emergent forms, as in a shape grammar. There are, however, certain implementational advantages. Set grammars can be represented more easily within a computer and conform more readily to the production-system formalism. They also match closely the use of a vocabulary of elements that must be configured.

We can describe objects in terms of facts about geometry with the name of the object and a list of points. A simple example of two objects that can be described as sets of connected points is shown in Figure 6.73; these objects are represented by the following facts:

Shape(Triangle(1), [1, 3] **and** [3, 1] **and** [1, 1] **and** Nil)

Shape(Triangle(2), [4, 5] **and** [2, 3] **and** [2, 5] **and** Nil)

The first argument in each case is the name by which the object is identified; the second, a list of points (x and y coordinate pairs). Other complex objects could be described as made up of these primitive objects.

A grammar rule is essentially a production rule (or transformation rule) with a left side and a right side. The facts on the left side of the rule facilitate matching with the current design description. When the rule is implemented those facts are deleted, and those on the right are asserted. The general form is

$$N : A \to B$$

where N is the name of the rule, A is the list of antecedents (those facts which are to be deleted), and B is a list of new facts that are to be asserted.

For non-geometrical object descriptions, the control for a production system described in Section 4.4.1 is sufficient. Control for rules involving the transformation of geometrical attributes must incorporate a mechanism for determining transformations. Such a mechanism is illustrated by considering the facts that describe the two objects in Figure 6.73. Although these objects are isomorphic, one is a rotation and translation of the other. It could be said that there is a match between the objects if there is a geometrical transformation that maps one object onto the other:

$$\mathrm{Match}(u, v, t) \Leftarrow \mathrm{Shape}(u, l_u) \wedge \mathrm{Shape}(v, l_v) \wedge \mathrm{Map}(l_u, l_v, t)$$

The third argument of the Match predicate is the transformation. Thus two shapes match if they map onto each other by transformation t. The definition of Map is as follows:

$\mathrm{Map}(\mathrm{Nil}, \mathrm{Nil}, t)$

$\mathrm{Map}([x_1, y_1] \textbf{ and } u, [x_2, y_2] \textbf{ and } v, \mathrm{RotTranslation}(r, t_x, t_y)) \Leftarrow$
   $\mathrm{Rotx}(r, x_1, x_2, y_2, t_x) \wedge \mathrm{Roty}(r, y_1, x_2, y_2, t_y) \wedge$

$\mathrm{Map}(u, v, \mathrm{RotTranslation}(r, t_x, t_y))$

The third argument of the Match predicate is the transformation, described as a rotation and a transformation (called RotTranslation). The first argument of RotTranslation is the rotation in degrees; the two other arguments are the translations in the x and y axes, respectively. The Rotx and Roty predicates are true by the following facts:

$$\text{Rotx}(0, x_1, x_2, y_2, x_1 - x_2)$$
$$\text{Rotx}(90, x_1, x_2, y_2, x_1 - y_2)$$
$$\text{Rotx}(180, x_1, x_2, y_2, x_1 + x_2)$$
$$\text{Rotx}(270, x_1, x_2, y_2, x_1 + y_2)$$

$$\text{Roty}(0, y_1, x_2, y_2, y_1 - y_2)$$
$$\text{Roty}(90, y_1, x_2, y_2, y_1 + x_2)$$
$$\text{Roty}(180, y_1, x_2, y_2, y_1 + y_2)$$
$$\text{Roty}(270, y_1, x_2, y_2, y_1 - x_2)$$

The predicate RotTranslation indicates a combined rotation and a translation. The goal

$$\exists t[\text{Match}(\text{Triangle}(1), \text{Triangle}(2), t)]$$

produces

$$t = \text{RotTranslation}(270, 6, -1)$$

which determines that object Triangle(2), when rotated by 270 degrees and translated by 6 and −1 along the x and y axes, respectively, provides a one-to-one mapping with shape Triangle(1). As defined earlier, rotations are constrained to matchings between orthogonally similar objects. Scaling, reflection, and other transformations are not taken into account. (This would make the control considerably more complicated.) The following goal can also be given:

$$\exists x \exists t[\text{Match}(\text{Triangle}(2), x, t)]$$

This asks, with what does shape Triangle(2) match and under what transformation? The goal is satisfied by

$$x = \text{Triangle}(1)$$
$$t = \text{RotTranslation}(90, 1, 6)$$

and

$$x = \text{Triangle}(2)$$
$$t = \text{RotTranslation}(0, 0, 0)$$

This states that object Triangle(2) matches Triangle(1) under a certain transformation and that it maps onto itself. (It is a simple matter to provide some sort of check to prevent the final, redundant match.)

Figure 6.74   A rule for transforming a triangle into a square.

The matching process involves comparing the lists of points of which the objects are comprised under the same rotations and translations. If there are many objects, then the process becomes computationally expensive.

Based on the preceding transformation rule schema, a rule for transforming a triangle into a square can be represented (Figure 6.74):

1: Triangle(1) $\rightarrow$ Square(1)

Assuming we have defined the shape Square(1), and employing the knowledge contained in the preceding program, a control mechanism can be devised that results in the replacement of Triangle(1) with Square(1) according to the following rule:

$$\text{Fire}(n) \Leftarrow (n : a \rightarrow b) \wedge \text{Match}(a, c, t) \wedge \text{Copy}(b, t) \wedge \text{Delete}(c)$$

This rule is analogous to that discussed in Section 4.4.1 for production systems in general, except that the predicates Match and Copy take account of transformations. We have not elaborated on the Copy predicate here, but it is similar to the definition of the Match predicate.

With this control logic we should be able to prove a goal, such as

Fire(1)

and in the process implement the changes due to rule 1 to the facts base describing the design. It is a simple matter to devise other mechanisms to ensure the exhaustive generation of all designs belonging to the space defined by the grammar, or to permit some kind of user interaction.

6.4.2   Producing Designs through Shape Generation

A simple grammar concerning the placement of rectangular elements is shown in Figure 6.75. Rule 1 matches an initial condition when there is merely a point labeled i, and results in a rectangle with a corner labeled p. Rule 5 displaces an element to one side. Rule 6 removes the labeled point p. (The labeled points assist in avoiding ambiguities in matching objects.) This results in a terminating condition in which no other rules can be applied. In order to avoid the possibility of two elements partially overlapping, the left sides of rules 2, 3, 4, and 5 contain the condition that no line segment exists where there is a dotted line. This condition can be stated as an extra argument on the left side of a transformation rule. (We would also require some changes to the control

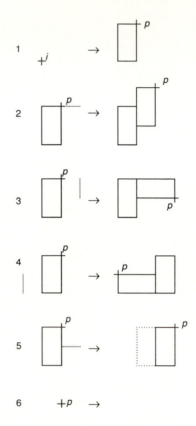

Figure 6.75 Rules for placing non-overlapping rectangles. Points i and p are labeled points. The dotted lines are segments that must not overlap existing lines.

mechanism to handle this.) The grammar rules are shown here graphically, though in program form they are represented in terms of points, as described earlier.

Figure 6.76 illustrates part of the tree of designs and partial designs generated by the rules of Figure 6.75. Not every rule is a candidate for implementation at each node in the tree. For example, certain rules cannot be implemented if they would result in the partial overlapping of elements. The computer output from an interactive program that incorporates this grammar is shown in Figure 6.77.

The design grammar in Figure 6.78 is also concerned with labeled points and lines. In this case the lines produce shapes resembling simple building components. The first seven rules establish a grid of points labeled d, which are the locations of columns, and points u, which form a secondary grid of mid-points between columns. Rule 8 establishes the locations of mid-points on the perimeter of the building grid, and changes their labels from u to w.

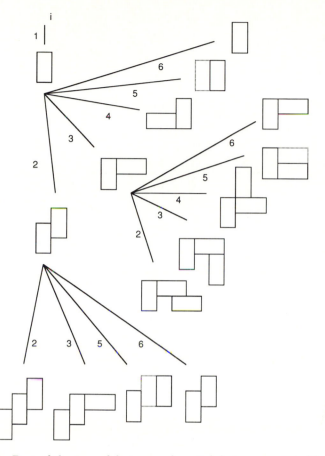

Figure 6.76 Part of the tree of designs and partial designs generated by the rules of Figure 6.75.

Rules 9 to 12 alter the labels at certain points to indicate the positioning of a front entrance, a rear entrance, north windows, and east windows. Rules 13 to 16 assert line segments into the design description to represent stylized building elements, such as a column, a wall panel, a front entrance, a semi-opaque wall element (window), and a rear entrance. Rules 18 to 25 simply delete labels. Rules 3, 8, 11, and 12 contain geometrical conditions that must be met and that are inferred from the design description by means of inference rules. Figure 6.79 illustrates the designs generated by this grammar. Unlike the previous example, the number of designs is finite.

Figure 6.80 shows that a control system can be devised that invokes the rules automatically, but that provides prompts so that a designer can intervene and cause the system to take a different path. In this case Rule 1 is followed by the implementation of rule 2 three times, then by rules 4 to 7, to produce

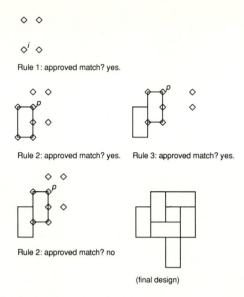

Rule 1: approved match? yes.

Rule 2: approved match? yes.   Rule 3: approved match? yes.

Rule 2: approved match? no

(final design)

Figure 6.77 Graphical output showing the automatic generation of designs; the option of accepting or rejecting the changes proposed by a rule is displayed on a computer screen. The design begins from the initial condition (the labeled point *i*). Rule 1 is applied. The corner points of the shape matched and the corners of the new shape to be added are indicated by diamonds. If the response is "yes," then the rule is implemented and the next applicable rule (rule 2) is called. The first rule that finds a match in the design description is called each time. In response to "no" the program calls the next available rule. The final design shown here results from the following rule sequence: 1, 2, 3, 3, 3, 4, 6.

the grid. Rule 8 labels the external grid lines w. After the firing of rules 9 to 12, the positions of the front entrance, the rear entrance, and the windows are fixed. Rules 13 to 17 draw the elements to produce a labeled plan. These labels are subsequently removed.

Figure 6.81 illustrates a set of design rules for generating simple building forms from wall elements and columns. The elements form rooms, each of which has a name and an orientation. A dot indicates the top of a room. The design space is depicted in Figure 6.82 as a tree, with the root node as the initial state (the fact *start*), and the ends of the branches as various end states, where no further rules can be applied. As in the preceding examples, the application of certain rules precludes the achievement of certain states. Thus each end state is different.

One of the rules, rule 2, is depicted in program form in Figure 6.83. In this grammar only facts about objects are to be matched, and there are no conditions. The rule has a name, a list of preconditions, and a list of consequents. Those facts that are not to be deleted appear on both sides of

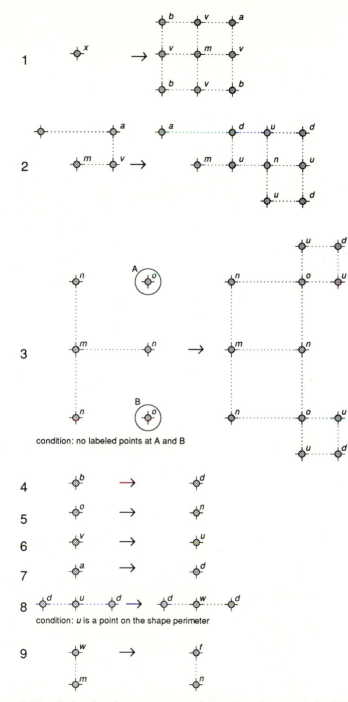

Figure 6.78 Rules for the generation of the plan of a simple building form.

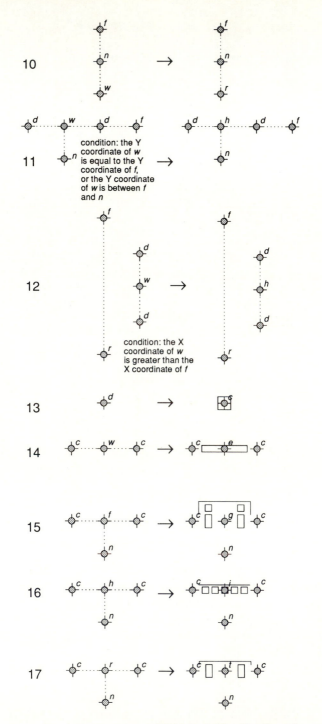

Figure 6.78 Continued.

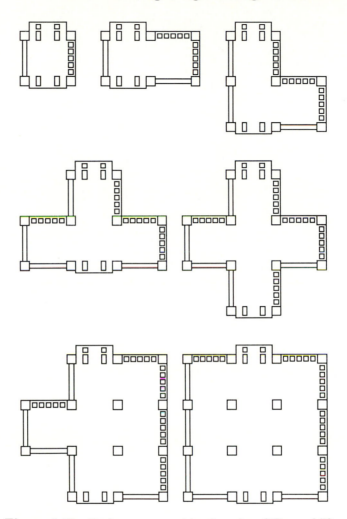

Figure 6.79   Designs generated by the rules of Figure 6.78.

the rule (for example, RoomA(1)). The objects to which the facts refer are described hierarchically, as shown earlier.

The rule is shown diagrammatically in Figure 6.84. A room is defined by four columns. For this rule to be fired, a pattern must be found in the current design state that matches the description of RoomA(1) and ShortWall(2). These objects are deleted from the state description and replaced with objects RoomA(1), again, RoomC(2), ShortWall(4), ShortWall(5), LongWall(5), and OpenWall(2). The arguments denote different instances of the same object.

RoomA(1) is a composite object made up of Column(1), Column(2), Column(3), and Column(4). These objects are represented hierarchically, as described in Sections 3.2 and 5.5.1, and consist of line segments that are defined

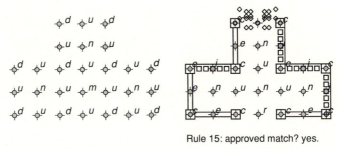

Rule 2: approved match? yes.

Rule 15: approved match? yes.

Figure 6.80 The progress of a design generated by the selective invocation of rules from the grammar of Figure 6.78.

by points expressed in homogeneous coordinate notation:

> Composite[RoomA(1), Column(1) **and** Column(2) **and**
>   Column(3) **and** Column(4)] Object[Column(1), L(1) **and** L(2) and
> L(3) **and** L(4)]
>
> ...
>
> Line[L(1), [P(1) **and** P(2)]
>
> ...
>
> Point[P(1), 260, 200, 1]
> Point[P(2), 260, 280, 1]

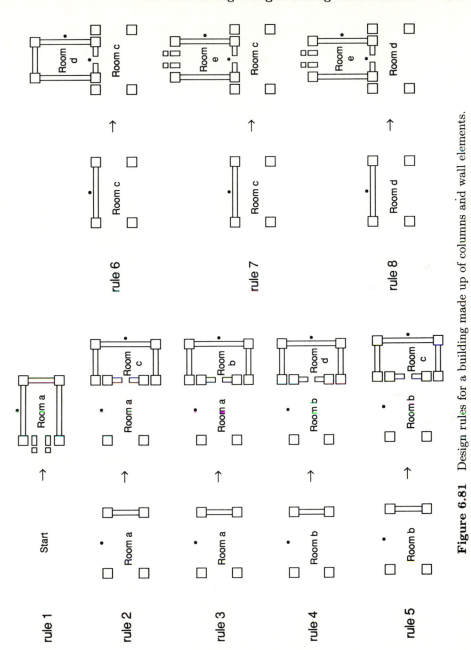

Figure 6.81 Design rules for a building made up of columns and wall elements.

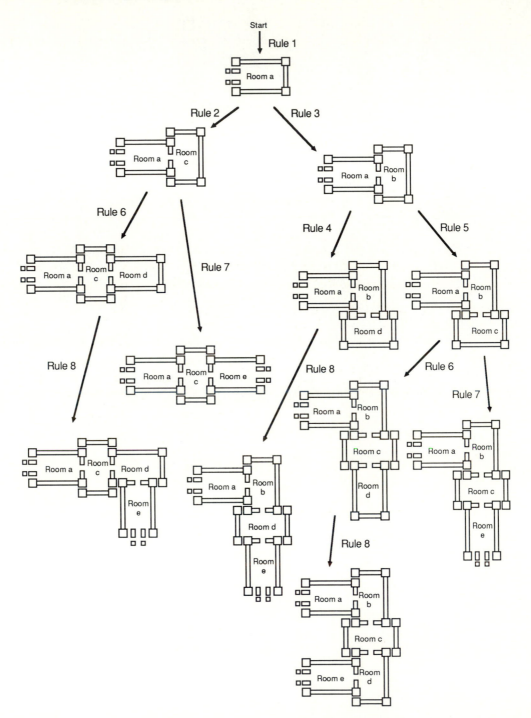

Figure 6.82   State space for the rules of Figure 6.81.

Figure 6.83 Rule 2 of the grammar of Figure 6.81 in program form.

Again, we require a more sophisticated control mechanism than the one described earlier in order to exploit the hierarchical nature of this description. It is possible to construct different rules from such object descriptions. A composite object in a rule matches the current design description if there exists a composite object described in the facts base that has the same predicate name but different instance number, and if both the composite object and the design description are made up of similar objects. The geometrical transformation that enables the object described in the rule to be mapped onto the object in the state description is calculated from the points defining the ends of line segments.

These systems deal only with the generation of geometrical shapes. They are valuable as a way to manipulate a vocabulary to produce rich and varied solutions starting from simple beginnings. They are useful because the generation of shapes in the geometric sense is one of the main concerns of design.

6.5   Producing Designs through Generate and Test

In this section we discuss the utility of generating designs from the space of designs defined by generative knowledge, employing interpretive knowledge

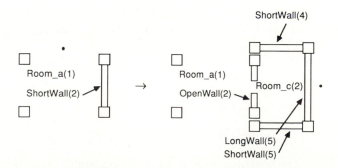

Figure 6.84 Graphical representation of the rule in Figure 6.83.

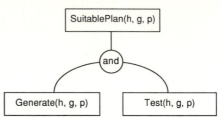

Figure 6.85   The logic of a simple generate-and-test system for the site-layout problem.

to guide the process. The simplest way to realize this idea is through the generate-and-test model of problem solving.

6.5.1   The Control of Generate and Test

One of the earliest expert systems to successfully exploit the idea of generate and test was DENDRAL (Buchanan and Feigenbaum, 1978; also described by Winston, 1984, and Hayes-Roth, Waterman, and Lenat, 1983). This expert system infers the structure of an organic molecule of known chemical formula from its spectrogram. There is a *generator* that exhaustively produces feasible molecule structures that conform to the formula, and a *tester* that produces a synthesized spectrogram of each of the feasible molecules (effectively sophisticated graphical interpretations) and compares them with the actual spectrogram. The number of possible structures generated, although large, does not represent an unacceptable computational burden. Because the generator starts with the molecular formula, it is not necessary to consider every possible chemical structure, only those with the given formula.

As discussed previously, the granularity of the generative mechanism is important. If the generative knowledge defines a small space, then, as in the case of DENDRAL, the task may not be too expensive. For a generate-and-test system to be useful, we require an effective generator that is efficient, complete, non-redundant, and succinct (Winston, 1984). The key to an effective generate-and-test system is that it be able to prune branches of the search tree as early as possible. Let us explore the generate-and-test idea with a spatial layout problem.

We refer to the simple example of Section 6.2.5, which has three elements to be located on a building site—a house, a garage, and a swimming pool. The logic of a generate-and-test system for this problem can be expressed simply as follows (Figure 6.85):

$$\text{SuitablePlan}(h, g, p) \Leftarrow \text{Generate}(h, g, p) \wedge \text{Test}(h, g, p)$$

The generator of such a system can simply be a list of candidate locations for elements. Its logic can be expressed as follows:

$$\text{Generate}(h, g, p) \Leftarrow \text{SiteCells}(s) \wedge \text{Member}(h, s) \wedge \text{Member}(g, s) \wedge$$
$$\text{Not}(h = s) \wedge \text{Member}(p, s) \wedge \text{Not}(p = s) \wedge \text{Not}(p = h)$$

The logic of a tester can be expressed as

$$\text{Test}(h, g, p) \Leftarrow \text{ExclusionZone}(z) \wedge \text{Not}[\text{Member}(h, z)] \wedge$$
$$\text{Not}[\text{Member}(g, z)] \wedge \text{Not}[\text{Member}(p, z)] \wedge$$
$$\text{NextTo}(h, g) \wedge \text{Not}[\text{NextTo}(p, g)] \wedge \text{NorthOf}(p, h) \wedge \text{NorthOf}(p, g)$$

Assuming we have a theorem prover, in which subgoals are proven in order from left to right, we have a system in which candidate locations for all three elements are generated, then tested. There are 16 positions and 3 elements to be located. This means that there are 3,360 ways in which these elements can be configured. We can see that there would be a definite advantage in early pruning—that is, reducing the size of the search tree by building some of the knowledge of the tester into the generator. There are several means at our disposal for doing this: (1) delete cells in the exclusion zone so that they are not even considered as candidate locations (this involves modifying the generator); (2) generate the location of two elements, then test for adjacency (NextTo) and orientation (NorthOf) before considering the third element; and (3) order the testing of the adjacency and orientation constraints so that the "most-constrained" element is tested first. We need not elaborate on this particular example further because the general idea is very simple. We would expect the approach to work here because the criteria we use for evaluating the end state are also entirely applicable to intermediate states of the design.

6.5.2   Generate and Test in Spatial Layout

We can pursue this idea with a slightly more realistic example in spatial layout. An interesting form of generative knowledge is that which takes dimensionless spaces as vocabulary elements and operates on them. The rules discussed here are similar to the set of rules described by Mitchell, Steadman, and Liggett (1976) for generating "rectangular dissections." Other rules for the same purpose are proposed by Earl (1977) and Flemming (1978), as discussed by Steadman (1983). The study of rectangular dissections is ostensibly concerned with enumerating the different ways in which a rectangle can be divided up into sub-rectangles, but it can also be applied to the generation of integrated circuitry and building layouts. Dimensioning is of only secondary importance in this context. (Dimensioning can be considered separately, and numerical techniques exist for handling this problem [Gero, 1977; Balachandran and Gero, 1987a].)

Rules by which dissections can be generated are considered here as a means of generating spatial layouts. These are shown schematically in Figure 6.86. Rule a adds a rectangle by appending it to the right side of an existing composition. Rule b appends a rectangle to the top of the configuration. Rules

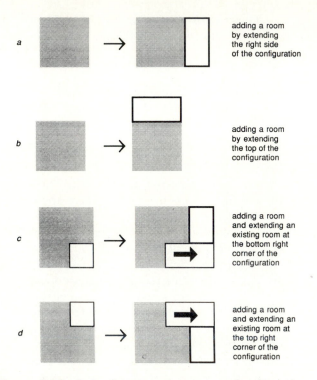

Figure 6.86 Rules for generating rectangular dissections.

c and d operate by extending some of the rectangles at the edge of the configuration and adding a new rectangle in the space thus formed. Other rules can also be created for adding rectangles in different ways.

The idea is to begin with a single rectangle (a room or space) that then undergoes successive transformations resulting in a complex composition of rectangles. After the operation of each rule, the composition retains its overall rectangular shape (as well as being composed only of rectangles). States in the development of a simple rectangular configuration are illustrated in Figure 6.87. The success of the grammar for generating building layouts relies on three assumptions. First, it is assumed that there is a small set of rules with which every possible arrangement of rectangular dissections can be generated. It is also assumed that any geometrically orthogonal arrangement of rectangular spaces can be derived by first creating a rectangular dissection (allowing for the deletion of spaces). Finally, it is assumed that the generation of rectangular dissections is best accomplished if every intermediate state in the generation process is also a rectangular dissection. We will not argue the validity of these assumptions here. It is assumed that any deficiencies in the rules proposed can be overcome by providing more rules of the same type, or refining the rules. Note that this rule set requires more complex geometrical operations than the grammars discussed in Section 6.4.2.

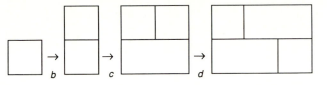

Figure 6.87 States in the development of a simple rectangular configuration.

Given a generator based on these rules, we can formulate some criteria for evaluating layouts (required performances):

a and b must be adjacent.

b and c must be adjacent.

c and d must be adjacent.

d and a must be adjacent.

c must be to the right of b.

Minimum width of a room is greater than or equal to W.

Width of arrangement is less than or equal to available area width.

Depth of arrangement is less than or equal to available area depth.

We can employ the rules of Figure 6.86 as generators, using these performances as criteria for testing partial states. As shown in Figure 6.88, this approach may be satisfactory for small numbers of rooms. With larger numbers of rooms we have less confidence in such criteria, and the process will require some search.

One application of this approach in floorplan generation is a system for the configuration of "loosely packed arrangements of rectangles" (Flemming et al., 1986). Loosely packed arrangements of rectangles are sets of rectangles that are pairwise non-overlapping and that have sides parallel to the axes of an orthogonal system of Cartesian coordinates. Such arrangements can contain "holes" and thus can be used to represent a wide range of rectilinear layouts. The system consists of a generator able to produce representations of all possible layouts, and a tester able to evaluate representation of layouts. The system is not limited to rectangular dissections, but also involves the location of equipment or furniture or preliminary layouts—any domain in which the details needed to completely specify a densely packed arrangement are not available.

The intention of the system is to act as a design assistant able to generate alternative layouts systematically within a broad range of constraints and criteria. Its aim is not to provide an optimum solution, but rather to generate and give a preliminary ranking to the possibilities, which are presented interactively to the designer for a selection. The basic components of the system are a generator, which generates orthogonal structures as representatives of loosely packed arrangements of rectangles; a tester, which evaluates these

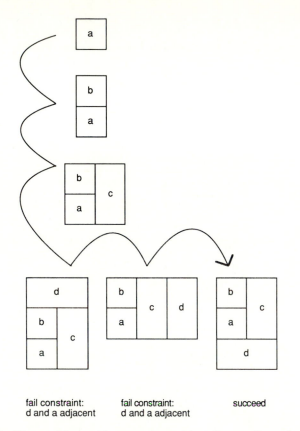

Figure 6.88 The problem with generate and test for small numbers of rooms.

structures within the context of a particular design problem; and a search strategy, which organizes the search for solution alternatives.

The generator accepts an orthogonal structure from the search strategy and expands it in all possible ways. It consists of a number of rewrite rules, where each rule is given by a left-hand side describing a configuration of vertices and arcs and a right-hand side describing how the new vertex can be added to this configuration. Figure 6.89 shows an example of an arrangement containing 12 rectangles and all of the arrangements that can be generated from this example by adding a new rectangle at the upper left corner. The tester takes in a structure produced by the generator and evaluates it. The tester sets up scores in three categories: (1) independent or dependent constraints reflecting geometric considerations, (2) strong criteria reflecting rules of good design or critical design requirements, and (3) weak criteria dealing with aspects deemed less critical than those of strong criteria.

Knowledge in each category is expressed in the form of interpretive rules.

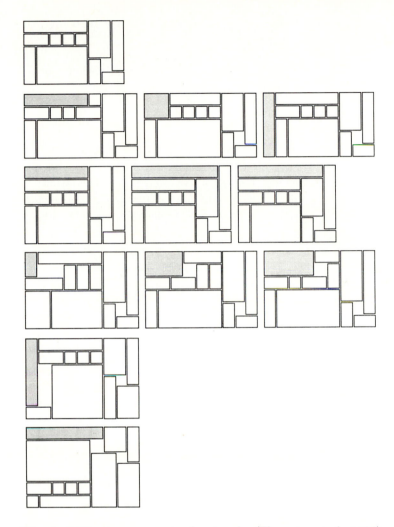

Figure 6.89 Arrangements of rectangles (Flemming et al., 1986).

An example of a rule evaluating a dependent constraint is

If   in the arrangement under evaluation, the sum of the minimum
        horizontal dimensions of any sequence of objects arranged
        from left to right exceeds the maximum horizontal dimension
        of the available area

then   update the score for constraints violated by the arrangement.

According to Flemming et al., (1986) experiments have shown that the preceding type of constraint is one of the most frequently violated dependent
constraints and should therefore be a very effective pruning device. This constraint demonstrates how dimensional knowledge can enter the search at an

early stage. In the domain of kitchen remodeling, the following rule enforces a strong criterion:

If   the arrangement under consideration contains a sink and
           plumbing outlets **and**
       the sink is not adjacent to the outlets
then   update the score for strong criteria violated by the arrangement.

The rule-based approach for the tester allows new knowledge to be formulated incrementally and interactively using intuitively appealing forms of encoding. The search strategy selects and passes the structures to be expanded to the generator, accepts the new structures, and passes them to the tester. It inspects the results to determine whether to terminate the process or reinitiate the cycle.

6.6   Producing Designs through Knowledge about Design Actions

In this section we discuss the formulation of an expert system that carries out actions to place elements in certain positions. The ideas are based on those developed in the XCON (or R1, as it was originally known) expert system for configuring DEC's VAX and PDP computers (McDermott, 1982). XCON is one of the earliest and most successful commercial applications of an expert system. First, however, we consider a simpler domain that highlights some of the features of such systems.

6.6.1   A Simple Allocation Problem

Our example concerns the problem of allocating staff members to rooms. We may go through various steps. First, we check that we have a complete list of staff members. Second, we locate the staff members who require the largest amount of space, paying particular attention to those who require a graphic workstation and a digitizer. Third, we locate those who require a medium amount of space, being careful to locate secretaries near the entry to the room. We are also careful not to put people who require only a medium amount of space together with those who require a large amount of space; we assume that they are generally somewhat incompatible. Fourth, we can locate those who require only a small amount of space; they can be located in any room with enough space. These rules are simple heuristics of the kind we might apply to such a problem.

Rules in this system are of the following general form:

if <step> **and** <performance> **then** <action>

The elements in this system, rooms and people, are to be configured as particular people in particular rooms. The main actions are simply to place people in rooms, though the rules are also concerned with other changes to the facts base relating more to control, such as moving on to the next step.

The approach demonstrated here is successful when relatively little consideration needs be given to interrelationships between elements. This assumption is not unreasonable in some cases—for example, when the interaction may be unknown at the outset. In the case of placing staff in rooms, we suppose that the criteria are size of workplace and general compatibilities among tasks. We intend to formulate general layouts that can be refined or revised subsequently.

We require an initial facts base. The system is to be carried out one step at a time. The current step is represented with a fact. We also keep a record of the rooms as they are "filled" and who has been allocated to them. There can be any number of rooms; we start with room 1 and proceed through the rooms in order. We also maintain a list of people who are not yet located. The initial state of the facts base looks like this:

step:	check staff list
room 1:	—
unlocated:	Stephen
	Cathy
	Mike
	Sally
	Fay
	Assistant

We need a set of information about people, their space and equipment needs, and their job title in the organization.

Person	Workstation Type	Space Requirement	Job Title
Stephen	text terminal	medium	system manager
Cathy	graphic workstation	large	research fellow
Mike	workstation and digitizer	large	research fellow
Sally	text terminal	medium	research student
Assistant	work desk	small	research assistant
Fay	text terminal and typewriter	medium	secretary

The rules are listed in the order we give them here. In this approach, rule order is important. Each rule contains a precondition to ensure that it is fired according to the step to which it applies.

Certain staff members tend to be accompanied by other, ancillary staff. This could be one of several rules for making sure the right number of staff are accommodated. Once the action has been implemented, we find that the following rule in the knowledge base can be applied:

> **If**   the step is check staff list
> **then**   change step to locate according to large space requirements.

This rule changes the step to "locate according to large space requirements." Only rules applicable to this step will be applied until the step is changed.

The next applicable rule is given here:

> **If**   the step is locate according to large space requirements **and**
>   there is a staff member with large space requirements to be
>   located **and**
>   there is a workstation and digitizer to be located **and**
>   there are fewer than four staff member with large space
>   requirements already in the room
> **then**   locate the workstation and digitizer in the room

This rule gives priority to staff members requiring a digitizer. (Because their space requirements are more complicated, it is helpful to locate them before other staff who also require a large amount of space.) Because rules are considered in order at each cycle, this rule is fired as many times as it applies until the following rule can be fired:

> **If**   the step is locate according to large space requirements **and**
>   there is a staff member with large space requirements to be
>   located **and**
>   there are fewer than four staff members with large space
>   requirements already in the room
> **then**   locate the staff member in the room.

This rule keeps locating staff members in the same room until we run out of staff members requiring a large amount of space or until the room has four such people. In the event that a room is full and there are still more staff to be located, we need another rule to start a new room:

> **If**   the step is locate according to large space requirements **and**
>   there is a staff member with large space requirements to be located
> **then**   start on a new room.

The following rule changes the step to "locate according to medium space requirements."

> **If**   the step is locate according to large space requirements
>
> **then**   change step to locate according to medium space requirements.

The state of the facts base is now as follows:

step:	locate according to medium space requirements
room 1:	Mike
	Cathy
unlocated:	Stephen
	Sally
	Fay
	Assistant

The following rule is applicable to this facts base:

> **If**   the step is locate according to medium space requirements **and**
> there is a staff member with medium space requirements to be
>   located **and**
> there is an empty room **or**
> a room containing only staff members with medium space
>   requirements **and**
> the room is not yet full **and**
> the position of the staff member is secretarial **and**
> the staff member is not located near the entry door
>
> **then**   locate staff member near an entry door.

This rule ensures that secretarial staff are located near the entrance to the room in which they are located. The following rule applies to the rest of the staff with medium space requirements:

> **If**   the step is locate according to medium space requirements **and**
> there is a staff member with medium space requirements to be
>   located **and**
> there is an empty room **or**
> a room containing only staff members with medium space
>   requirements **and**
> the room is not yet full
>
> **then**   locate staff member in that room.

The following two rules are similar to those about large space requirements:

If   the step is locate according to medium space requirements **and**
there is a staff member with medium space requirements to be
located
then   start on a new room.

If   the step is locate according to medium space requirements
then   change step to locate according to small space requirements.

They produce the following state of the facts base:

    step: locate according to small space requirements
    room 1:   Mike
                 Cathy
    room 2:   Stephen
                 Sally
                 Fay (near entry door)
    unlocated: Assistant

The following rules are for locating staff members with only small space
requirements; because their roles partly overlap, it is best to spread them
around. According to these rules, staff members with small space requirements may be located in rooms occupied by either staff with large or medium
space requirements, provided they are not in with a secretary.

If   the step is locate according to small space requirements **and**
there is a staff member with small space requirements **and**
there is a room that is not yet full **and**
the room does not contain a secretary
then   locate the staff member in that room.

The rest of the rules are described here and require no explanation:

If   the step is locate according to small space requirements **and**
there is a staff member with small space requirements **and**
there is a room that is not yet full
then   locate the staff member in that room.

If   the step is locate according to small space requirements **and**
there is a staff member with small space requirements
then   start on a new room.

One rule is required to terminate the process satisfactorily:

> **If**   the step is locate according to small space requirements
> **then**   change step to stop.

This rule changes the context to "stop." The final state of the facts base is as follows:

step:	stop
room 1:	Mike
	Cathy
	Assistant
room 2:	Stephen
	Sally
	Fay (near entry door)
unlocated:	—

Much of the knowledge in this system is contained within the strategy for selecting between competing actions, in this case partitioning into steps and ordering rules. Here we have looked at allocating people to rooms. We would expect this approach to apply to any packing problem, such as locating furniture in rooms, storing, and placing computer components in cabinets. Although the system does not optimize the packing, it might be a useful first cut at a problem in which optimal packing is required.

The approach could also be applied to spatial allocation, although this would be difficult if there were a large number of proximity constraints—for example, if the secretary must be adjacent to the two research fellows and next to the window; if the staff are to be located to minimize space; or if travel distances between people who work together are to be minimized.

6.6.2   Configuring Computer Systems

We are now in a position to describe XCON. Given a customer's order for a computer system, XCON checks to see whether all necessary components have been included. If not, the missing components are added to the order. XCON produces diagrams showing how the various components of the order are to be associated. A typical VAX computer may have more than 100 different components in its final form. These are selected from a large set of components, many of them bundles of more basic components. All these individual items must be placed in cabinets that will fit into the available floor space. The components must be interconnected properly; power supplies must be provided with sufficient capacity to supply adequate voltage and current to all components for all conditions of operation. Each cabinet space must accommodate the assigned components. Many peripheral devices can be connected to the system. XCON has sufficient knowledge of the configuration domain and of the peculiarities of the various configuration constraints that at each step in the configuration process, it simply recognizes what to do. Consequently, little search and no backtracking are required.

XCON is a large and powerful design system. Beginning as R1 in 1981, it grew from 420 parts and 777 rules to more than 5,000 parts and 3,300 rules in 1984 (Bachant and McDermott, 1984). XCON performs a number of sub-tasks in the process of configuring a system. Some of the tasks the system is required to carry out are determining whether the order received is a reasonable system and is complete, and that all the components are compatible; determining the optional module sequence; selecting a backplane; assigning a backplane to a box; when the module is a multiplexer, verifying that there is sufficient panel space in the cabinet. Other tasks are to configure unibus adaptors, calculate unibus length, and assign boxes to cabinets and unibuses. If insufficient space remains in a box or cabinet, then additional components must be added to the list.

XCON is implemented in the OPS language (Forgy, 1981), based on a forward-chaining technique in which the production conditions are checked to see which productions can be triggered—that is, which productions have all their preconditions satisfied. A recognize-select-act control cycle is used. After all the productions triggered are recognized, one is selected using OPS's selection strategy and is then fired—that is, its actions are carried out. English translations of two of XCON's production rules follow:

If   the most current active context is distributing massbus devices **and**
there is a single-port disk drive that has not been assigned to a
massbus **and**
there are no unassigned dual port disk drives **and**
the number of devices that each massbus should support is
known **and**
there is a massbus that has been assigned at least one disk
drive **and**
the massbus should support two additional disk drives **and**
the type of cable needed to connect the disk drive to the previous
device on the massbus is known
then   assign the disk drive to the massbus.

If   the most active context is selecting a box and a module to put
in it **and**
the next module in the optimal sequence is known **and**
at least that much space is available in some box **and**
that box does not contain more modules than some other box
on a different unibus
then   try to put that module in that box.

A database holds the details about each of the components likely to be used.

Each entry in the database contains the name of the component, its type, its classification, and other relevant information. The database also holds information about cabinet types and sizes so that components can be configured to fit within a specific cabinet.

Given the nature of the rules in XCON and the problems of configuration, it is quite possible that several rules will be triggered at each cycle, giving rise to the possibility of obtaining several alternative configurations, each satisfying the requirements. The strategy adopted by XCON is slightly different from that just demonstrated in the staff-allocation problem. XCON contains a "special-case" strategy for selecting the configuration that contains the greatest number of elements most specialized for the particular requirements.

6.6.3 Construction Detailing

Let us now apply this approach of generating designs through knowledge about actions to another configurational problem. We consider the design of a building detail, in this case the junction between roof and walls in house construction—the eaves. The precise domain of this example is the design of eaves detail to match Australian construction around the years 1910 to 1920 (Mitchell and Radford, 1986), given a design for the external wall construction of a building. Examples of actions include the following (Figure 6.90):

> Place wall plate over single brick wall.
>
> Notch rafters over wall plate.
>
> Add battens to underside of rafters.

Along with other actions, these actions define a generative language for producing detail designs. To be useful, the actions should be described geometrically. They make changes to a graphical database, and thus the results can be seen as a graphics model. In this example we do not consider the imposition of constraints other than those required for an action to be implemented. This is appropriate in certain design tasks—for example, generating and testing. If

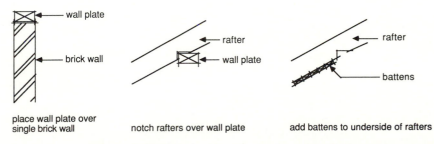

Figure 6.90 Actions for configuring eave details.

we want different designs, we may change the actions or generate alternatives. As with XCON, detail design can be organized into certain steps:

fixing of roof to the walls
trimming rafter ends
fascia
eaves lining
gutter
roofing

We can use this information to control the application of actions through the design.

Attributes are modeled as facts about the object. The most important attribute is shape. An object's geometric description is modeled in a graphics database referenced by the name of the object. Part of this geometric description is an origin for the local coordinate system in which the vertices of the shape of the component are defined. The state of the design as it develops is modeled as a list of its component objects. The spatial configuration of adjacent objects is predefined in the actions. Therefore a list of components is not merely an inventory of parts but also implies a particular spatial organization sufficient to draw the detail.

Rules are based on one of two types of conditions.

1. In the arrangement of two elements, check the presence of a first element before the second element can be added—for example, a gutter can be fixed only to a vertical fascia or a vertically cut joist.

2. In the current step being undertaken, which simplifies an otherwise extended list of individual relationships, permit or negate the next transformation of the design—for example, if the "lined eaves" step (boarding on the top of joists for the eaves extent) has been taken, then soffits (eave liners) are not applicable. This makes it easy to categorize and test a set of design conditions.

Because there are no explicit goals, we can consider the control to be simply based on the following form:

if <step> **and** <preconditions> **then** <action>

The design process is modeled as a stage/state process, with the design progressing from an initial state (the external wall) to a final state (the complete detail) through a sequence of partial design states (Figure 6.91). At each stage, there is at least one and generally a set of alternative rules by which the current state of the design can be transformed into the state appropriate for the next stage. Designs are generated by the serial "firing" of a list of rules. The final state of the design is reached once all rules have been tested. An updated

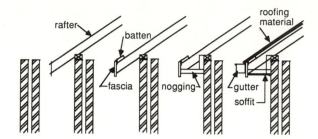

Figure 6.91   States in the development of an eave detail.

fact records the progressive application of classes of operations, and execution of a rule is conditional upon certain facts. For example

If   the rafter end has been configured **and**
      the fascia fixing has not been configured **and**
      the rafter has a square end
then place the batten on the rafter.

Satisfaction of the first condition determines that the "rafter end" stage has been completed; of the second condition, that no other rule in the "fascia-fixing" stage has been applied; and of the third condition, that the state of the design includes a "square end." A rule also moves the drawing origin to a marker position associated with the "square end" element. If these conditions are met, then the element "batten" is added to the state of the design and the current mode of the design is updated to "fascia fixing." No further rules of this stage will then apply. The later rules are more complex because the elements cannot be serially added to the last element, but still use the location of an existing element in the design state to reference the position of a new element. Part of the network of actions and the implicit links between them by means of conditions is shown in Figure 6.92. The heavy lines in that figure show the dependencies between rules in the design of one of the eave details, shown in Figure 6.93. Only the knowledge base and the graphics database are peculiar to eave detailing; the rest of the system can be applied to other details.

It is informative to consider the utility of this approach when combined with generate and test. In the eave-detailing example and the examples using shape grammars, the generation process is purely syntactical; this process is illustrated in Figure 6.94. As pointed out, this blind process may produce an unacceptable number of legally correct solutions. By incorporating interpretive knowledge, we can reduce the number of designs generated by the system. In this process a design is generated, syntactically as before, but is then interpreted. This interpretation or performance is matched to the performance requirements and, if satisfactory, the design so generated is accepted. If the

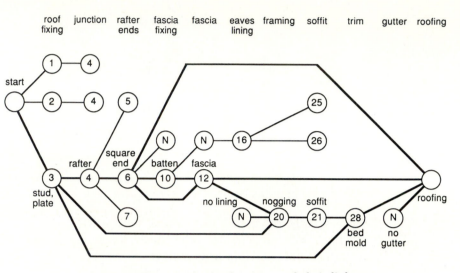

Figure 6.92   A network of actions and their linkages.

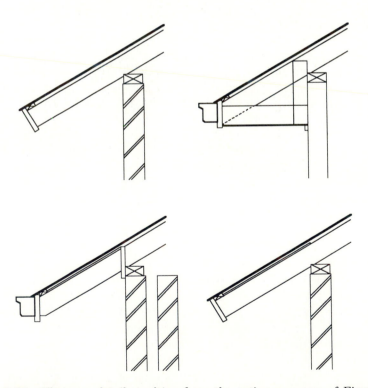

Figure 6.93   The eave detail resulting from the action sequence of Figure 6.92 (top left) and three other eave details.

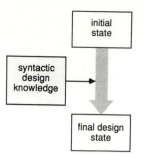

Figure 6.94   Generating designs using syntactic knowledge.

interpretation is unsatisfactory another design is generated—that is, different syntactic rules are applied. The generate-and-test process is illustrated in Figure 6.95.

There are two approaches to this latter method. In the first, a complete solution is generated and then tested. If it is acceptable the process stops; if not, then a new solution must be generated. This may involve simple chronological backtracking or some other, more efficient control technique. This approach suffers from the fact that much effort may be expended in generating large numbers of complete solutions that are rejected. The second, more efficient approach is to test at each stage of the generation by applying knowledge in the form of constraints and criteria and determining whether a potential

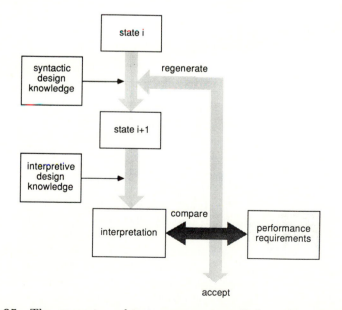

Figure 6.95   The generate-and-test process as applied to the eave-generating system.

action will produce a feasible and/or desired state. If at any stage any partial solution produced violates some constraint or seems no better than some other partial solution already generated, then there is no need to proceed along that path. For this strategy to work, there must be good evaluators of partial states. In some cases the evaluator of the end state can be used for intermediate states. For example, "there is to be no eaves lining" simply means that any intermediate state that has an eaves lining is ruled out. Or goals may be decomposable into subgoals that can be checked at intermediate states. For example,

If the goal is to produce a detail that admits ventilation
        to the building
then (ensure that there is no eaves lining **or**
        eaves lining is above rafters) **and**
        no trim above top plate **and**
        there is insect screen above top plate.

The second approach is that taken by stage-state optimization techniques such as implicit enumeration, branch-and-bound, heuristic search, and dynamic programming. In general, at each stage the process is at some state and the partial solution is expanded by considering all new states at the next stage that can be produced from that state. The optimization function is used to decide with which of the alternative expansions to proceed so as to arrive most efficiently at an overall optimal solution. Optimization is applicable only if all the values of all the variables can be expressed numerically.

6.7   Summary

In this chapter we have considered the production of designs from knowledge about semantics and knowledge about syntax. These issues can be translated into concerns about goal satisfaction and the configuration of design elements.

We discussed the problems inherent in beginning with statements about desired interpretations, or performances, and proceeding to design descriptions. Our commonsense understanding tells us that we can reason from a description to an interpretation. In design, the problem is that we begin with the desired interpretation and must arrive at a description—a process we can label "abduction." We discussed the issues of the closed-world assumption, conflict resolution, and one-to-many mappings that make abductive knowledge difficult to formalize. We can overcome these problems by limiting our knowledge so that these issues do not arise. We demonstrated one example in which we were able to reason about design specifications. Instead of treating the performance as an interpretation of the design, we considered the design to be an interpretation of the desired performance—in other words, the specification for the project.

In interpreting a design specification, we bring knowledge to bear in building up a description of a design. At some stages in this process we may simply be turning the specifications into a form that is more "legible" to other knowledge. We demonstrated the idea of interpreting design specifications to produce a problem formulation suitable for an optimization algorithm.

We then adopted the approach of attaching plausibility factors to abductive design rules as a means of deciding among competing design descriptions. We followed a MYCIN-like example, treating performance requirements as evidence contributing to competing design decisions. We then showed the utility of adopting fuzzy reasoning as a means of expressing varying degrees of commitment within abductive rules.

We tackled the issue of constraint propagation, where the intention is to determine the bounds on a design before actually making any design commitments or choices. The challenge is to define the constraints in such a way that we can gain an appreciation of the nature of the design space, and possibly produce a design from that bounded space. We also considered the role of assumption-based reasoning, in which we reason about contradiction in abductive rules. This latter approach represented a serious attempt to tackle the problems of inconsistency and conflict inherent in abductive reasoning.

From the point of view of syntax, we considered the key role of control as a means of producing designs that conform to a set of intended performances. We discussed the implementation of several simple control techniques, including generate and test; the use of performance/action mappings; generate, test and modify; prototype selection and refinement; and hierarchical control abstractions. We then focused on several examples that demonstrated different approaches to design by using syntactic knowledge. These included simple generation, in which we discussed shape and set grammars; generate and test, in which we considered spatial layout; and the production of designs through knowledge about design actions, in which we discussed a simple spatial allocation problem, computer configuration, and construction detailing.

In all of these examples we wished to demonstrate the roles of knowledge about semantics and syntax without clouding the discussion with too much emphasis on control. In order to understand knowledge-based design systems better, we must now graduate to more complex control mechanisms. In the next chapter we will consider reasoning about design processes in greater detail.

6.8   Further Reading

Material on producing designs is widely scattered in the literature. The following books all provide either techniques or examples of knowledge-based design: Coyne (1988), Dym (1985), Gero (1985a, 1987a, 1988a), Holden (1987),

Kostem and Maher (1986), Latombe (1978), Maher (1987), Rychener (1988), Smith (1986), Sriram and Adey (1986a, 1986b, 1987a, 1987b), ten Hagen and Tomiyama (1987), and Yoshikawa and Warman (1987). Further material on generative grammars can be found in March and Stiny (1985).

Exercises

6.1   What is the difference between the semantic and the syntactic approaches to producing designs?

6.2   What are the difficulties in goal satisfaction? How do we get around these difficulties? Do you think the assumptions we make in this chapter about formulating design systems are well-founded?

6.3   What is the difference between interpreting a design to determine its performances and interpreting a specification to produce a design? What assumptions do we make when we formulate a design system to produce design descriptions by interpreting specifications?

6.4   List several design tasks in which plausible inference might have a role to play. What features of these tasks lend themselves to this approach? What are the strengths and weaknesses of this approach?

6.5   Plausible inference is often applied to diagnostic reasoning. What is the difference between diagnosis and design? Does diagnosis have a role to play in design systems?

6.6   Devise a simple system for reasoning with plausible inference.

6.7   What is the essential difference between plausible reasoning and fuzzy reasoning? What role can each play in design systems?

6.8   Extend the ideas for a fuzzy reasoning system developed in Section 6.2.4. How could we avoid having to supply numerical values for "termite protection," "slope," and "tiled floor preference"?

6.9   What is the difference between constraint checking and constraint propagation? What role can each play in design systems?

6.10   What are the advantages of the constraint propagation approach to producing design descriptions? What makes constraint propagation difficult? What kind of knowledge is required for constraint propagation to work effectively in a design system?

6.11   What are the advantages of the assumption-based reasoning approach to producing design descriptions? What makes assumption-based reasoning difficult? What kind of knowledge is required for assumption-based reasoning to work effectively in a design system?

6.12   Summarize the different methods available for controlling generative systems. How could each be expected to operate in a knowledge-based design system?

6.13 The results of automated shape generation are often impressive. How useful is this method to capture design knowledge and operate with it?

6.14 What are the strengths and weaknesses of generate and test? How effective is this approach as the basis of a knowledge-based design system?

6.15 What makes the XCON approach and the task of configuring computer systems so well-suited to the rules as actions approach? Discuss other design tasks for which this approach could be effective.

6.16 Implement the allocation rules of Section 6.6.1 as a production system (in LISP, PROLOG, or OPS5, or using some expert system shell).

CHAPTER 7

Design Processes

In this chapter we concentrate on design processes, which we consider to be concerned with the decisions that must be made during the operation of a design system. Process can also be characterized as concerned with the "temporal" aspect of design knowledge—controlling, selecting, and ordering operators to achieve some end. In knowledge-based systems the operators are items of knowledge. Following the model of design developed in Chapter 2, process is concerned with reasoning about interpretive knowledge and about design actions. For example, at any state in the system it may be necessary to decide which interpretive rules to bring into play or which actions to implement.

Design process encompasses a wide range of activities. In most of our discussion in this chapter we assume that process is concerned with making decisions about design actions: we have a set of design elements and bring knowledge to bear in their configuration, as actions. We are interested in controlling the generative process. We employ generation as a vehicle for discussing process, but when appropriate we expand our discussion to other aspects of design, such as reasoning with constraints.

We wish to consider two kinds of control. Although they represent extremes along a continuum, they can be discussed separately, and any design system may incorporate both approaches. In the first kind of control, we make decisions immediately before carrying them out. This kind of decision making can be demonstrated by considering a car trip. We may embark on the trip and make decisions from the range of choices presented to us at each intersection, on the basis of clues such as signposts and orientation. This is a kind of regulatory decision-making. In the second approach, we sit down with a map and plan the trip. In planning we reason about actions before implementing them. This may mean "acting out" the decision-making process in an environment in which the costs are small—that is, on a map. But it may also involve other kinds of reasoning such as chaining through the operators in certain ways. We argue that with planning it is possible to make control knowledge explicit and operable in much the same way that we make knowledge about form explicit and operable. It is possible to formulate design systems so that knowledge about process can be made explicit in much the same way that we make interpretive and generative knowledge—that is, knowledge about form—explicit.

This chapter is organized into five major sections. First, we discuss issues in reasoning about design actions, including control as regulating and planning, the value of different abstraction levels, and the resolution of conflict between actions. Second, we discuss the blackboard system of control, a relatively straightforward model of regulatory control that has considerable utility. We discuss two examples in which this approach is useful: interpreting a simple scene (demonstrating the control of actions other than those that produce configurations of elements) and generating site-plan layouts (in this case, design constraints are treated as objects). Third, we focus on planning with design actions. After reviewing the different planning approaches, we present an example of a system for producing spatial layouts. In this system, conflicts among competing actions are resolved by reasoning about sequences of actions before they are implemented. Fourth, we consider reasoning about design tasks. If actions are concerned with configuring design elements, then tasks are "meta-actions" by which we "configure" design actions. In this discussion we also consider tasks that are described procedurally. Finally, we present an overview of how design knowledge can be organized to take account of different abstraction levels.

Much of this chapter explores ill-defined territory in knowledge-based design systems, pointing the way rather than describing what current systems offer.

7.1   Issues in Reasoning about Design Actions

In this section we look at three important issues in formulating knowledge about design process, as exemplified in the control of design actions. We define the difference between regulating and planning and discuss the uses of each. The essential feature of the planning approach is in the use of appropriate abstractions. Design systems can be formulated to exploit hierarchies of abstraction. Because abstraction hierarchies can be realized in several ways, we consider the three major abstraction hierarchies with which we are concerned in detail. The major advantage afforded by the use of action abstractions is that we can exploit the advantages of decomposition. Finally, we discuss conflict and decomposition.

7.1.1 Regulating and Planning

Regulating is best exemplified by the mechanism by which we decide which action to fire next in a production system. By *planning* we mean reasoning about design actions independently of their implementation. Thus we decide on a sequence of actions before executing them. Although we can readily see the place of planning in organizing a car trip, it might be difficult to see what place planning has in design. In fact, because the idea of planning may run counter to our intuitions about design, it is helpful to pursue the value of these two approaches by considering an example in some detail.

We take a straightforward problem, that of designing a crossing for a shallow river. The goal is to reach the other side. We can approach this problem with various degrees of sophistication, from that of simply walking and wading across to designing a structure to facilitate a dry passage. We start with walking and wading. In the regulating model we negotiate rocks and sand bars, with possibly a little wading, until we reach the other side. We make each decision from where we happen to be, and we consider only the choices immediately in front of us (Figure 7.1) Of course we may employ some heuristics that take account of possible future states, such as selecting rocks as stepping-stones that are closest to the destination, or keeping to shallow water.

In the planning model we get into a position where the entire search space is before us, perhaps from the top of a cliff. In planning we work out all the steps required to cross the river first, perhaps by mentally rehearsing the process before committing ourselves to the task (Figure 7.2). The difference between regulating and planning is quite clear in this case.

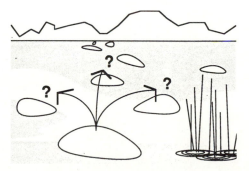

Figure 7.1   Making choices from where we happen to be.

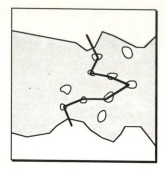

Figure 7.2 Planning a river crossing.

We now consider the task of making a bridge to get across the river, assuming that we have some rocks and planks at our disposal. By "make" we mean some process that involves designing and constructing—the sort of activity that is possibly carried out by a child building sand castles on the beach or a beaver building a dam. In the regulating approach we may start by spanning planks across rocks from the shore. For the wide stretches we may install extra rocks to support the planks (Figure 7.3). In planning we first work out the appropriate construction steps in order to span the river before laying down any planks (Figure 7.4).

A further level of sophistication is to construct a bridge to a design. Here the goal is to achieve some known configuration of rocks and planks to get across the river. The difficulty is in getting the construction sequence right. In the regulating approach we may make decisions on the basis of some heuristics (Figure 7.5). For example, initially we may start from the bottom—that is, placing rocks to form pylons—then work our way up to positioning the planks. For complex structures, such as those involving suspension cables, it would be possible to achieve the goal only with planning (Figure 7.6).

Finally, we can consider the task of designing a bridge, or at least configuring one, using pencil and paper. Here we are not immediately concerned

Figure 7.3 Making a bridge—the regulating model of control.

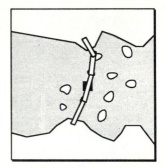

Figure 7.4 Planning before making a bridge.

with how the bridge will be manufactured. We can characterize the problem
as one in which actions are concerned with arranging planks and rocks spatially, with no regard for how things should be ordered during construction.
Thus we may begin by drawing planks end to end at some distance above the
river, then decide how they might be supported (Figure 7.7). The regulating

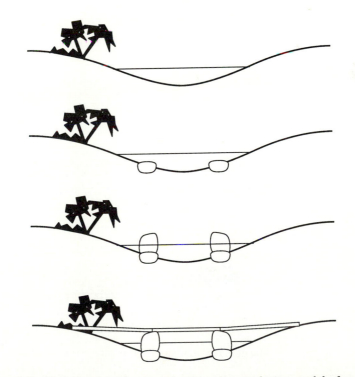

Figure 7.5 Building a bridge to a design—the regulating model of control.

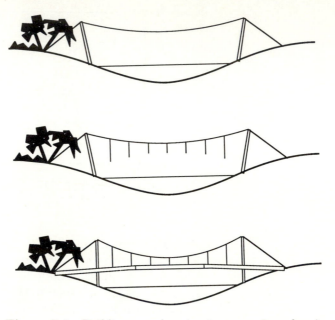

Figure 7.6   Building complex structures requires planning.

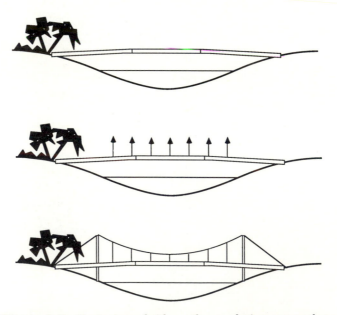

Figure 7.7   Designing a bridge—the regulating approach to control.

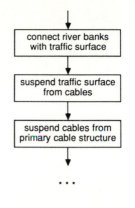

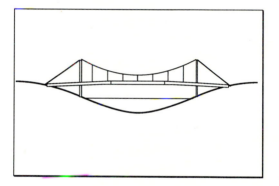

Figure 7.8   Planning in the design of a bridge.

approach involves searching the space of states for one that matches the goal state: a structure spanning the river.

Whereas the role of the planning approach is fairly obvious in the case of construction, it is less so in the case of design. In planning we decide first on the sequence of design actions, by whatever means are available, before committing ourselves to pencil and paper. When re-enacted on paper, these actions result in a design description (Figure 7.8). We should notice that the idea of planning design actions does not readily map onto our intuitions about human design activity. As we show later, however, planning does show promise in providing frameworks for representing certain kinds of design knowledge.

7.1.2 Abstraction Levels

Before pursuing the role of planning in design in greater detail, let us consider the role of abstractions. It is important to distinguish among three kinds of abstraction hierarchies, each of which can be exploited to some degree in formalizing knowledge about design processes: abstractions of design descriptions, abstractions of design actions, and abstractions of control.

Design Description Abstractions

There are obvious economies in carrying out design activity at appropriate levels of abstraction. For example, it is less expensive to design buildings on paper, by manipulating drawn lines representing walls and spaces, than to design with bricks and concrete on site (Figure 7.9). It is also generally less expensive to operate on highly abstract descriptions of objects as facts in a computer system than to operate on complicated graphical representations. There may be several levels of abstraction that can be considered. There are advantages in choosing a system of describing designs and partial designs so that the cost of searching the space of possible states is low. For example, we may explore the space of designs and partial designs by considering only certain attributes of the elements being arranged. Thus if we are designing a bridge made up of rocks and planks of wood, then we may choose to describe states only in terms of certain key attributes and relationships, such as what is positioned on what, rather than the geometry of the elements.

There are at least two kinds of description abstraction that we may consider. First is a hierarchy of generality; we may describe a design in terms of line drawings in which much of the geometry of the design is made explicit. Then there are sketches that are less precise, showing the relationships among elements, their shapes, and their relative sizes. A further level of abstraction

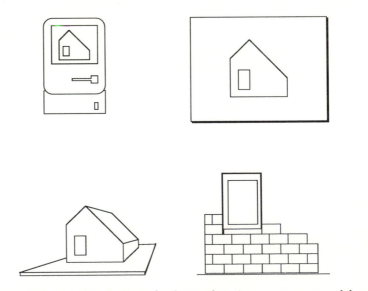

Figure 7.9 Different abstractions of a design domain: a computer model, a drawing, a cardboard model, bricks and mortar.

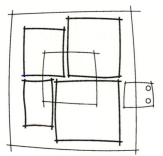

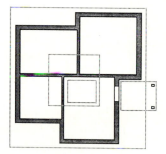

Figure 7.10 Some abstractions used in design.

may be portrayed as "bubble diagrams" that show relationships and relative sizes only (Figure 7.10). In each case the design is represented at a different level of generality. The bubble diagram is the most general representation. A further system of description hierarchy is to consider levels of decomposition (Figure 7.11). In the case of a bridge, we can describe a design in terms of pylons and spanning members. But the pylons can perhaps be further described in terms of base, shaft, and brackets.

We can exploit the idea of different description abstractions in design systems by searching a design space at one level of abstraction, then progressing through more complex abstractions. Thus we may design with bubble diagrams first, then proceed to sketches and then to precise line drawings. Or we may design a bridge in terms of pylons and spanning members, then resolve

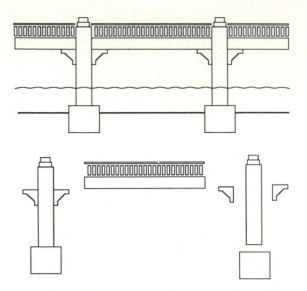

Figure 7.11 A decomposition hierarchy for bridge design.

the design by considering the components of each of those elements. We may use the product of the process at one abstraction level to guide the process at another level of abstraction (Figure 7.12). What form this guidance may take is discussed in Section 7.3.4.

Action Abstractions

In addition to abstractions of design description, we may consider different abstractions of design actions. Again, these can be categorized in terms of generality (or class) and decomposition. Different levels of generality might include actions such as "locate A next to B", "locate A touching B," and

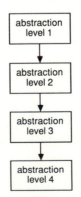

Figure 7.12 The relationship between different abstractions.

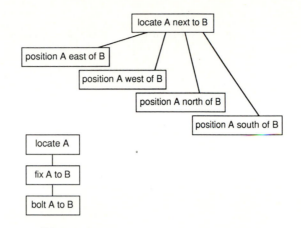

Figure 7.13 Class hierarchies of actions.

"position A east of B." We may proceed from general statements about the placement of elements to more specific actions. A further system of abstraction might include actions such as "locate," "fix in place," and "bolt." The abstraction levels proceed from general actions concerned with how elements go together to specific actions about types of fixing (Figure 7.13).

We can also look at levels of abstraction in terms of the decomposition of actions (Figure 7.14)—for example, the action that A is to be bolted to B. This action can be "expanded" to the actions "put A on B," "place a hole in each so that they are aligned," and "place a bolt in the hole." Certain design actions can be broken down into other actions, the cumulative effect of which is to achieve the major action.

As before, we can exploit the idea of different action abstractions in a design system by searching a design space, considering actions at one level of abstraction, then progressing through more complex abstractions. We may use the product of the process at one abstraction level to guide the process at another level of abstraction.

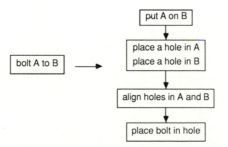

Figure 7.14 Decomposition hierarchies of actions.

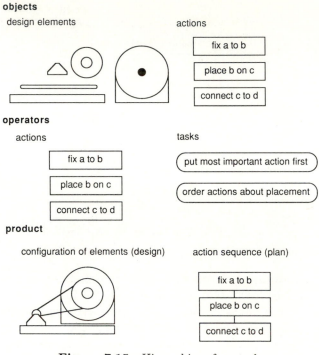

Figure 7.15 Hierarchies of control.

Control/Process Abstraction

The fact that we can have different abstractions of design actions suggests that actions can be treated in much the same way that we treat descriptions of designs. As with design elements, actions can be organized to reflect hierarchies of generality and decomposition. In principle, it is possible to consider the issue of control as simply another abstraction of the design process. Design actions can be treated as objects; relational statements about actions can be treated as facts. Control knowledge is therefore that which defines spaces of possible action sequences. We may further suppose that there is knowledge pertaining to the interpretation of action sequences and knowledge for generating action sequences. More generally, we can say that design knowledge can pertain to the interpretation and generation of form, as well as to the interpretation and generation of process. There may also be other, higher abstractions of the design process.

Although we maintain that control abstraction levels apply to both regulatory and planning control, these levels are realized most easily when we talk about planning (Figure 7.15). We can formulate hierarchies of control to exploit this idea. We have referred to the operators that configure design elements as *actions*. The operators that "configure" actions may be called *tasks*, while those that "arrange" tasks can be called *strategies*. There may be further levels of control above these operators. This formulation is only one of several possible ways of describing the idea of hierarchies of control (and defining the terminology of actions, tasks and strategies), but it has some intuitive appeal. We may suppose that designers possess knowledge (actions) about the configuration of elements, knowledge (tasks) about the appropriate selection and ordering of actions, and knowledge (strategies) pertaining to the selection and ordering of tasks. Later in this chapter we discuss this system of organization in greater detail as a means of structuring design knowledge.

7.1.3   Conflict and Decomposition

The question arises as to why we would want to institute such systems of organization. If they appeal to our intuitions about design knowledge, perhaps this is sufficient reason. There are also operational advantages, however, in exploiting control abstractions. Here we pursue the advantages afforded by decomposition.

As illustrated in the previous chapter, in some limited domains the production of the space of designs that can be generated by the design actions, and that conform to a desired interpretation, is relatively straightforward. In such cases, there is a direct mapping between interpretations and actions. Such control knowledge may take the following form:

> **If**   the design is to exhibit interpretation X
> **then**   implement action Y.

Although it is convenient to formulate a system in this way, decisions cannot always be made independently. In essence, actions conflict with one another.

Let us demonstrate why the order in which actions are implemented generally matters. Consider the procedure of stacking three objects on top of each other in a particular order. These objects might be modular components in some very simple building system: a sleeping module (A), a living module (B), and a service module (C). There is only one action, "place x on y," where x and y are variables that can be any of the three objects A, B, or C.

Because there is enough room for only one object on any other object, and an object can be moved only if there is nothing on top of it, the preconditions of the action are that both objects x and y are clear on top and that x is on something. The consequent of the action is that x is still clear, but y is no longer clear; object x is now on top of y, and whatever x was on is now clear.

If the goals are that A is to be on top of B and B is to be on top of C, the order in which an attempt is made to satisfy the goals matters. An initial state shows the three objects sitting on a surface (the ground) (Figure 7.16). If we start with the action "place A on B," it does not appear possible to achieve the second goal without undoing the last action performed. Starting with the second goal by placing B on C, however, permits the first goal to be

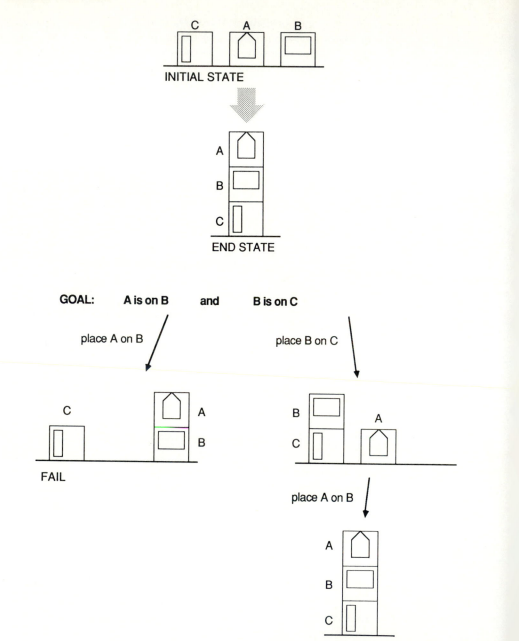

Figure 7.16 Two paths in which an attempt is made to proceed from an initial state to a known end state. The first path starts by attempting to satisfy the first goal. The second path starts from the second goal. The first attempt fails because it is necessary to undo what has already been accomplished in order to satisfy the second goal.

satisfied by the action "place A on B." In some situations, simply reversing the goals does not work—see Figure 7.17, in which C is already on A in the initial state. In this case the paths starting with either goal reach a state at which it is not possible to proceed without undoing some action already performed. Yet there must be a simple way to satisfy both goals without performing

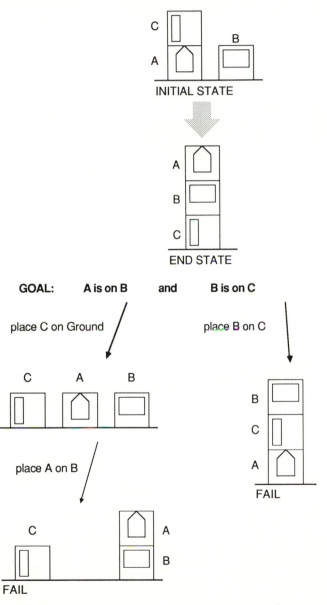

Figure 7.17 Two paths in which an attempt is made to achieve an end state. Both result in failure.

redundant actions. Although this example is a gross characterization, it has something to say about design. We would expect designers to bring knowledge to bear on the effective selection and ordering of actions to produce designs that obviate goal conflicts.

In Section 6.3.1 we discussed the issue of decomposition in design generation. Even when it is impossible at an abstraction level in which design elements are treated as objects, decomposition can be exhibited in various other representational abstractions. This can be shown in the way the design process is organized, as opposed to the way elements are organized. For example, in the context of urban design, we may say that even though the spatial form of a city is not decomposable such that one part cannot be changed without affecting another, the processes that bring about that form may exhibit

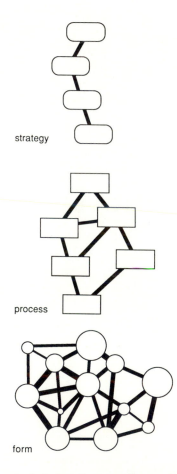

Figure 7.18 Different control abstractions can be employed to take advantage of decomposition.

a certain compositional independence. If not the processes, then the strategies that order the processes may be decomposable, and so on. The idea that the process can be decomposed—even though the artifact cannot be—can be exploited to advantage in knowledge-based design systems (Figure 7.18).

Knowledge about design is often represented in the form of prototypical decomposition hierarchies. The role of a fairly detailed strategy based on decomposition (Figure 7.19) can be demonstrated simply. The decomposition may result in the definition of tasks such as (1) concentrate on ranking objects in terms of their spatial importance, (2) without locating anything, decide how each object should be placed so that it is adjacent to an object with which it interacts and which is above it on the ranking list, (3) then decide what each object should be adjacent to. There are then further sub-tasks to be considered. The tasks can be regarded as relatively independent in that the outcome of one task provides information that can be utilized in the next; the system can be formulated to minimize backtracking.

If a particular body of generative knowledge exhibits this property of being decomposable, then it should not be necessary to undo the effects of any action (that is, backtracking is unnecessary). Each action firing brings the design closer to a resolution (Nilsson, 1982).

This approach represents an ideal. Rarely is the behavior of a system entirely predictable, especially when the system models some aspect of design. Thus a system may have to utilize a range of control strategies to generate a range of alternatives.

The idea that design can be modeled as a system in which there is some kind of interaction between levels of abstraction is attractive. What is passed from one abstraction level to another? One level may provide a partial design

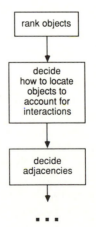

Figure 7.19 Decomposing a design task.

description that will be expanded at another abstraction level. One level may provide a gross configuration of the design, the components of which can be treated as independent sub-problems. In Section 7.3, we consider whether one level of abstraction provides a sequence of operators, a plan, for the next level. First, however, we consider the kind of refinements possible with the regulatory method of control.

7.2   Control by Regulating Design Actions

In this section we discuss control by regulating design actions. This approach has already been presented in previous chapters, most simply as generate and test. Here we demonstrate some of the refinements that can be employed with this approach, allowing us to make explicit the knowledge pertaining to different abstractions of design descriptions.

7.2.1   The Blackboard Control Model

First we discuss the control of what might be termed a complex interpretation task. Our concern is not actions for producing design descriptions, but rather actions for producing facts that constitute interpretations of an existing design. We then transfer the approach to the production of design descriptions.

In this section we consider a useful approach to control in design systems, that afforded by *blackboard systems*. This approach embodies many of the ideas discussed in previous chapters, particularly scheduling and focus of attention, discussed in Chapter 4. As discussed in Chapter 5, several important issues exist in the control of interpretive systems. Here we focus on four areas that an effective interpretation system must handle: (1) the combinatorial problem, (2) multiple search strategies, (3) non-uniform knowledge representations, and (4) contradictory and conflicting knowledge.

The example of interpreting a CAD database (Section 4.4) could be described in these terms. The design description may be considerably larger than that illustrated in Figure 4.13; many hundreds, if not thousands, of facts could describe a design. If many interpretations can be inferred from the description, then there will be many interpretive rules. Some of the knowledge by which interpretations are made, and that a designer may use to interpret a drawing of a building design, can be expressed in terms of strategies, such as the following:

If   attempting to identifying structural members
then   look for elements that conform to some system of repetitive
         organization.

If   attempting to identify office spaces
then   look first for rooms that are placed along perimeter walls.

If   calculating area
then   focus on the largest units of subdivision in the design.

This heuristic knowledge, separate from that by which we actually infer new facts about the design, "suggests" how we might go about making the right interpretation more efficiently. Blackboard systems provide a framework for representing and using this type of knowledge.

In interpreting certain features of the design, it may also be desirable to employ non-uniform methods of knowledge representation. For example, highly efficient procedures exist for computing areas of polygons and for calculating energy loads in buildings. It should be possible to employ this knowledge in interpretation tasks without having to convert it into a less convenient form, such as rules or frames.

There may, however, be conflicting interpretations that could be placed on a design feature. For example, certain configurations of lines in a drawing might suggest a circular structural column. Other knowledge may result in the interpretation that the feature is a light fitting. The configuration of lines may be entirely consistent with the knowledge for inferring each of these interpretations. Thus we appear to require further knowledge to resolve these conflicting interpretations.

These problems are also exemplified in the automated interpretation of natural language statements. In natural language understanding, conflicting interpretations can occur, each constituting a rival hypothesis. The *blackboard model of control*, initially developed to provide expedient interpretations of statements within a limited language, gained currency with the development of the HEARSAY-II speech-understanding project (Erman et al., 1980). This led to the development of HEARSAY-III, a generalized system appropriate to other domains (Balzer et al., 1980; Erman et al., 1981), and to the AGE system (Nii and Aiello, 1979) and the BB1 system (Hayes-Roth et al., 1986). These latter three systems can be regarded as computer tools for experimenting with and creating rule-based problem-solvers; they are also described by Hayes-Roth et al. (1983). Hayes-Roth and Hayes-Roth (1979) have demonstrated the applicability of blackboard systems for modeling cognitive planning activity.

The family of automated problem-solving systems bearing the "blackboard" label sometimes appears to be associated very loosely. The key property of a blackboard system is its ability to direct its attention to different tasks according to certain explicit control heuristics. The "blackboard" label, however, refers to an analogy with the cooperative behavior of human experts. There is an autonomous medium through which the experts communicate, called a *blackboard*. By changing the pattern, or facts, on the board, each independent expert can respond to whatever happens to be on the blackboard at any moment. Each change prompts another expert to contribute to the contents of the board. So far this description sounds little different from the way in which we expect a production system to operate. The rules, or actions, are essentially autonomous in the same way. The model, however, has refinements. The blackboard can be *partitioned*; the units of knowledge, which we have been calling rules or actions, are termed *knowledge sources*. This label emphasizes that the knowledge need not be represented with any uniformity.

The partitioning is intended to facilitate control, allowing the system to focus attention on parts of the facts base that most warrant attention. The focusing is determined by heuristic control knowledge. This ability to control and direct its own attention is the key ingredient of a blackboard system. In summary, a blackboard system embodies the following important features (Figure 7.20).

1. System organization. Blackboards are a development of the production system idea. There is a global database, a facts base. As described in Section 4.4, control is concerned with the relationship between knowledge statements, knowledge sources, and the facts base. Uniformity in the representation of knowledge is not as important as in rule chaining.

2. Partitioning of the facts base. The facts base can be partitioned in various ways. Basically, the attention of the system can be directed to different partitions (see Section 4.4.4). Thus the partitioning must be "meaningful" in terms of the specific task. Typically, the partitioning reflects various systems of hierarchical organization among facts. For example, facts regarded as primitive, in that they contribute to the establishment of other facts, may occupy a low partition in the hierarchy. There may be several overlapping hierarchies. As demonstrated later, the partitioning can also be "spatial" in the sense that facts can be partitioned according to the location in a coordinate system with which they deal. How the facts base is partitioned bears on the tractability of the interpretation task.

3. Control. As introduced in Section 4.4.4, in a blackboard system knowledge sources are triggered prior to firing. Knowledge sources are selected for firing on the basis of the outcome of each triggering. Choices must be made regarding (1) the operator to be fired and (2) the part of the facts base to be operated upon. The basis of selection is essentially heuristic. If there is no basis for selecting among knowledge sources, the blackboard formulation provides little advantage.

In general, more than one action applies—that is, more than one knowledge source could be activated given the current state of the system. *Schedul-*

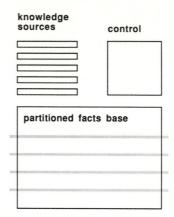

Figure 7.20 Major features of a blackboard system.

ing knowledge sources decide which knowledge source to activate based on scheduling heuristics, the information recorded on part of the facts base called a *scheduling blackboard,* and a *domain blackboard.* Hayes-Roth and Lesser (1977) have proposed a set of selection criteria for use in the HEARSAY speech-understanding system. Examples of such scheduling rules follow:

If   one of the competing tasks is known to make the others unnecessary
then   select that one.

If   a knowledge source is operating on a part of the blackboard
      known to contain more reliable information,
then   select that knowledge source.

If   a knowledge source is operating on a part of the blackboard
      known to contain more important information
then   give it priority.

If   a knowledge source is known to be more efficient
then   select that one.

If   a knowledge source is known to be more likely to lead
      to a final state that matches the goals of the system
then   give it priority.

A weighting system may be applied to select among competing knowledge sources. The first element from the ordered list maintained by the scheduler is executed. Control is generally implemented through the following cycle:

trigger
evaluate
focus attention
fire

As with a scheduling system (see Section 4.4.3), temporary records are kept of the effect on the knowledge base of the activation of each knowledge source. Heuristic control knowledge is brought to bear in comparing these records, and attention is focused on a particular partition of the facts base. Knowledge sources may be fired according to the partition determined to be of most interest; this may also influence the partition in which triggering takes place.

A significant feature of this approach is that it facilitates simulation of two different search models, backward chaining and forward chaining.

Let us consider an example in which the blackboard model is applied to the task of interpreting a facts base of line primitives. The design to be interpreted is depicted in Figure 7.21. The configuration is the rudimentary floorplan of a house, though it could be any geometrical configuration, including a block diagram for a circuit layout. We consider how a facts base can be organized, the nature of the knowledge base, and an appropriate control cycle.

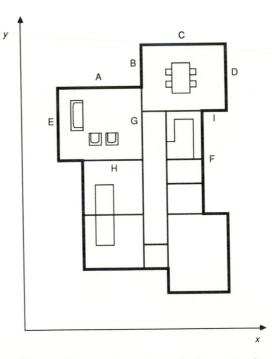

Figure 7.21   A floorplan to be interpreted.

Organization of the Facts Base

We choose to partition the facts base both hierarchically and spatially. We can consider several hierarchical abstraction levels organized on the basis of dependence between facts. The truth of higher level facts depends on those at a lower level. The facts with which the system must deal can be categorized according to the levels to which they belong:

level 0	primitive line segments
level 1	composite lines
level 2	simple shapes made up of lines (Rectangles)
level 3	simple properties of shapes and relationships between shapes (Area, Adjacent, Inside, North, East)
level 4	comparisons and more complex relationships between shapes complex objects described by function (BiggerThan, SmallerThan, SameSize, West, South)

It should be noted that these divisions are "correct" only to the extent that they help focus attention. The hierarchical partitioning is illustrated in Figure 7.22.

The facts base is also organized spatially: the facts base is divided into a spatial grid. Here we consider this partitioning to be relevant only to prim-

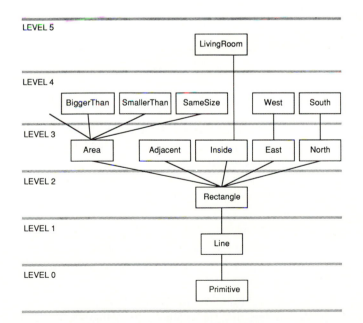

Figure 7.22   The hierarchical partitioning of a facts base.

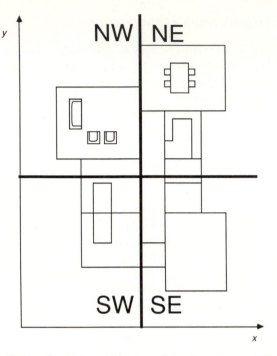

Figure 7.23   Spatial partitioning of the facts base of Figure 7.21.

itives, lines, shapes (rectangles), and areas because facts about comparisons are difficult to locate spatially. The spatial partitioning is illustrated in Figure 7.23. To keep the example simple, the design is divided into four quadrants labeled NW, NE, SE, and SW. If an element falls within a quadrant, then it belongs to that quadrant, which means that there is some overlap and that an element may belong to more than one quadrant. It should be noted that spatial partitioning can be achieved in many other ways—for example, by selecting concentric zones arranged around the center of the design. The partitioning might also be dynamic, changing according to the dictates of the control knowledge. Again, the utility of the spatial division depends on the nature of the knowledge sources and should facilitate focusing of attention.

Some of the facts describing Figure 7.21 include the following:

Primitive(A, 12.0, 72.0, 40.0, 72.0)

Primitive(B, 40.0, 72.0, 40.0, 80.0)

Primitive(C, 40.0, 80.0, 68.0, 80.0)

where the first argument is a label attached to the element and the other arguments are the x and y coordinates of the end points of the elements (lines), respectively.

Knowledge Sources

As indicated, in a blackboard system it is normal to employ knowledge sources that are not uniform in structure. For simplicity and in order to make effective comparisons with the control models described elsewhere in this book, we restrict this example to knowledge sources as rules or actions. In tabular form they appear as follows:

Knowledge Source	Preconditions	Consequent
1	$\text{Area}(a,c)$ $\text{Area}(b,d)$ $c > d$	$\text{BiggerThan}(a,b)$
2	$\text{Area}(a,c)$ $\text{Area}(b,d)$ $c < d$	$\text{SmallerThan}(a,b)$
3	$\text{Area}(a,c)$ $\text{Area}(b,d)$ $\text{Not}(a=b)$ $c = d$	$\text{SameSize}(a,b)$
4	$\text{SameSize}(a,b)$	$\text{SameSize}(b,a)$
5	$\text{Rectangle}(a,n_a,s_a,e_a,w_a)$ $\text{Rectangle}(b,n_b,s_b,e_b,w_b)$ $\text{Not}(a=b)$ $n_a < n_b, s_a > s_b$ $e_a < e_b, w_a > w_b$	$\text{Inside}(a,b)$
6	$\text{East}(b,a)$	$\text{West}(a,b)$
7	$\text{North}(b,a)$	$\text{South}(a,b)$
8	$\text{Rectangle}(a,n,s,e,w)$ b is $(n-s) \times (e-w)$	$\text{Area}(a,b)$
9	$\text{Rectangle}(a,n_a,s_a,e_a,w_a)$ $\text{Rectangle}(b,n_b,s_b,e_b,w_b)$ $s_a >= n_b$	$\text{North}(a,b)$
10	$\text{Rectangle}(a,n_a,s_a,e_a,w_a)$ $\text{Rectangle}(b,n_b,s_b,e_b,w_b)$ $w_a >= e_b$	$\text{East}(a,b)$

Knowledge Source	Preconditions	Consequent
11	$\text{Rectangle}(a, x, s_a, e_a, w_a)$ $\text{Rectangle}(b, n_b, x, e_b, w_b)$ $\text{Overlap}(e_a, w_a, e_b, w_b)$	$\text{Adjacent}(a, b)$
12	$\text{Rectangle}(a, n_a, x, e_a, w_a)$ $\text{Rectangle}(b, x, s_b, e_b, w_b)$ $\text{Overlap}(e_a, w_a, e_b, w_b)$	$\text{Adjacent}(a, b)$
13	$\text{Rectangle}(a, n_a, s_a, x, w_a)$ $\text{Rectangle}(b, n_b, s_b, e_b, x)$ $\text{Overlap}(n_a, s_a, n_b, s_b)$	$\text{Adjacent}(a, b)$
14	$\text{Rectangle}(a, n_a, s_a, e_a, x)$ $\text{Rectangle}(b, n_b, s_b, x, w_b)$ $\text{Overlap}(n_a, s_a, n_b, s_b)$	$\text{Adjacent}(a, b)$
15	$\text{Line}(w_a, w, s, w, n), n > s$ $\text{Line}(e_a, e, s, e, n)$ $e > w$ $\text{Line}(s_a, w, s, e, s)$ $\text{Line}(n_a, w, n, e, n)$	$\text{Rectangle}([n_a, s_a, e_a, w_a],$ $\quad n, s, e, w)$
16	$\text{Line}(a, x_2, y, x_3, y)$ $\text{Line}(b, x_1, y, x_2, y)$ $\text{Append}(a, b, c)$	$\text{Line}(c, x_1, y, x_3, y)$
17	$\text{Line}(a, x, y_2, x, y_3)$ $\text{Line}(b, x, y_1, x, y_2)$ $\text{Append}(a, b, c)$	$\text{Line}(c, x, y_1, x, y_3)$
18	$\text{Primitive}(a, x_1, y_2, x_2, y_2)$	$\text{Line}([a], x_1, y_2, x_2, y_2)$

The facts can be tagged according to the abstraction levels to which they apply:

level 4	BiggerThan
	SmallerThan
	SameSize
	South
	West
level 3	Area
	North
	East
	Inside
level 2	Rectangle
	Overlap
level 1	Line
level 0	Primitive

Propositions with no abstraction-level numbers assigned to them are conditions that are to be satisfied but that do not appear as assertions in the facts base, such as the proposition that 6 is greater than 4.

Control Cycle

The control mechanism can be demonstrated with a simple interpretation task: to interpret the configuration of lines in Figure 7.21 that constitutes a living room. This is essentially a recognition problem. In order to use the formulation described so far, we require two kinds of information: (1) a definition of what constitutes a living room and (2) some heuristics to guide the search process. The definition may be along the lines of how a living room can be recognized. A very simple definition appropriate to our example is to say that a space is a living room if it contains three items of furniture—that is, three other rectangles (Figure 7.24). This can be represented as just another knowledge source in the same manner as those shown previously.

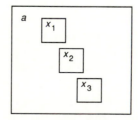

Figure 7.24   The definition of a living room.

Knowledge Source	*Preconditions*	*Consequent*
19	$\text{Inside}(x_1, a)$	$\text{Function}(a, \text{LivingRoom})$
	$\text{Inside}(x_2, a), \text{Not}(x_2 = x_1)$	
	$\text{Inside}(x_3, a), \text{Not}(x_3 = x_1)$	
	$\text{Not}(x_3 = x_2)$	

We will assign facts about Function to level 5.

We require the system to assert new facts about the design (interpretations). Because many facts could be asserted, however, we are most interested in facts that have some bearing on function. We need some means of directing the interpretation process. This is even more important for a facts base describing a drawing with many line segments.

Because living rooms in the southern hemisphere are often located to take advantage of the sun, an appropriate item of knowledge to aid search might be to consider northeasterly spatial partitions first (followed by the northwest, southwest and southeast partitions). We might also decide that it is appropriate to proceed from lower to higher abstractions in a breath first manner. We can summarize the control as follows:

1. Evaluate description: to see whether the current state of the design description contains a description that matches the goal—that is, corresponds to a description of a living room.

2. Trigger: this involves matching each knowledge or action source against the facts base and keeping a temporary record of the effect of each rule and each event or instantiation.

3. Evaluate events: this involves using search heuristics to select among events according to the partition affected and assigning the events to a priority list.

4. Focus attention: knowledge sources that operate on a particular hierarchical level are selected. Spatial partitioning can be implemented by ignoring all triggered facts that are outside a particular cell.

5. Fire: execute the event that has the highest priority.

In this example we would expect the system to produce new facts about the design in a breadth-first manner (initially relating to level 0), beginning with facts pertaining to the interpretation of the northeasterly quadrant of the layout. Because this does not produce a description corresponding to a living room, the system then proceeds to develop the facts pertaining to the northwest quadrant until the desired description is achieved.

What we have described is a relatively simple application of the blackboard idea, allowing communication through different abstractions of design descriptions to produce an interpretation of a design of interest:

Function([A, G, H, E], LivingRoom)

In the process other facts have been asserted, some of them irrelevant to the function of the room bounded by primitives A, G, H, and E. We should note that the scheduling heuristics do not have to be "right"; the objective is to guide search.

In this example none of the knowledge sources will lead to contradictory interpretations of the design. Where competing knowledge sources produce contradictory facts, the scheduling knowledge must incorporate some means of producing the most plausible interpretation, as in the HEARSAY system.

7.2.2   A Blackboard Approach to Producing Designs

Here we describe SIGHTPLAN, a design system that uses the blackboard approach for representing and exploiting knowledge about the generation of construction-site layouts—the arrangement of cranes, site huts, and materials storage around a building site (Tommelein et al., 1987). The system described here is concerned with actions that operate on constraints as objects. There are also classes of constraints, and the system makes use of some of the ideas of object-oriented programming discussed in Section 4.6. The system utilizes the BB1 blackboard control system (Hayes-Roth et al., 1986). The task is to configure elements within a bounded, gridded site as a two-dimensional spatial arrangement. It is necessary to assign every element to a location on site to minimize the number of conflicts.

The position of elements is constrained by their behavior. For example, the reach of the crane constrains the distance between the crane and the building under construction. Each of these constraints is represented by a frame with slots for the design elements. Individual constraints are treated as objects (Figure 7.25):

Object	$NextTo_{15}$	
Link	name:	Involves
	value:	$Crane_1$ $Building_1$
Link	name:	Exemplifies
	value:	NextTo
Object	$MustZone_7$	
Link	name:	Involves
	value:	$Crane_1$ $Zone_1$
Link	name:	Exemplifies
	value:	MustZone

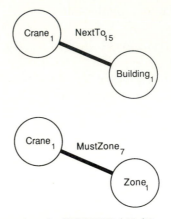

Figure 7.25   Constraints in SIGHTPLAN (Tommelein et al., 1987).

The Link slot is a general-purpose slot provided within the BB1 system. The first frame can be interpreted as follows: there is a constraint instance $NextTo_{15}$ that involves $Crane_1$ and $Building_1$. This constraint belongs to the class of NextTo constraints. In the second frame: there is a constraint instance $MustZone_7$, which involves $Crane_1$ and $Zone_1$, stating that $Crane_1$ must be inside $Zone_1$. This constraint belongs to the the class of MustZone constraints. Other frames relate to other constraints, such as PreferZone, which as its name suggests is a constraint that a design element should be located within a particular zone. In the event that there is a conflict with the satisfaction of other constraints, however, PreferZone is not mandatory. A fully working system would also have to take account of time constraints and locational constraints.

Further object descriptions relate to design elements, such as a particular crane, $Crane_1$ (Figure 7.26):

Object		$Crane_1$
Attribute	name:	Dimensions
	value:	12 18
Attribute	name:	Area
	value:	218
Link	name:	Exemplifies
	value:	Crane
Link	name:	InvolvedBy
	value:	$NextTo_{15}$ $MustZone_7$
Link	name:	Plays
	value:	Anchoree

Here the design element $Crane_1$ belongs to—that is, exemplifies—the Crane

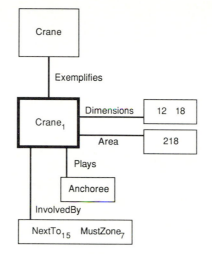

Figure 7.26 An object in SIGHTPLAN.

class of elements. It inherits the attributes of that class. This object is also involved in the constraints $NextTo_{15}$ and $MustZone_7$, defined previously. $Crane_1$ also plays the role of an "Anchoree," meaning that it is fixed in place relative to other elements whose locations are fixed (Anchors).

The site is divided into three zones. $Zone_1$ is the area immediately around and including the "footprint" of the building under construction; $Zone_2$ is the space skirting $Zone_1$; and $Zone_3$ is the remainder of the site.

Design knowledge or knowledge sources about construction-site planning pertain to design elements (construction site equipment and structures), their geometrical and other attributes, and the relationships among elements. The overall control strategy is to develop a partial arrangement on the site by locating the most appropriate anchors. One anchor will be the site. The building under construction will be another.

In addition to the domain blackboard, on which is described the construction-site layout, there is a separate *control blackboard* for reasoning about the design process. A general control-knowledge source operates on the control blackboard, and the site-planning knowledge source operates on the domain blackboard. All applicable knowledge sources are triggered and compete for scheduling priority.

The execution cycle is as follows.

1. The execution of an instance of a knowledge source on a blackboard causes changes to the blackboard. These changes constitute events.

2. After an event, certain knowledge sources are triggered to produce knowledge source activation records, which appear on the control blackboard. A control knowledge source determines a focus of attention, such as "position large equipment in an important zone." Several knowledge sources,

which operate on the domain blackboard, may be in competition—for example, the knowledge sources (or actions) anchor crane in $Zone_1$; anchor concrete pump in $Zone_1$; append concrete truck.

3. An agenda of executable events is maintained.

4. The event to execute is determined according to priorities, which are determined according to the extent to which the events accord with the current focus of attention. For example, according to the focus of attention just determined ("position large equipment in an important zone"). the knowledge sources "anchor crane in $Zone_1$" and "anchor concrete pump in $Zone_1$" will rank higher than the knowledge source "append concrete truck." The priority rating also takes account of the importance and size of design elements, as well as the importance of any constraints associated with the elements affected by the knowledge sources. This information is provided *a priori* or interactively by the system user.

The cycle is repeated until a "design" for the site layout is produced (Figure 7.27).

The layout task is treated as decomposable in that the locations of design elements can be treated with a fair degree of independence. Proximity is more

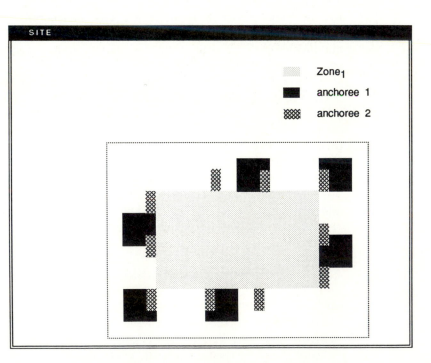

Figure 7.27   A design produced by SIGHTPLAN.

important than adjacency and orientation. The system assumes there is a fair margin for "error" and loose fit in a final layout.

In summary, SIGHTPLAN develops its designs incrementally and makes decisions opportunistically based on control knowledge. It determines which action to take next, positioning design elements at a limited number of possible locations while satisfying the constraints about relationships among elements. The result is not necessarily the "best" design, but simply a workable design that gives the designer a starting point to explore alternatives in response to changes to the knowledge base. A fully developed system, intended for preliminary construction-site layouts, would assist project managers on site by ensuring that all important factors are taken into account.

We return to the blackboard approach in Section 7.3.6 in the context of planning.

7.3   Planning Design Actions

Planning can be considered the effective selection and ordering of actions before they are implemented. In this section we discuss the various approaches to planning with design actions: forward chaining, backward chaining, reordering sub-goals, the use of abstraction hierarchies, the hierarchical refinement of plans, scheduling planning knowledge, and meta-planning.

In isolation, none of these approaches provides a wholly satisfactory model of design processes. In order to model design effectively, we would need a combination of approaches. Here, however, we discuss them independently.

7.3.1   Forward Chaining

We discussed forward chaining in the context of interpretive knowledge in Section 4.3.2. In the realm of actions it is convenient to consider the example of planning a car trip. The actions to be performed are "turn right," "turn left,"' and "go straight ahead." A map represents the entire search space, where intersections are problem states (Figure 7.28). The objective is to get from one city to another. In forward search we begin at city A and trace along all the paths emanating from A until we reach destination B (the goal). We may use some system of exhaustive search, such as trying each action at each intersection and backtracking when we reach the edge of the map without reaching B (Figure 7.29). The end product of this process is a route marked out on the map (a plan), which can then be implemented in the real landscape of roads, cities, and cars (Figure 7.30). The expensive business of search and

Figure 7.28 A problem space as a map.

backtracking has been carried out at a level of abstraction at which the costs are relatively minor.

Design tasks, of course, are much more complicated than suggested by this example. We can characterize certain aspects of the design process, however, as a case of choosing a suitable abstraction in which to work (an appropriately scaled map). We can then generate partial designs by applying actions to transform states. If we represent states as pencil lines on paper, then backtracking is somewhat similar to using an eraser to revert to earlier states. The end product of this particular process may be a set of actions that can be fed

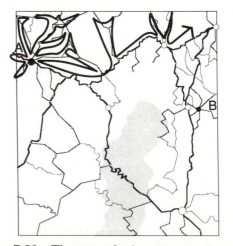

Figure 7.29 The start of exhaustive search on a map.

Figure 7.30 A route from A to B.

into a different abstraction of the design process, producing a more complete design description.

Because of the general nature of the conflicting-actions problem and the simplicity of a blocks world (such as that described in Section 7.1.3) in modeling the problem, it is often used as a "bench test" for systems that reason about actions. This domain serves to demonstrate the general utility of different reasoning approaches in satisfying multiple goals. A design action can be represented in predicate calculus as Puton(x, y)—that is, put x on y. Its preconditions are Clear(x), Clear(y), and On(x, z). The consequents are Clear(x), Clear(z), and On(x, y). A suitable abstraction for an initial states can be described simply as On(C, A), On(A, Ground), On(B, Ground), Clear(C), Clear(B). In the forward-search approach, the action is applied to the current state by finding a suitable match for the precondition facts and their variables. These facts are then deleted from the current state and replaced by the consequents. The action is applied repeatedly until a state that matches the goals is achieved. If a dead end is encountered, then the procedure is to backtrack. (A dead end is a state in which it is not possible to proceed without undoing an action already performed.) The state to which a transformation is applied is therefore simply a list of facts in predicate calculus. The search procedure is shown in Figure 7.31; for clarity, states are shown graphically, rather than as lists of facts.

This approach provides the advantage that it is a relatively simple matter to evaluate states against any goal set. If the knowledge is made available by which the necessary inferences can be made, it is possible to match the goal—for example, "the structure is to contain three stories with the service module (C) at ground level." In this case the knowledge must also be available for interpreting design descriptions.

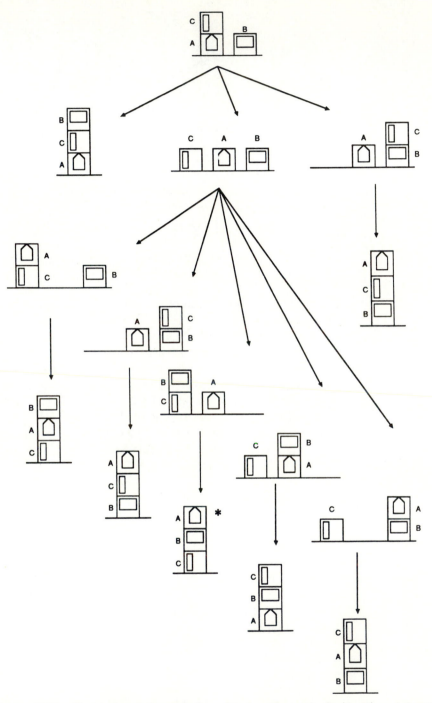

Figure 7.31 Forward search graph for achieving the goals On(A, B) and On(B, C). The goal state is indicated by *.

7.3.2 Backward Chaining

In this approach we begin tracing our car trip from the destination to the city of origin (an operation that is possible only using an abstraction such as a map). As far as maps are concerned, in practice this is little different from the forward-chaining approach (Figure 7.32). In other domains, however, there are advantages in starting with a set of requirements and working out the sequence of actions necessary to get there. We can ask, what action is necessary in order for a state with this performance to be achieved? That action has certain preconditions. We can then ask, what action must be performed in order for a state with the required preconditions to be achieved? Of course, complexities arise when there are multiple goals, and the satisfaction of one goal negates the satisfaction of another. Backward chaining, essentially a reversal of the forward-chaining model, is described formally by Nilsson (1982). Such a system is described here.

The search process can be constrained by employing the goals as a starting point and working backwards to see how they can be best achieved. This is behind STRIPS, a planning system developed by Fikes and Nilsson (1971). A problem is formulated so that the major component of state descriptions is a stack of goals, the members of which are added and removed one by one according to the conditions in the current description of the world. The system can also be described conceptually as operating by matching the first goal against the consequent of an action. This is the same as asking what action is necessary in order to achieve a particular goal. The preconditions of this action are then inspected. These preconditions become sub-goals, and the system attempts to find what actions are necessary to achieve them. This process of chaining through the actions continues until a set of preconditions is found that match the initial state of the world. When that state is reached, the

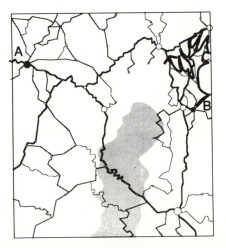

Figure 7.32 Backward chaining on a map.

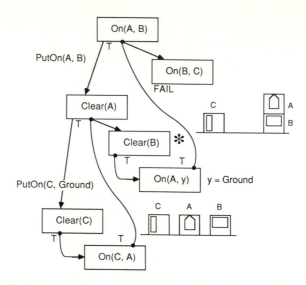

Figure 7.33 Backward search graph to satisfy the goal On(A, B) and On(B, C), resulting in failure.

actions that have just been investigated are implemented in order to transform the state of the world.

The second goal is now investigated. If it cannot be satisfied, either by the current state or by some non-redundant action, then the system backtracks to an earlier state and attempts to resatisfy the first goal in an alternative way. This is indicated in Figures 7.33 and 7.34 for the blocks problem, where the flow of attention from sub-goal to sub-goal is shown by means of arrows. The letter T indicates that a subgoal is true because it matches the current state description. The * symbol in Figure 7.33 indicates the sub-goal to which the system backtracks in order to produce the graph of Figure 7.34. The sequence of actions is Puton(C, Ground), Puton(B, C), Puton(A, B). In this manner goals are propagated by chaining through the rules. Sub-goals may be considered as constraints; they are propagated through the system by chaining. The backward-chaining approach therefore still involves searching and backtracking. It has the disadvantage that, if the goals are not represented in terms of the consequents of rules, it is necessary to furnish the knowledge by which the link can be made.

The STRIPS approach is also amenable to the introduction of certain efficiencies, which are discussed next. An early problem solving system, GPS (an acronym for General Problem Solver), was developed by Newell and Simon (1972). When there are a large number of possible actions by which goals can be achieved, GPS reduces the amount of search by the use of *means-end analysis*. It identifies appropriate actions by calculating the "difference" between the current world state and the goals, then selecting an action relevant

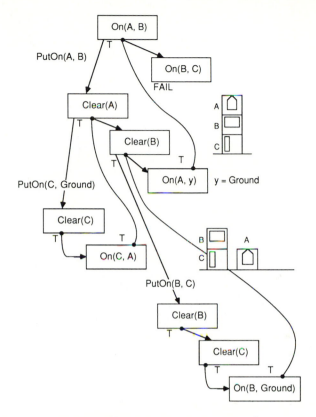

Figure 7.34 Continuation of the search graph of Figure 7.33.

to reducing the difference. This requires the formulation of a "difference table", in which actions are ordered according to their ability to reduce certain differences. The search process for an appropriate chain of actions therefore becomes more efficient. The cost is that the difference table must be given to the system explicitly.

7.3.3 Reordering Sub-goals

In the preceding approaches, goal conflict is primarily handled by first detecting conflicts and then backtracking to earlier states to produce alternative, conflict-free solutions. This entails the inefficient operation of undoing work that has already been performed.

In certain cases, the order in which we consider multiple goals bears on the tractability of a problem. In the case of planning a car trip, if the destination is not one but several towns, then the order in which we attempt to connect the towns will have a bearing on the ease with which we find an efficient

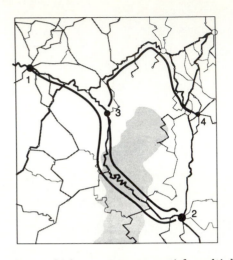

Figure 7.35   The order in which we attempt to satisfy multiple goals has a bearing on the ease with which we solve the problem (demonstrated here with a map).

path (Figure 7.35). In design, the order in which we attempt to satisfy certain requirements bears on the resolution of conflicts between requirements. For example, if the most constraining goals can be identified, it is generally worth attempting to satisfy them first to reduce the amount of search necessary.

An early computer system that demonstrated the usefulness of reordering goals was HACKER, developed by Sussman (1975). A more sophisticated approach is exhibited in the INTERPLAN system (Tate, 1975) and the WARPLAN system (Warren, 1974), which is similar to a system developed by Waldinger (1977). WARPLAN is similar in principle to the STRIPS system, but is structured to handle the problem of interacting subgoals more efficiently.

Although primarily a "learning system," HACKER also addresses the issue of goal conflicts. It formulates plans that may be defective and then uses various debugging procedures to correct them. For the block-stacking problem, the strategy is to reorder the initial goals. If the ordering means that the first goal is violated by the second, then it attempts to solve the second before the first. This approach works in some cases, as in the problem in Figure 7.16, but is not general enough to handle the situation of Figure 7.17.

WARPLAN and INTERPLAN, on the other hand, can reorder goals at various levels in the plan. The basic idea is that a plan is constructed by attempting to solve each goal in order, checking that each goal does not interfere with the solution for the previous goal. When the solution to a goal violates a previous goal, that solution is ignored and the goal is moved back to an earlier place in the plan where its solution will not cause any violations. Although this process may still involve search and backtracking, it enhances efficiency for some domains.

7.3.4   Abstraction Hierarchies

In planning a car trip between two cities we may consider several abstractions of the domain (Figure 7.36). We may trace our path by first considering regions rather than cities, and major connecting highways rather than the entire road system. The space of possible paths connecting two regions is less than that required when all the roads are considered. Having resolved the problem at that level of detail, we can then consider secondary roads, and sub-regions within regions; there may be additional levels of detail. In the case of design, planning through abstraction hierarchies is similar to the idea of resolving a problem with rough sketches, then refining the design at the abstraction of careful line drawings before considering further abstractions of scale and detail.

The approach here is to enhance the efficiency of the goal-chaining model of STRIPS by distinguishing between factors that are most important in determining the shape of the plan and those that are less important. In order to do this, the planner operates at various levels of abstraction. At a high abstraction level, only a few of the attributes of the blocks world are considered. When a plan is formulated considering the most important factors, the system is operating at the highest level of abstraction. When the least consequential factors are being considered, it is operating at the lowest level. In effect, the plan worked out at the highest level of abstraction is a skeletal or partial plan (that is, it is correct, but incomplete); the details are filled in at the next level down in the hierarchy. Goal conflict is therefore handled by resolving conflicts at a higher level of abstraction before the details of the plan are worked out.

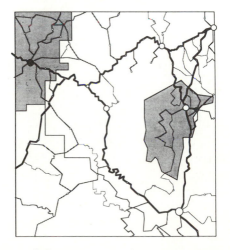

Figure 7.36   The use of abstractions in planning a car trip. Here the major road system (shown in black) is considered, along with regions (shown in gray), rather than towns.

This approach introduces efficiencies by relatively simple means. The difficulty lies in assigning degrees of importance to the preconditions of the actions. These levels are not intrinsic to the problem domain, but must be derived experimentally or intuitively—that is, by making a good guess. The hierarchical approach has been developed in the ABSTRIPS planning system by Sacerdoti (1974).

7.3.5 Hierarchical Refinement of Plans

Returning to the map analogy, in the hierarchical refinement approach we begin with the city of origin and the destination marked (Figure 7.37). We may then join them with a straight line, which is probably a path not represented on the map, but, it is hoped, one that can be transformed into a legitimate route. One approach is to find towns close to this connecting line and join them with straight lines. We can then join intervening villages and eventually road intersections and curves in the roads; eventually the route connecting the cities corresponds to the road system. We thereby produce a path by a process of incremental refinement. Another way of looking at this process is as one in which we progress through different action abstractions. A route on a map is a graphical representation of a sequence of actions. We begin with a general action, perhaps "proceed south-southeast", then conclude with a plan of very specific actions—turn right, turn left, and so on.

In the NOAH planning system (Sacerdoti, 1977), operators called *expansion rules* and *critics* are employed to transform partial plan states consisting of "loosely formed" sequences of general actions to linear plans consisting of more specific actions. *Critics*, which effectively "observe" what is happening to states, are triggered into action when something occurs that they are designed to respond to (Sussman, 1975; Dietterich and Buchanan, 1981). In this

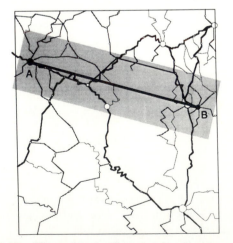

Figure 7.37 Hierarchical refinement in planning a car trip.

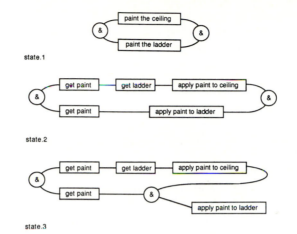

state.1

state.2

state.3

Figure 7.38 States in the development of a procedural network (after Sacerdoti, 1977).

formulation a planning state contains explicit information about actions and how they are related to one another. Actions can be related serially—that is, they can be laid out in the order in which they are to be executed—or groups of actions may be related in some parallel configuration. It is therefore possible to represent a partial ordering of actions. The *procedural network* is a convenient structure for representing these relationships. The initial state consists of a parallel arrangement of a few, highly abstract actions, derived from the goals, and the final state consists of a serial arrangement of low-level actions that can be executed by a design system.

States are transformed by means of expansion rules, which substitute low-level actions for high-level actions in the network, and critics are knowledge sources, rules, actions, or meta-actions that incorporate knowledge about resolving conflicts among actions. Figure 7.38 shows a simple example given by Sacerdoti that illustrates the major principles behind NOAH. A goal is represented by the high-level actions paint the ladder and paint the ceiling, which are represented in parallel in the network (there is no commitment to order as yet). When the network is expanded, it becomes clear that certain actions conflict. (It is not possible to "get" something that has just been painted.) The order of the actions matters, and it becomes necessary to apply one of the critics (the Resolve Conflicts critic) to adjust the network. The network can then be expanded further if necessary, and further conflicts resolved by the critics as they appear.

This approach ensures that partial plans are refined by critics before they become too large. The system therefore proceeds from a less detailed to a more detailed plan, the critics ensuring that the actions make some sort of global sense before the network is expanded further. NOAH can also be seen as a hierarchical approach in which attempts are made to create simple, correct, high-level plans before they are expanded to greater detail, when the costs

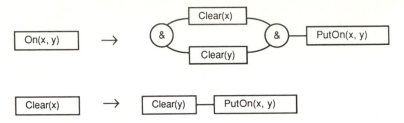

Figure 7.39   Expansion rules for the blocks world.

of rectifying plans are greater. This approach means that backtracking is not necessary. The cost is that knowledge about conflict resolution must be made explicit.

The initial state is a conjunction of high-level actions, depicted as a procedural network, which are expanded to form a plan that can be executed. The components of the end state are not specified, but it is achieved when no operators can be applied. The end state is expected to contain a substantial serial component.

Critics and expansion rules represent two classes of operators in NOAH: (1) a hierarchical mapping between actions (called expansion rules here) and (2) the critic. Expansion rules contain knowledge about how individual actions are achieved—for example, that painting a ceiling involves getting a ladder, getting paint, and applying paint to the ceiling. The procedure for handling this type of operator is to substitute the actions on the right of a rule for the actions on the left. The critics are procedural modules of specialized knowledge that enable them to detect certain interactions among actions, and to make changes to procedural networks. Because each critic may have a different procedure, it is not always possible to represent the knowledge they contain in a uniform way.

Two expansion rules are shown graphically in Figure 7.39. Three critics which are of use in this domain bear the following names:

Resolve Conflicts

Remove Redundant Preconditions

Use Existing Objects

The control system is very simple because the power of this approach lies within the critics. The procedure is to cycle through each expansion rule in turn and, after each expansion, cycle through the critics. There is no need to consider backtracking (although the system, as described by Sacerdoti, retains a record of earlier states to assist in monitoring the execution of robot actions). Figure 7.40 shows how the goals, On(C, B) and On(A, C), are achieved. State.1 can be seen as the conjunction of two high-level actions/goals, On(C, B) and On(A, C). This is expanded to state.2. The Resolve Conflicts critic detects that the action, Puton(A, C, deletes (or denies) the action, Clear(C). In order

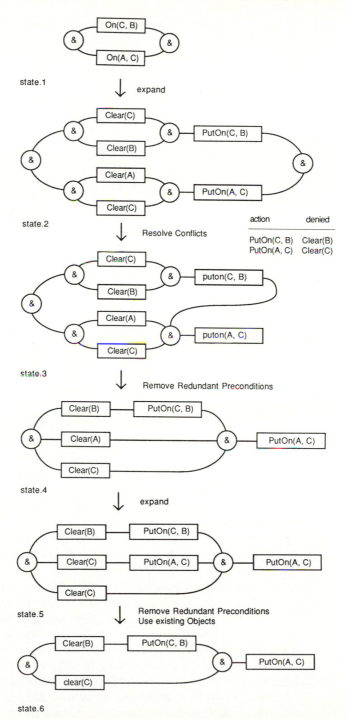

Figure 7.40 States in the development of a procedural network for a plan satisfying the goals On(C, B) and On(A, C).

to resolve the conflict, the critic adjusts the network so that the sequence of actions of which Clear(C) is a part occurs before Puton(A, C). This is shown in state.3. The Remove Redundant Preconditions critic produces state.4, and this is expanded to state.5. The Use Existing Objects critic instantiates the values of any variables. This critic plus the Remove Redundant Preconditions critic produce state.6. The nodes, Clear(B) and Clear(C), exist in the initial state and so can be ignored. The plan is therefore Puton(C, B), Puton(A, C).

An alternative way of visualizing states is to regard them as partially formed search graphs indicating sub-goal chainings. Critics are therefore designed to construct an end state that has been pruned. The procedural network formulation of states, however, makes it easier to represent partial orderings than branched search trees.

The expansion rules and critics constitute a production system that is decomposable (Section 7.1.3)—that is, it always produces the same end state irrespective of the order in which the operators are applied. There are no failed states, and changing the ordering of operators (expansions or critics) has a bearing only on the efficiency with which an end state is achieved. The system also has the major advantage that the intermediate states are "understandable" to a greater degree than the goal stacks of the systems described earlier. The system therefore lends itself to being understood by human observers while the plan is generated.

7.3.6   Scheduling Planning Knowledge

In Section 7.2 we introduced the blackboard idea. Here we consider the application of this approach to planning. The idea of blackboards was developed as a structure for handling complex problems that rely on a rich body of knowledge for their solution. Most design and planning tasks fall into this category.

Returning to the map example, there are a multitude of ways that we could formulate the task as a blackboard problem. One is to consider that we have several experts working on the task. One expert has knowledge pertaining to the intricacies of the major highway system, another knows about the way to bypass congested cities en route, and yet another knows about scenic routes and pleasant places to stop. Each expert is an independent "knowledge source," who may contribute changes to the plan as it is being developed (Figure 7.41). There will be inevitable conflicts during the development of the plan as the experts propose alternative routes. We require some system of scheduling the contributions of the different experts to give priority to the most effective changes. Scheduling in the context of planning sequences of actions can be illustrated with the NOAH example described previously.

Even though the example is very simple, and can readily be solved without scheduling the operators, there are choices to be made at each state: the expansion operators could be applied to different elements in the network; it

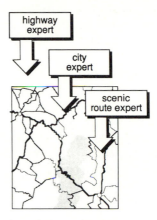

Figure 7.41 Independent knowledge sources contributing to the development of a plan for a car trip.

may be possible to apply the critics in a different order. An example of two scheduling rules for such a system might be as follows:

> **If** a critic and an expansion rule are competing
> **then** give the critic higher priority.

> **If** several critics are competing
> **then** give priority to those that reduce the size of the network.

Blackboard systems form the basis of the Hayes-Roth and Hayes-Roth (1979) model of human planning activity. This approach makes it possible to reproduce the opportunistic way in which, it is maintained, people make plans. Planning is treated as a multi-directional process, in which the process of formulating plans actually leads people to change and adapt strategies as they think through a problem. Human planning activity operates at different levels of abstraction in both a top-down and a bottom-up fashion, and it involves both forward search and backward chaining. The blackboard approach provides a loosely structured, knowledge-dependent mechanism for simulating such behavior.

Provided the problem has been formulated such that the knowledge or action sources operate as a decomposable system, the scheduling approach is reasonably sound—even an incorrect ordering of actions will lead eventually to the solution. For non-decomposable systems, it is more important that the heuristics that guide the scheduling process are effective in producing a solution.

This approach provides the advantage that knowledge that would otherwise have been part of a control procedure is made explicit in the form of scheduling rules. This provides scope for creating a system in which control knowledge is explicit and legible, although perhaps at the expense of efficiency.

7.3.7   Meta-Planning

Some knowledge-based systems have been proposed that contain components called *meta-planners*. Such systems include MOLGEN, a system for planning steps in laboratory experiments (Stefik, 1981b), the Hayes-Roth and Hayes-Roth (1979) model, and a general system developed by Wilensky (1981). The term meta-planning is generally employed in two different senses, which overlap somewhat. First, the term is employed in the sense of describing a particular level of control in a planning system, usually the top level. A meta-planner is therefore a control mechanism that contains knowledge about the formulation of plans within a particular domain. In this sense the scheduling knowledge in the preceding example is meta-planning knowledge. In the map example, a meta-planner would "know" about the different experts and how to assign priority to their activities.

The second view concentrates on the contents of that knowledge. Meta-planning knowledge is seen as completely independent of any particular problem domain. It concerns universal principles generally adopted in decision making. Wilensky outlines the kinds of *meta-themes* or goals that might guide the planning process:

Conserve resources.

Achieve as many goals as possible.

Maximize the value of the goals achieved.

Avoid impossible goals.

Do not violate desirable states.

It can be seen that these same kinds of principles guided the formulation of the various planning systems described earlier. If the knowledge by which these goals are achieved can be made explicit, as in a scheduling system, then it is possible to devise a completely general problem-solving system responsive to any domain. No system has yet been devised that effectively exploits this idea.

7.3.8   Reasoning about Actions for Spatial Layout

In this section we demonstrate the utility of the planning approach. While utilizing a flexible and layered control environment, the SIGHTPLAN system (Section 7.2.2) does not directly address the issue of conflicting actions. Here we turn to a control structure in which an attempt is made to address this issue directly. We describe an approach to generating spatial layouts based on some of the ideas discussed in earlier subsections about reasoning about actions, particularly the hierarchical refinement of plans. The system uses the kinds of actions discussed in Section 6.5.2 for generating rectangular dissections. Actions are given names and can be represented in the following general form:

Put(a, *dir*, b)

Put(a, *dir*$_1$, b, *dir*$_2$, c)

where a, b, and c represent the names of spaces; *dir*, *dir*$_1$, and *dir*$_2$ are directions (NorthOf, SouthOf, EastOf and WestOf), and Put is a descriptive predicate indicating that the spaces should be put in place. As well as carrying a descriptive name, actions contain a set of conditions that must be matched against the facts base describing the current state of the design, and a consequent that describes the changes to be made to the facts base. Rules in Figure 7.42 therefore bear the following names:

Put(a, EastOf, b)

Put(a, EastOf, b, NorthOf, c)

where a, b, and c are variables representing spaces. The actions

Put(Bathroom, EastOf, BedRoom)

Put(DiningRoom, EastOf, Kitchen, NorthOf, LivingRoom)

are therefore specific instances of these rules. The relationship between these actions is important. The final state of the system is to consist of a sequence of actions that, when implemented, generate a spatial layout.

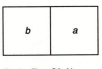

Put(*a*, EastOf, *b*)

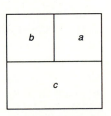

Put(*a*, EastOf, *b*, NorthOf, *c*)

Figure 7.42 Rules for locating a in a rectangular dissection.

As discussed in Section 7.3.5, there are two useful relationships between actions that can be considered: serial and parallel relationships. In *serial* relationships actions are configured in the order in which they are to be executed. In *parallel* relationships actions are arranged in groups as conjunctions and disjunctions. It is therefore possible to represent actions with varying degrees of commitment to order. These relationships can be represented in a procedural network that undergoes a series of state transformations according to the actions of critics and expansion rules. Although the use of critics is intended primarily to resolve conflicts between actions, the device can be extended to incorporate diverse sorts of knowledge, including heuristics that also help to avoid conflicts.

Actions bear a relationship to each other. In the context of design, the action Put(DiningRoom, EastOf, Kitchen) is also a specific instance of the action Put(DiningRoom, NextTo, Kitchen), which is a subclass of the action Put(DiningRoom, Near, Kitchen). Sets of such actions eventually map onto high-level actions such as DesignHouse. An action system can therefore be described that takes as its initial condition some high-level action (or actions), and produces a predominantly serial configuration of executable actions as an end state. Intermediate states consist of actions at different levels arranged with various degrees of commitment to order. The transition is from a general set of actions to an executable set of actions, and from a set of actions in which there is relatively little commitment to order, to a set of actions with a more complete commitment to order. Mappings between actions can be employed to transform sets of actions. These mappings can be utilized by knowledge about actions, as a kind of expansion rule, which replaces general actions with highly specific, executable actions.

As illustrated with the example about the conflict in painting a ladder and painting a ceiling (Section 7.3.5), simply substituting executable actions for higher-level actions is insufficient for producing appropriate sequential plans. Actions interact with one another. Other knowledge is required, concerned with interactions between actions. In the NOAH system, critics tend to resolve conflicts within a network, but the general idea can be extended to other operations on sets of actions. An example of a rule appropriate to the domain of layout planning is one that states

> **If** the procedural network contains a set of actions concerned
> with locating spaces **and**
> these actions are arranged in parallel
> (that is, there is no commitment to order)
> **then** order the actions according to the importance of the spaces
> on which they operate.

This can be interpreted as follows: if there are several spaces to be located and the desirable interconnections between spaces are known, then locate the space with the most interconnections first. Although not entirely general, this is a simple heuristic of the type that a designer might employ in beginning a layout task.

Figure 7.43 shows a simple example of states in a domain about actions in the design of a building. State.1 contains the single action DesignHouse. Given that there is a simple mapping between this action and the actions of locating the components of House, State.2 can be produced by a kind of planning rule that expands the plan. State.3 is produced by applying the planning rule about locating the most important element first (described earlier). The application of further planning rules may expand the network into more specific details and take account of more complex mappings between actions. To bring about the transformations of Figure 7.43, more information about building components is required than evident from the mappings described.

As illustrated by the simple example of Figure 7.43, it might be expected that designers readily make use of this sort of knowledge about ordering actions. Also, this formulation can be constructed as a decomposable system. It is possible that the continuation of the sequence of states in Figure 7.43 would produce a single end state, and that the order in which the rules are applied affects only the efficiency with which the end state is reached. This is not to say that only a single building layout is produced; the end state may consist of many sequences of actions in disjunction. When executed, the sequence of actions produced by this system constitutes a set of operations that

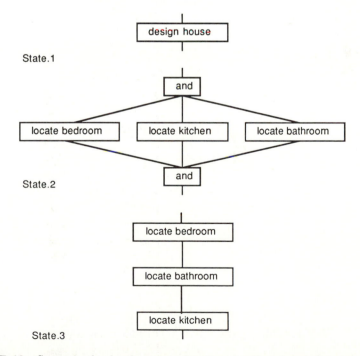

Figure 7.43   States in the development of a plan of actions for configuring a house.

generates design descriptions. The sequences may result in the generation of many different descriptions.

Nothing has yet been said in this discussion about goals—the intended performances—of the design. The knowledge for transforming states of the plan of actions can be formulated so that it responds to a set of design requirements. The general form of such planning knowledge is therefore

if <performance> **and** <plan description> **then** <plan action>

Intended performances for the spatial synthesis task may therefore be concerned with factors such as spatial relationships, the hierarchical relationship among parts and relative room sizes.

A set of facts constituting a set of desired performances is represented in predicate calculus, and is depicted graphically in Figure 7.44. The first fact is that a particular building (called Building(X)) is composed of seven spaces, six of which are rooms and the last of which is a void (that is, a courtyard or other open space). The desired relationships between the spaces are depicted in an adjacency network. The requirements constitute a type of input to the system, but may themselves be a product of another similarly formulated design system. Although it is not entirely realistic to assume that such relationships will remain static throughout the design process, they are assumed static for this exercise. In this example the relative sizes of some of the rooms are also considered important. This serves to constrain—or in this example guide—the selection of plan actions. The spatial configuration should be such that these relative sizes can be achieved when the rooms are dimensioned, even though the relative size differences are not actually evident from the dimensionless configuration.

This set of facts therefore constitutes an explicit, high-level description of the final artifact. Also part of the context is the initial state of the planning

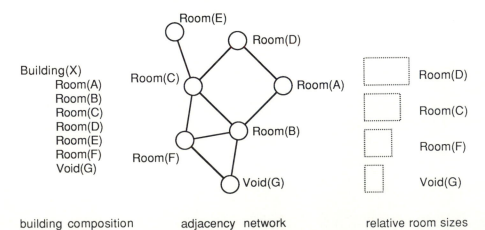

building composition   adjacency network   relative room sizes

Figure 7.44   Requirements for a spatial layout task.

process. Here it is the action, Configure(Building(MyHouse))—that is, the task is to arrange the components of building(MyHouse). This represents the high-level design goal that must be expanded and shaped into a sequence of executable actions.

Some actions for spatial synthesis are represented schematically in Figure 7.45. (These rules are simplified. Coyne (1986) represents them more precisely than illustrated here in predicate logic and describes them more fully.) The expressions in the boxes are predicates, where uppercase characters are variables. As a rule is executed, its left side is matched against the current state of the facts representing the state of the plan. The variables are instantiated to corresponding values in the matched facts, and the terms on the right side of the rule are substituted in the current state.

Rule (i) is an expansion rule. It means that the action to configure an object should be expanded into the conjunction of several actions placing the components of that object. Rule (ii) orders a set of such actions according to the valencies of the rooms on the adjacency network. This assumes that the placement of rooms with the most interconnections is somehow critical, and that these rooms should be placed first. Other rules could be devised that order on some other basis, such as size. Rule (iii) simply says to anchor the first room on the "work plane."

Rule (iv) states, if there is a room b and, according to the set of adjacency relationships it should be linked to room a, and room a has already been

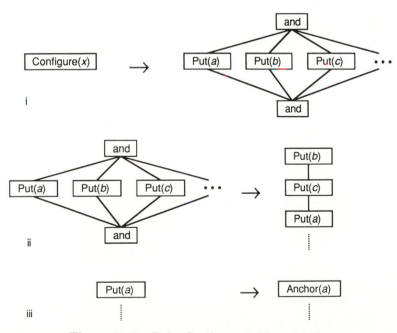

Figure 7.45   Rules for the spatial layout task.

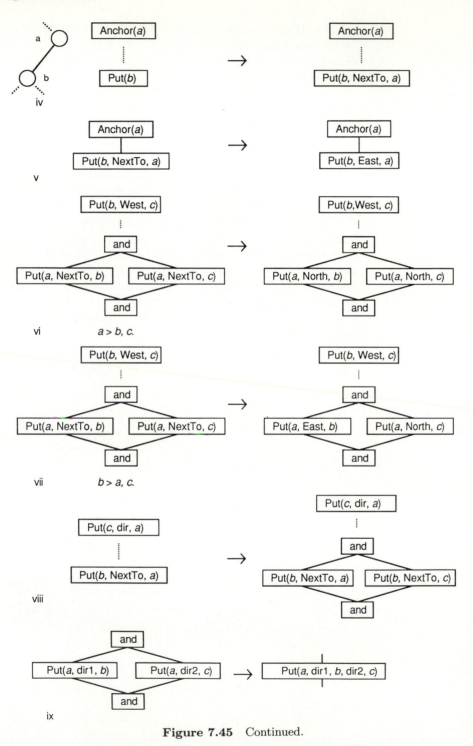

Figure 7.45   Continued.

anchored, then substitute Put(b, NextTo, a) for Put(b). This is equivalent to placing objects adjacent to objects in place and with which they should be linked. Rule (v) arbitrarily locates the second object to the east of the first object. This is a simple expedient; it is considered that other configurations can be handled by the rotation and reflection of the entire configuration. Rules (vi) and (vii) are examples of rules for orienting the rest of the rooms. They take the NextTo relationships and refine them into more specific relationships, depending on the relative sizes of the rooms. Rule (vi) therefore states that if room a is to be NextTo room b and also room c and the action for putting b west of c has already been established, and a is to be the larger of the three rooms, then place a north of b and north of c (provided nothing has already been placed there). Rule g applies if b is to be larger than a and c. The effects of these alternative rules on the eventual spatial configuration are shown graphically in Figure 7.45. Although the rules for achieving this and other orientations can be represented in a single rule, they are separated here for clarity.

Rule (viii) states that if there are no other rules to determine the placement of room b, the action Put(b, NextTo, a) can be expanded to the conjunction of two actions as shown. Room b is simply located adjacent to c and the room next to which c is already positioned. Rule (ix) replaces the conjunction of two actions with a single action. Implicit in these rules is the general strategy of proceeding from less specific to more specific actions (that is, NextTo relationships to North, South, East, and West directional relationships), as well as the strategy of using triangulation for fixing objects in place. That is, objects are placed in relation to at least two objects already in place.

These are examples of the types of rules that can be employed in manipulating plans of simple design actions. The rules are intended to reduce the likelihood of conflicts between actions by means of simple heuristics. When conflicts occur, they can be handled by means of rules after the fashion of the critics of the NOAH system, which adjust the network by re-ordering actions. Rules that specifically handle conflict resolution have not been illustrated here, but they operate in the same way as the other planning rules.

Some of the rules are independent of any domain and therefore seem to be widely applicable; others might be regarded as idiosyncratic and of limited applicability. The programming environment in which this example is implemented is such that the system is not entirely dependent on the appropriateness of the rules. The knowledge base (rules) can be enhanced and developed without detriment to the overall framework. An example of the development of a plan of actions employing the rules of Figure 7.45 is shown in Figure 7.46.

Six states in the development of the plan are shown. State.1, the initial state, is expanded to the conjunction of actions in State.2 derived from rule (i) of Figure 7.45. State.3 depicts the ordering produced by rule (ii). It arbitrarily selects one of the two spaces with equivalent highest valencies to be positioned first. State.4, which results from the application of rule (iii) and the repeated application of rules (iv) and (v), therefore depicts a sequence of actions starting with the placement of the most important space. Subsequent actions locate rooms next to other rooms already in place and with which they are to be linked. State.5 is a plan produced by the repeated application of rules (vi), (vii), and (viii). The repeated application of rule (ix) produces the end, State.6. When this sequence of actions is implemented, it results in the states depicted in Figure 7.47. The final state is a floorplan. Further rules can be applied to delete voids, to draw openings between spaces that are linked, and to position windows.

Figure 7.48 shows a computer graphics environment devised primarily for experimenting with planning knowledge for spatial synthesis. The computer

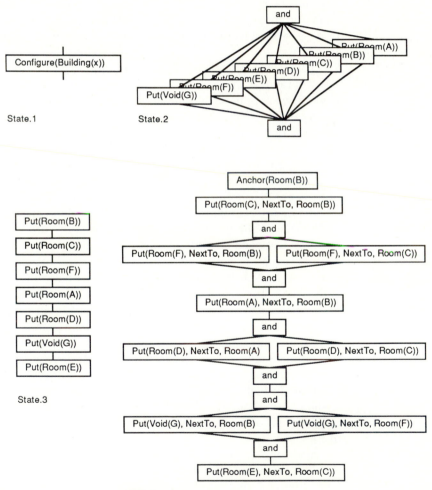

Figure 7.46   States in the development of a plan of actions.

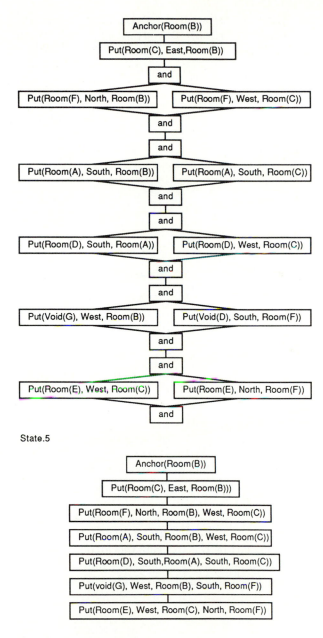

State.5

State.6

Figure 7.46 Continued.

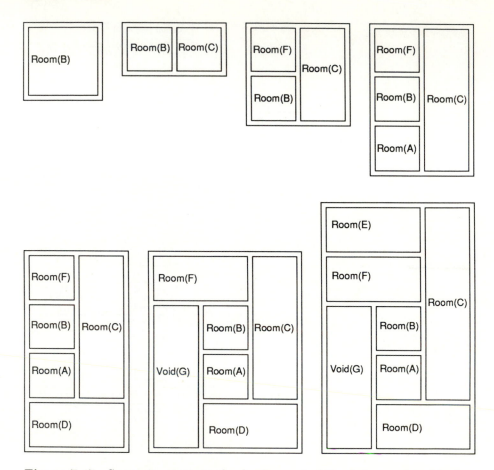

Figure 7.47 Successive states in the development of the design produced by implementing the final action sequence of Figure 7.46.

screen is divided into windows displaying information about design requirements: an adjacency graph, schedules of planning rules, plans, and the development of the form of the artifact. The intention is to create a flexible environment in which design requirements, knowledge, and system processes are visible and amenable to modification and development.

7.4   Reasoning about Design Tasks

In this section we talk about the organization of design tasks as a control abstraction above that of actions. This also enables us to discuss the control of design procedures. We are interested, however, in the knowledge by which

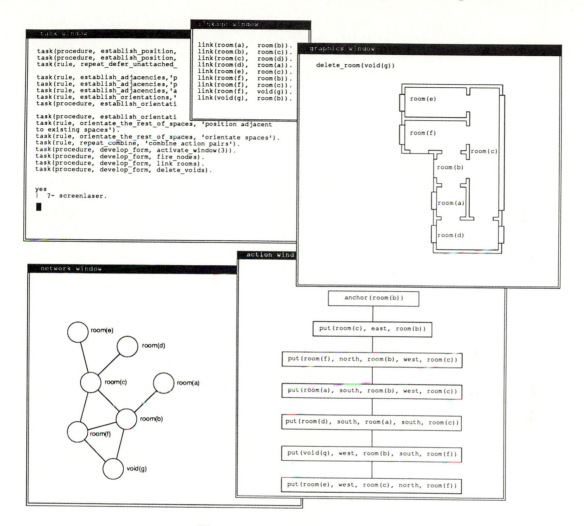

Figure 7.48 Screen display of a planning system.

we can control the use of different procedures. Here we demonstrate how this knowledge can be made explicit in knowledge-based systems.

For certain design problems there are well-established activities that can loosely be called "tasks," "methods," and "plans." For example, in the design of a reinforced concrete beam the "tasks" might be to determine the load distribution, calculate the cross section of the beam on the basis of bending moment, calculate resistance to shear stress, compute the deflection, and calculate the amount of reinforcement. There are various "methods" for carrying out these tasks, such as the use of bending moment, shear and deflection diagrams, look-up tables, and formulas. The "plan" generally involves some well-established sequence of tasks—start with bending moment, then shear then deflection, and so on. (Here "plan" does not mean only "a sequence of actions," but any sequence of activities, including tasks.) The whole design process involves generating design descriptions using well-established methods and checking the design using other methods for interpretation and evaluation. If the design fails the checking process at any stage, then some modification is made to parameters whose values were decided earlier in the process. For example, if there is excessive deflection, we might go back and make the beam a little deeper, or stiffen the flange. Various strategies can be applied.

While the methods of determining cross sections from bending-moment diagrams and of calculating the deflection of a beam are well-established, much depends on the expertise of the designer in knowing what to check, what to adjust, and what methods to revise in deriving a satisfactory design. This applies to many domains. We must be able to capture the knowledge by which designers navigate through different tasks, methods, and plans, and make adjustments to parameters.

A highly specialized but interesting example of a design problem for which there are known "methods" is the design of the paper transport, or feeder mechanism, for duplicating machines. An expert system called PRIDE (Mittal, Dym, and Morjacia, 1986) was developed to assist in the automation of this task. The transport mechanism of a photocopier moves paper from one part of the machine to another (Figure 7.49). The aim of the design process is to produce a complete specification of a paper-handling system: the number of rollers, their position, their type, and the geometry of the paper path between the rollers.

The features of the PRIDE system that interest us most here are the use of tasks, methods, and plans. As with the beam example given earlier, the totality of the design problem can be decomposed into tasks. These tasks can be organized hierarchically, and they have some degree of independence. Figure 7.50 shows a decomposition tree of some of these tasks. Returning to our analogy with beam design, it is similar to saying that the task "determine the beam cross section" can be decomposed into the tasks "determine load conditions," "determine support conditions," and so forth. A sequence of tasks is a "plan."

In addition to tasks, there are "methods," four types of which are used in PRIDE.

1. Design (value) generators provide a value for one (or a small number of related) design parameters at increments within a specified range and assign it to the parameter. Heuristic knowledge can be attached to the method for making a good first guess. The generator may be as simple as looking up a database of standard values.

2. Calculators calculate a value for a parameter given other values. Some

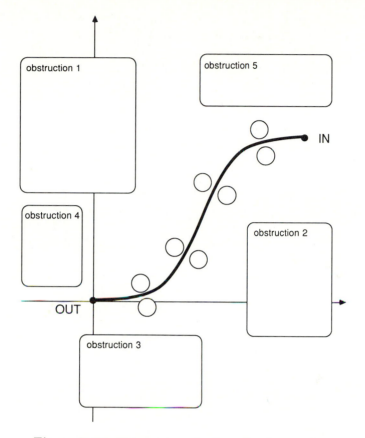

Figure 7.49   Transport mechanism of a photocopier.

parts need complex calculations that cannot be represented as declarative knowledge.

3. Procedures describe procedural knowledge for establishing a value, represented as a closed piece of code that cannot be reasoned about.

4. Task-decomposition methods decompose the single task into a set of subtasks. A default sequence of tasks is given at the outset, the actual sequence being decided in the course of executing the tasks based on the information available at the time. The approach essentially says that the best way to achieve a task is to break it up into simpler tasks.

Information about tasks and methods is represented with frames. For example, the frame for the task "decide number and location of roll stations" has slots for the tasks that must already be accomplished, the parameters required, the parameters that will be determined, possible design methods to

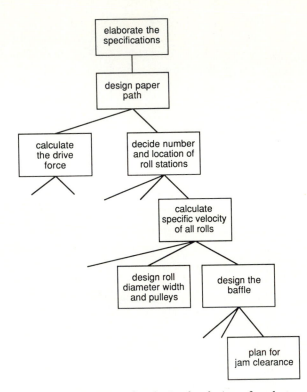

Figure 7.50 Decomposition tree of tasks in the design of a photocopier transport mechanism (Mittal, Dym, and Morjacia, 1986).

achieve the task, and constraints that affect the operation of the task:

frame	
Slot	*Value*
type	Simple task
name	Goal 5
descriptor	Decide number and location of roll stations
status	initial
antecedent goals	Design paper path
input parameters	Paper path, length of paper path
output parameters	Number of roll stations, location of all rolls
design methods	subtasks
	Decide minimum number of roll stations
	Decide abstract placing
	...
constraints	First station <= 100 millimeters
	Distance between adjacent stations
	<= 160 millimeters
	...

All designs derive from an initial plan of tasks. The tasks in the plan are not pre-ordered. Control knowledge defines (1) the tasks to be carried out, (2) information about how to order the tasks (based on the availability of input and the need for output parameters), (3) how to carry out each task (via particular design methods), (4) how to detect failures in the state of the design as it develops (through design validation), and (5) suggestions about how to fix the failures (using heuristic knowledge).

The plan is structured around the design tasks and methods. One type of design method is the decomposition of a task into simpler steps with appropriate sub-tasks. Each task is associated with designing or redesigning a small set of parameters that describe some part or aspect of the transport system. A typical design session may involve the implementation of several hundred tasks. The design methods attached to a task contain all the different ways a decision can be made about the design parameters under consideration. The methods are responsible for generating only a plausible design, not necessarily the best or even a correct design.

The PRIDE system also incorporates information about design constraints in the form of frames. Constraints serve as a means of testing the outcome of a task. An example of a frame for a constraint is as follows:

frame
Slot	*Value*
type	The type of constraint—for example, a "single constraint"
name	The name or number of the constraint
description	For example, "idler width $>= 1.2$ times the driver width"
apply when	Whether "always" or dependent on some other condition
must satisfy	Can the constraint be relaxed if it cannot be satisfied?
parameters constrained	List of the dependant parameters involved
test expression	An expression by which the satisfaction of the constraint can be determined
advice	Advice on what actions to take if the constraint fails—for example, for the preceding frame description "increase idler width" and "decrease driver width"
....	....

If a constraint fails, the system backtracks by undoing some of the decisions already made. The backtracking is not a simple reversal through the last goal or the last sequence of goals, but is directed by knowledge about the dependencies among components of the design. Some of this knowledge is captured in the "advice" slot in the frame just shown. We need not pursue this example here. For our purposes it serves to demonstrate the utility of capturing knowledge about problem-solving procedures in order to control them.

A further example is provided by Brown and Chandrasekaran (1985) in the context of the design of air cylinders. An air cylinder is a mechanical device intended to ensure accurate, reliable backward and forward movement of some component. In such a cylinder, compressed air forces a piston back into a tube against a spring. When the air pressure drops, the spring returns the piston and the attached load to the original position (Figure 7.51).

The goal is to produce a complete design in which all the attributes are assigned consistent values. The components, general configuration, and certain attributes of the design are given in advance; the requirement is to produce a set of values for the attributes sufficient to define a unique instance of this general description. This process is an example of prototype refinement. Design knowledge is formulated as tasks for making a single design decision about a single attribute. For example, one task might deal with a component's material; another task, with its thickness. Furthermore, the design knowledge is concerned with choosing from among a limited number of options for each parameter. The order in which the options should be selected is also given.

The task consists of first obtaining required information, then doing any necessary calculations, then making the decision. Figure 7.52 shows a task to decide on the width of a seat for a piston seal. To make this decision, the operations under "to do" are required. First, values for the attributes—piston thickness, piston material, and spring depth—must be determined, and a minimum thicknesss for the piston material must be decided. Then there is the means for making the decision: calculating the available space for a seal seat and checking whether the available space is greater than a postulated standard width. Also represented in the task is design knowledge about what to do if no decision is possible: increase the piston thickness. This knowledge would be useful to a "piston seal seat width failure handler" if redesign was

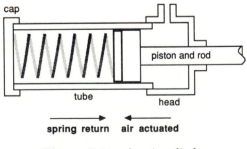

Figure 7.51 An air cylinder.

```
STEP
    NAME                        Piston Seal Seat Width
    USED BY                     Piston Seal
    COMMENT                     Written by DCB, Sept. 83
    ATTRIBUTE NAME              Seal Seat Width
    FAILURE HANDLER
      FOR DECISION FAILURE      Piston Seal Seat Width FH
    FAILURE SUGGESTION          INCREASE Piston Thickness
    REDESIGN                    NOT POSSIBLE
    TO DO
        KNOWNS          FETCH Piston Thickness
                        FETCH Piston Material
                        FETCH Minimum Thickness
                           OF Piston Material
                        FETCH Spring Seat Depth
        DECISION        Available IS
                            (Piston Thickness
                            MINUS DOUBLE Minimum Thickness)
                        Seal Seat Width IS 0.156
                        COMMENT Using one size only
                        TEST Available > Seal Seat Width?
                        STORE Seal Seat Width
```

Figure 7.52   Deciding on the width of a seat for a piston seal.

possible. (It is labeled "not possible" at this stage in the development of the system.)

The design is represented as a two-level plan, where a plan consists of a set of tasks, some of which may be executed in parallel, and each task consists of a set of sub-tasks. A sub-task is concerned with making a single design decision; the piston-seal-width task of Figure 7.52 is used by the piston-seal task. The plan for the design of an air cylinder shown in Figure 7.53 lists five tasks "to do." One of the tasks is to check a constraint that the head and spring are compatible. "Air cylinder," "spring," "head," and "rest" are the names of procedures that will carry out the rough design tasks.

A further example of organizing procedural knowledge is provided by the HI-RISE structural-design system (Maher and Fenves, 1985). The design process is divided into two major tasks, the design of the two major functional systems. These functional systems are designed in a fixed order—first the lateral load-resisting system is designed, then the gravity load-resisting system. Results from the gravity system—that is, the type, depth, and weight of the floor system—are needed to design the lateral system. This information is estimated within the lateral-system design task.

Each of the two major tasks is decomposed into a set of sub-tasks, the first of which is to synthesize a set of alternatives for the functional system under consideration. This is accomplished by carrying out a depth-first search through a tree of prototypical sub-systems stored in the knowledge base. The analysis sub-task evaluates the feasibility of an alternative using feasibility

```
PLAN
    NAME                    Air Cylinder Design Plan
    TYPE                    Design
    USED BY                 Air Cylinder SPECIALIST
    USES                    Spring Head Rest SPECIALISTS
    QUALITY                 Reliable BUT Expensive
    FINAL CONSTRAINTS       Design details OK?
    TO DO
       Validate and Process Requirements
       ROUGH DESIGN  Air Cylinder
       PARALLEL DESIGN  Spring AND Head
       TEST  Head and Spring Compatible?
       DESIGN  Rest
```

Figure 7.53   Plan for the design of an air cylinder.

constraints. The parameters of the system are initially selected using heuristics. These initial parameters are employed to evaluate the feasibility constraints. If a constraint is violated, "heuristic recovery rules" are applied to revise the parameters. HI-RISE presents all structurally feasible systems to the user, indicating which system it has selected as the "best" given an evaluation function.

7.5   Reasoning about Design Procedures

In addition to the approaches discussed so far, a space of designs can be delineated by means of generative procedures. In this book we have said little about the representation of design knowledge by means of procedures because this topic is well covered in the computer-aided design and design-methods literature—for example, Radford and Gero, 1988. We are interested, however, in the knowledge by which we might decide to adopt one procedure as opposed to another. In this section we demonstrate how this knowledge can be made explicit in knowledge-based systems.

We consider the knowledge for selecting an appropriate optimization algorithm given a problem formulation, as discussed in Section 6.2.2. It would be useful to identify certain features within the problem formulation that suggested the appropriate algorithm to solve the problem. The criteria we should use for selecting an algorithm are (1) the nature and type of design variables, (2) the nature and type of objective functions, (3) the nature and type of design constraints, and (4) the number of objective functions. We can describe some of this knowledge in the form of interpretive rules:

If   all the variables are of continuous type **and**
all the constraints are linear **and**
the objective function is linear
then   conclude that the model is linear programming **and**
execute linear programming algorithm.

If   all the variables are of continuous type **and**
all the constraints are linear **and**
the objective function is nonlinear and quadratic
then   conclude that the model is quadratic programming **and**
execute quadratic programming algorithm.

If   all the variables are of continuous type **and**
all the constraints are posylinear **and**
the objective function is posylinear
then   conclude that the model is geometric programming **and**
execute geometric programming algorithm.

If   all the variables are of binary type **and**
all the constraints are linear **and**
the objective function is linear
then   conclude that the model is zero one linear programming **and**
execute zero one linear programming algorithm.

If   all the variables are of integer type **and**
all the constraints are linear **and**
the objective function is linear
then   conclude that the model is integer linear programming **and**
execute integer linear programming algorithm.

If   all the variables are of integer and continuous type **and**
all the constraints are linear **and**
the objective function is linear
then   conclude that the model is mixed integer linear programming **and**
execute mixed integer linear programming algorithm.

If   all the variables are of continuous type **and**
the constraints are nonlinear **or**
the objective function is nonlinear
then   conclude that the model is nonlinear programming **and**
execute nonlinear programming algorithm.

If all the variables are of continuous type **and**
all the constraints are linear **and**
the number of objectives is greater than 1 **and**
all the objective functions are linear

then conclude that the model is multiobjective linear programming **and**
execute multiobjective linear programming algorithm.

An expert system making use of these rules can act as a "front-end consultant" to an optimization package that contains a set of algorithms to solve various optimization problems (Balachandran and Gero, 1987b). Knowledge of this kind is quite straightforward. What is also interesting is the knowledge by which we determine that an algebraic expression is linear. A compound expression is linear if all of its terms are of the linear form. The logic of this idea can be represented with the following statements:

$$\text{Expression}(a + b, \text{Linear}) \Leftarrow \text{Term}(b, \text{Linear}), \text{Expression}(a, \text{Linear})$$

$$\text{Expression}(a - b, \text{Linear}) \Leftarrow \text{Term}(b, \text{Linear}), \text{Expression}(a, \text{Linear})$$

The first rule simply states that an expression of the form $a + b$, in which b represents its last term, is linear if both b and the rest of the expression a are linear. The second rule deals with the case in which b is a negative term. These two rules apply only when the expression has at least two terms. Hence we must define further rules to handle expressions that have only one term. The following rules are added for this purpose:

$$\text{Expression}(a, \text{Linear}) \Leftarrow \text{Term}(a, \text{Linear})$$

$$\text{Expression}(-a, \text{Linear}) \Leftarrow \text{Term}(a, \text{Linear})$$

These rules simply define that an Expression that has a single term is linear if that term is linear.

Now we must define the necessary rules to determine whether a given Term is linear. An algebraic term is linear if it is represented by a variable symbol or by a product of a variable symbol and a number. The following rules encode these conditions:

$$\text{Term}(a, \text{Linear}) \Leftarrow \text{Symbol}(a)$$

$$\text{Term}(a \times b, \text{Linear}) \Leftarrow \text{Term}(a, \text{Constant}) \wedge \text{Term }(b, \text{Linear})$$

$$\text{Term}(a, \text{Constant}) \Leftarrow \text{Integer}(a) \vee \text{Float}(a)$$

Integer, Symbol, and Float constitute a check on the kind of term we are dealing with; they are extra-logical predicates. We now have a theorem prover able to process these statements with the goal

$$\text{Expression}(X_1 + X_2 \times 3 + X_3 \times 12 + 4.6, \text{Linear})$$

We expect it to return "true." If any of the terms fails to match the conditions of the preceding rules—as when we have a term raised to a power—then the expression will fail to be linear. It is possible to devise similar rules to determine that an expression is quadratic, posynomial, or nonlinear.

This interpretation is then used in the knowledge to determine which procedure or algorithm is satisfactory. We can generalize this idea to encode the knowledge needed to ensure that any specific procedure is being applied appropriately. Balachandran and Gero (1988) present such a system, built around a graphical interface, and apply it to structural design.

7.6 Summary

We began this chapter with a discussion of two approaches to the design process— regulating and planning. With both approaches we can formulate design systems to exploit different abstraction levels. We discussed the blackboard model of control and the way in which knowledge can be brought to bear in facilitating communication between different abstraction levels to produce designs.

We considered planning in greater detail, discussing different techniques. We began by concentrating on design elements as objects, considering the design process as concerned with selecting and ordering actions for configuring design elements. This led us to consider different levels of control. Planning operates at an abstraction of the design process at which actions are treated as objects. We can also consider levels of control above this, where the operators that select and order actions are controlled by a meta-planner. We then discussed design "tasks" as objects. A design system can be formulated so that any "component" can be an object, including actions, tasks, methods, plans, goals, and constraints. Although we cannot fully demonstrate the applicability of this idea, we can speculate on its realization.

A knowledge-based design system may take account of some hierarchical view of control. Many systems of organization might be brought to bear on a design task, including those that control the operations of other systems. The knowledge by which plans are produced may be formalized as planning tasks, such as expanding actions and ordering actions. Tasks such as these can be organized according to their relationships with one another, and as groups of tasks that are related to other tasks.

An advantage of this formulation is that the importance of domain-independent principles becomes apparent. Knowledge for manipulating tasks could be based on general strategies for problem solving, such as consolidation, diversification, convergence, and divergence. The type of knowledge that one wishes to encapsulate in order to control a set of actions comes under the category of "meta-planning" knowledge, or control knowledge. The objective is to capture domain-independent principles employed in decision making.

A strategy to consolidate might be expressed as a rule that takes as its precondition the relative sizes of the states produced by each planning task competing to be executed, and that consequently selects the task that tends to reduce the size of the facts base. Alternatively, such a strategy may select the task that produces a state with the smallest "conflict set" (set of competing tasks). A rule about convergence may select the task which operates on the most recently affected fact. This may cause the system to concentrate on particular design elements. Rules about convergence may also operate by selecting planning tasks that operate on the most important elements first. It could be maintained, for example, that important elements are those that occur earliest on a procedural network.

The control strategy for this system could be one in which only a single scheduling rule is adopted, or there might be an overriding strategy which states, for example

Attempt to consolidate or converge attention (whichever is applicable at any given state). If, after a particular period of time, no satisfactory end state is reached, then attempt to diversify.

It is possible to conceive of a meta-controller operating above the planning system to facilitate the representation of this sort of knowledge.

In this discussion we have assumed the independence of the hierarchical levels—that instructions are passed from one level to another. Also, however, some "dynamic communication" between levels may be necessary. Certain important attributes of a partial design may come to light only when a plan of action is executed. This suggests that we need some kind of integration of the regulating and the planning approaches. A better understanding of design processes, and of how we can model them, is needed. Whatever systems of knowledge organization are adopted to account for design processes, they must relate to how designers see their knowledge, and they must render that knowledge operable.

7.7   Further Reading

Coyne (1988) places the notions of planning in a design context. Two important and readable books on planning as a process are Sacerdoti (1977) and Wilensky (1983). More details and some of the theory concerned with reasoning about actions can be found in Genesereth and Nilsson (1987) and Nilsson (1982). Georgeff and Lansky (1987) contains an up-to-date set of research papers. Engelmore and Morgan (1988) contains a state-of-the-art set of presentations on blackboard systems. Blackboard systems in a design context are the focus of a special issue of the journal *Artificial Intelligence in Engineering* (Volume 1, Number 2, October 1986).

Exercises

7.1 What is the difference between the regulating and the planning models of control? What are the advantages and disadvantages of each model as a method of control in design systems?

7.2 How can regulating and planning be combined in design systems? Why would we want to combine these approaches? What conceptual difficulties are involved?

7.3 How do blackboard systems appeal to our intuitions about the way designers work?

7.4 Describe how the careful use of abstraction levels can be exploited in design systems.

CHAPTER 8

Learning and Creativity in Design Systems

The preceding chapters outlined models and methods for the formulation and use of knowledge-based design systems. Do such systems adequately account for all the knowledge that a designer brings to bear on design tasks? Do such systems adequately capture all that is creative in the designer's art? In this chapter we extend the discussion of design systems to some areas of expertise that, while difficult to capture, are essential ingredients of intelligent design behavior.

We argue that learning plays a key role in creative design. As a design progresses, designers not only increase their store of information in the form of facts, but also appear to change the way information is organized. Over time designers do things better, and this cannot be attributed simply to the acquisition of more and more facts. When a system learns, it undergoes structural change. The system itself is seen as dynamic. Learning may also bring about changes in the control of a system and the formulation of goals and constraints. In addition, design appears to involve learning as the design progresses: knowledge undergoes change. A full discussion of the implications of these assertions is beyond the scope of this book. Here we review some machine-learning techniques in the context of design. We discuss learning as an isolated phenomenon before considering how it may contribute to creativity in design systems.

A discussion of learning provides an introduction to the overall question of creativity in design systems. Can design systems operating along the lines outlined in this book be regarded as truly creative? If not, what is lacking from such systems? The chapter concludes with a discussion of some of the requirements for creative knowledge-based design systems.

8.1   Issues in Learning about Designs

Learning is said to occur when a system constructs or modifies its representations (McCarthy, 1968). Using this definition, the acquisition of any kind of information may be seen as a form of learning. Here we are not interested in the type of learning concerned with acquiring facts.

Learning is closely associated with intelligence. People are regarded as more intelligent if they learn easily. One sign of intelligence is that if we have learned a lesson from a bad experience, we avoid repeating that experience. Learning seems to involve generalization and abstraction. We do not merely extract the relations that are explicit in a particular situation, but generalize and abstract concepts so that we can apply them in different but analogous situations. For example, in Aesop's fable of the tortoise and the hare, the lesson is not merely that tortoises beat hares in races, but that "slow and steady wins the race."

Why is it important to model learning in design systems? First, it is desirable that a system improve its performance over time. Without learning, a system can respond only to those situations for which it has been provided explicit knowledge. Upon encountering a situation similar to but differing in some degree from those about which it has information, a system without a learning capability will probably falter. While many situations can be accommodated in a system without learning, design of real-life objects demands explicit knowledge for producing a large number of designs. What is required is the ability to construct general principles from knowledge about specifics and to apply them as required.

Second, the articulation of design knowledge is labor intensive. It would be desirable if a system could reason from general principles to produce knowledge that is more specific to the task at hand, thereby decreasing human effort. Over time the system would acquire knowledge that enhances its own performance and provides useful explanations for its actions. Third, learning is important in design because it is essential to the design process. Learning is an integral part of design. A designer, or a design system, learns about the tasks and steps required to produce a design.

In the following subsections we look at issues in learning about design in general terms before considering different knowledge-acquisition techniques in Section 8.2.

8.1.1   Sources of Design Knowledge

As discussed in Chapter 3, spaces of designs can be represented in various ways—as knowledge in the form of rules, as procedures, as frames, and as generic descriptions. Design knowledge can be characterized in general terms as prototypical knowledge. Prototypes can be embodied in any system of knowledge representation. (They are represented most explicitly in generalized descriptions.) When we view a single design in the form of a prototype, we bring into play knowledge to define a space of designs—knowledge about acceptable variations from the description before us.

We may employ several individual designs to define spaces of designs. These designs may serve to define a "language of designs." We require a new design that is similar in some sense to the existing ones. In knowledge acquisition we generally refer to such existing designs as "positive examples": designs exhibiting some characteristics we wish to emulate in the new design.

There are two major sources for design knowledge. A system can learn from its own activities and also from the activities of other systems. In the examples presented here we consider only how a system can acquire knowledge from observing the results of other systems—that is, by observing existing designs.

8.1.2   The Form of Acquired Design Knowledge

Initially we consider how knowledge can be acquired as rules. One of the shortcomings of this approach is the difficulty we have in tailoring the extracted knowledge to the new design task. The knowledge we acquire should relate to what we want to do with it. In a later section we explore how the operation of extracting knowledge as rules can be circumvented. The body of existing designs constitutes the knowledge base, and we reason with these designs analogically.

8.1.3   Categories of Acquired Design Knowledge

A design system could be required to acquire knowledge pertinent to any control abstraction—about form and about process. We could also consider the acquisition of knowledge pertinent to interpretation or generation. In Section 8.2 we limit the discussion to the acquisition of interpretive and generative knowledge.

8.1.4   Learning as Generalization

A kind of naive learning can be demonstrated by considering knowledge in the form of simple rules: knowing that A is a fact contained in a design description and that B is an interpretation that can be placed on the design, we might infer the following statement:

If   A
then   B

We have inferred a rule—a mapping between two facts. In reality, learning is much more complicated. Other factors may impinge on the fact that B follows from A, and it may not always be the case that B follows from A. For interpretive knowledge to be useful, it should be a generalization that accounts for different statements about designs and their interpretations. If the rules contain only explicit mappings between every description encountered and every possible interpretation, then no generalization has taken place. A system for inferring interpretive knowledge must be able to derive rules that account for cases that have not been encountered previously. In the case of interpretive knowledge, a good test for a learning system is to present it with new examples and see whether the acquired knowledge accounts for their performance.

8.1.5   The Validity of Acquired Design Knowledge

In most cases, generalizations so derived cannot be verified with absolute certainty; rather, they are falsity preserving (Michalski, 1983). Strictly speaking, a body of knowledge constitutes a hypothesis, and the learning process is one in which a hypothesis undergoes development and refinement over time.

Human beings, however, do not seem to require absolute certainty in order to reason. Learning is a continuous process. Whereas in the theoretical world generalizing from a few instances is considered unacceptable, people appear prepared to generalize from even a single instance. The first experience serves as the generalization (rightly or wrongly), remaining as such until replaced by future contradictory experiences. New experiences that repeat previous ones reinforce the generalization, while new experiences that differ in some way lead to either the generalization being extended if possible, with the differences stored as exceptions, or the generalization being revised. In addition, we take notice of the kind of experience we encounter. A bad experience, such as being sued for professional negligence—for example, failing to provide adequate waterproofing in a sensitive area—usually leads to an appropriate generalization—for example, if a similar situation ever takes place again, adequate waterproofing must be provided. A good experience such as the warm acceptance of a design usually leads to a generalization that, in future similar situations, the same sort of design should be proposed.

8.2   The Acquisition of Interpretive Knowledge

Although learning appears to be integral with the design process, the acquisition of design knowledge can be discussed as an independent issue. In this section we consider knowledge acquisition as an isolated phenomenon, refer-

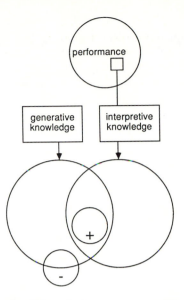

Figure 8.1 A learning set within a design space as positive and negative examples.

ring to two specific knowledge-acquisition techniques, before returning to the general role of learning in design in Section 8.3.

In Sections 8.2.1 and 8.2.2 we consider the task of acquiring interpretive knowledge, for which the source of the knowledge is other systems. Specifically, we can narrow the source of the knowledge to existing designs. Knowledge acquisition can be modeled simply by considering the descriptions of designs within the spaces designated + and − in Figure 8.1. These spaces are positive and negative examples, respectively—that is, designs that exhibit a particular interpretation and those that do not. The task is to infer the interpretive knowledge that defines a space that includes the positive examples and excludes the negative examples.

8.2.1 Learning through Successive Discriminations of Attributes

A simple algorithm for producing generalizations was devised by Hunt, Martin, and Stone (1966) under the title CLS (Concept Learning System). Later automated-learning systems followed a similar approach, notably ID3 (Quinlan, 1979) and two commercially available programs, EXPERT-EASE and RULEMASTER (Michie et al., 1984). Programs following the CLS technique are intended specifically for the derivation of expert knowledge from large numbers of examples. We demonstrate a simple, artificial application of a CLS-type learning system in the derivation of a few rules—in this case, rules about the setback of buildings from boundaries in residential design.

Figure 8.2 shows a learning set (a set of examples about which we wish

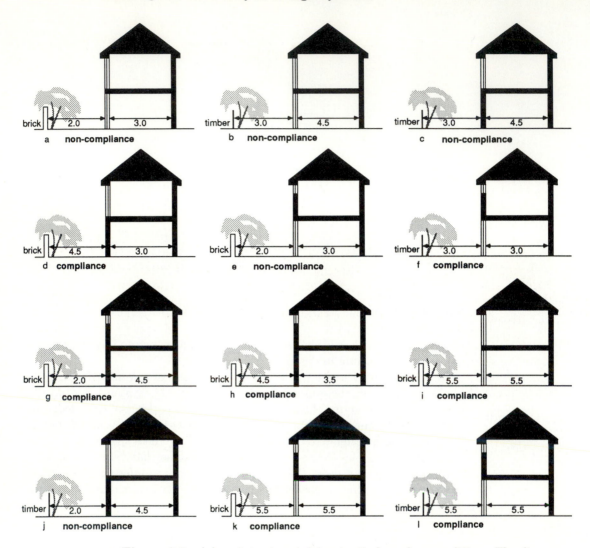

Figure 8.2 A learning set pertaining to site-boundary conditions. The diagrams, cross sections of buildings near the property line, show the materials of the boundary walls, distances from boundaries, room depths, and window conditions for two-story structures.

to generalize) that consists of positive and negative examples of setback conditions. The task of the learning system is to discover the essential features of the designs that contribute to their compliance or non-compliance, and to construct general rules to account for the important factors. The learning system is "told" about the five attributes represented explicitly on the diagrams, as well as the various values the attributes may take (Figure 8.3). Each example is described in terms of the attributes and their values. The system is

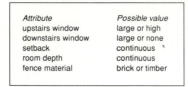

Attribute	Possible value
upstairs window	large or high
downstairs window	large or none
setback	continuous
room depth	continuous
fence material	brick or timber

Figure 8.3 Table of attributes and corresponding values.

also told which examples comply and which do not (see the "classes" column in Figure 8.4).

The CLS algorithm does not directly produce an interpretive rule, but rather a procedural *decision tree*. The decision tree produced for the learning set of Figure 8.2 is illustrated in Figure 8.5. The branch ends of the tree represent the classes (compliance and non-compliance); the other nodes are various attributes that have a bearing on the classes. The arcs of the tree are the values of the attributes.

Learning set

	Up-stairs front window	Down stairs front window	Set-back	Room depth	Fence material	Class
a	large	large	2.0	3.0	brick	non-compliance
b	large	large	3.0	4.5	timber	non-compliance
c	large	none	3.0	4.5	timber	non-compliance
d	large	none	4.5	3.0	brick	compliance
e	high	large	2.0	3.0	brick	non-compliance
f	high	large	3.0	3.0	timber	compliance
g	high	none	2.0	4.5	brick	compliance
h	high	none	4.5	3.5	brick	compliance
i	large	large	5.5	5.5	brick	compliance
j	large	none	2.0	4.5	timber	non-compliance
k	high	large	5.5	5.5	brick	compliance
l	high	large	5.5	5.5	timber	compliance

Learning set divided by class

	Compliance							Non-compliance				
d	large	none	4.5	3.0	brick		a	large	large	2.0	3.0	brick
f	high	large	3.0	3.0	timber		b	large	large	3.0	4.5	timber
g	high	none	2.0	4.5	brick		c	large	none	3.0	4.5	timber
h	high	none	4.5	3.5	brick		e	high	large	2.0	3.0	brick
i	large	large	5.5	5.5	brick		j	large	none	2.0	4.5	timber
k	high	large	5.5	5.5	brick							
l	high	large	5.5	5.5	timber							

Figure 8.4 Formal representation of the learning set of Figure 8.2.

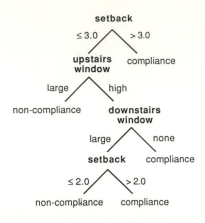

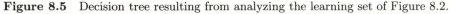

Figure 8.5   Decision tree resulting from analyzing the learning set of Figure 8.2.

Figure 8.6 gives the general idea of how the tree was produced. The objective is to successively sub-divide lists of examples according to attributes and values. With each division we try to discriminate between classes until divisions are achieved that exclude all of one class. Initially, we can divide the learning set into two lists, those that comply and those that do not. We try to find an attribute that acts as a means of discriminating among the classes. If we study these lists, we see that examples with setbacks greater than 3.0 (meters) always comply, though not only these examples comply. This is the basis of the second division in Figure 8.6. The learning set is divided into two sets, those that have a setback less than or equal to 3.0, and those that have one greater than 3.0. We have indicated the examples that comply and those that do not. If we then study the set of examples in which the setback is less than or equal to 3.0, we see that those examples for which the upstairs window is large do not comply. The attribute "upstairs window" provides the basis of the next division. Of those examples in which the upstairs window is high, we find that if there is no downstairs window then they comply. Of the examples that contain a large upstairs window, we can discriminate on the basis of setback: those examples that are less than or equal to 2.0 comply; those greater than 2.0 do not comply.

Thus the tree of Figure 8.5 represents a summary of this entire set of discriminations: if the setback is greater than 3.0, then the design complies with the standard; if it is less than or equal to 3.0 and the upstairs window is large, then it does not comply; and so on. The tree representation can be readily converted to a set of if-then rules:

    If   (setback $>$ 3.0) **or** n_1
  then   complies.

    If   (upstairs window = high) **and** n_2
  then n_1.

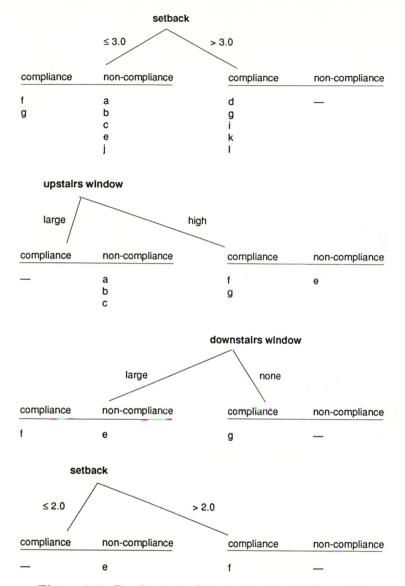

Figure 8.6   Development of the decision tree of Figure 8.5.

If   (downstairs window = none) **or** n_3
then n_2.

If   (downstairs window = large) **and** (setback > 2.0)
then n_3.

This conversion was accomplished "manually" and is not part of the CLS algorithm. The node n_1, n_2, and n_3 simply serve to split a large rule with

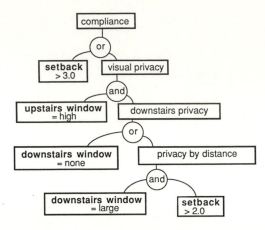

Figure 8.7   Inference network derived from the decision tree of Figure 8.5.

many nested conjunctions and disjunctions into four rules that are relatively simple to read. These nodes, however, are of some interest.

The CLS approach tells us nothing about "concepts" that fall between classes and discriminatory attributes. For example, it would be valuable to know that large downstairs windows and a setback greater than 2.0 mean that we had achieved "privacy by distance." This proposition or "no downstairs windows" means there is "downstairs privacy." A "high upstairs window" and "downstairs privacy" means "visual privacy." Thus we could enhance the preceding rules by substituting the terms "visual privacy," "downstairs privacy," and "privacy by distance" for the nodes n_1, n_2, and n_3. The result is the inference tree of Figure 8.7. Of course, the CLS approach cannot produce such intermediate concepts. To do so, we would have to provide extra knowledge and enhance the method.

One of the greatest difficulties with the CLS method is deciding on what basis to make divisions. Clearly, we require a set of divisions that produces the smallest tree possible. It is noteworthy that in this example we have been able to discern that the wall material and room depth do not have any bearing on whether the design complies with the standard. Another problem is that decision trees can be very "bushy." If the decision tree has as many decision nodes as there are examples, then no generalization has taken place. Considerable research effort has been directed to efficient pruning of decision trees, particularly when the example set is "noisy"—that is, when it contains inconsistencies (Quinlan, 1986).

We should also note that the success of the CLS method is dependent on the choice of attributes as accurate descriptors of class differences. For example, if the major determinant of compliance or non-compliance is the height of the building, then the resultant decision tree would be entirely misleading.

The method also cannot discern structural relationships between attributes—for example, that "the room depth must be half the setback," or that "the upstairs window must be the same as the downstairs window." In order to gain confidence in the decision tree and in the choice of attributes and values, we would have to test the decision tree against other examples that did not originally contribute to the formation of the decision tree (Quinlan, 1979).

The CLS method proves effective with learning sets much larger than the one just demonstrated, including applications outside design such as analyzing medical records (Quinlan, 1987) to produce rules for diagnosis. The method was used by McLaughlin and Gero (1987) to learn abductive design rules in the domain of window-wall design in buildings. The learning set was composed of design decisions in the form of design descriptions and design performances. In this case the design performances were all optimal. In the next section we consider an alternative approach to the acquisition of design knowledge that takes account of "background knowledge."

8.2.2 Learning through Knowledge about Generalization

When dealing with interpretive knowledge, the learning task can be characterized as one in which we produce the knowledge by which certain interpretations can be made—in other words, we discover mappings between design descriptions and interpretations. We may assume that such a process takes place within the context of other knowledge, often termed *background knowledge*. The task is therefore to modify or supplement links in an existing inference network, as shown in Figure 8.8.

It is possible to characterize a simple learning system, one that operates in the realm of interpretation, as a kind of production system. We can formulate

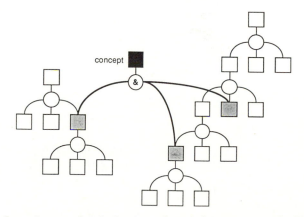

Figure 8.8 Learning as the derivation of an inference network. Essentially the purpose of the "concept" learning task is to discover the links indicated in bold lines.

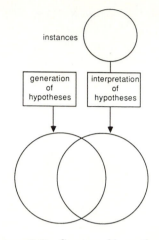

Figure 8.9   Spaces of hypotheses.

learning as a further abstraction of the design process. The knowledge to be
derived constitutes a hypothesis, for which there is generative and interpretive
knowledge (Figure 8.9). The interpretation of a hypothesis is simply the space
of designs—instances—that it defines. We can identify the components of such
a system:

Example set. Examples are available for which interpretations are known.
In the simplest case, the examples can be divided into two sets (as shown
previously in the CLS approach): those that display a particular interpretation and those that do not.

Concepts. In acquiring interpretive knowledge it is necessary to focus on
particular interpretations. These interpretations are sometimes referred
to as "concepts". We are therefore interested in inferring the mappings
for one or more concepts. (Concepts are equivalent to the "classes" of the
CLS approach.)

Description language. There has to be some way to describe the examples
—a description language. If the interpretation can be made in terms of
the primitives of the description language, as in the CLS example, then
the task of generalizing is simplified. For example, if building designs
are described in terms of their perimeter shapes, then the acquisition
of interpretive knowledge for determining whether a plan is "compact"
will be relatively straightforward. If designs are described in terms of
line segments, however, it may be necessary to provide the knowledge for
inferring shapes before the knowledge about compactness can be acquired.

Background knowledge. This is the knowledge by which various interpretations, (other than the concepts under study) can be made from the

design descriptions. We may call these interpretations and descriptions "facts about the designs."

Hypothesis. The hypothesis is some statement about the links between the interpretations and the facts about the design, generally in the form of a rule. It undergoes transformations during the learning process. Initially, it may be the null hypothesis. It may contain conjunctions or disjunctions of facts about the design. A hypothesis may develop in several ways, including from the general to the specific or from very specific statements to more general ones. The various approaches are reviewed by Langley and Carbonell (1984). Ultimately it should be possible to test the hypothesis by seeing whether it can account for interpretations of designs other than those presented for the purposes of learning.

Generalization knowledge. This is the knowledge that brings about transformations to the hypothesis. It can be modeled with production rules. Such rules respond to the state of the hypothesis, and to the examples, to modify the hypothesis.

Control knowledge. Various control strategies could be employed in such a system, including the simultaneous consideration of the entire example set. A simple strategy is to inspect the examples in turn and implement each of the generalizing rules exhaustively as they are applicable. It is desirable for the system to be formulated such that there is a direct mapping between the knowledge for interpreting a hypothesis and that for generating the hypothesis. However, knowledge may be required to control search. The space of hypotheses may be very large. The form of the final hypothesis may also depend on the order in which the generalizing rules are applied.

We can investigate a simple example of the acquisition of interpretive knowledge in these terms. A set of designs is displayed in Figure 8.10. The concept for which we wish to acquire interpretive knowledge is fairly intangible—perhaps it is simply that certain examples are liked or disliked. In the figure they are simply labeled as positive and negative examples. The knowledge-acquisition task involves finding out what the positive examples have in common that accounts for their grouping together that the negative examples (one in this case) do not exhibit. How we describe each of these examples has a bearing on the kind of generalizations we make. From Figure 8.10 we can see that it is possible to describe the examples in terms of "objects"; this is the specific information we provide to the learning system to describe the examples. Other facts may be inferred from this information; we present the system with the knowledge by which these facts could be inferred. This knowledge, similar to that shown in Section 4.3.1, constitutes the background knowledge.

A useful general procedure for this learning system is to consider each of the examples in turn and apply generalizing rules exhaustively. Some of the

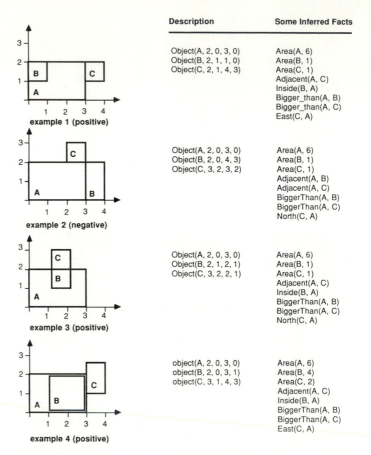

Description	Some Inferred Facts
Object(A, 2, 0, 3, 0) Object(B, 2, 1, 1, 0) Object(C, 2, 1, 4, 3)	Area(A, 6) Area(B, 1) Area(C, 1) Adjacent(A, C) Inside(B, A) Bigger_than(A, B) Bigger_than(A, C) East(C, A)

example 1 (positive)

| Object(A, 2, 0, 3, 0)
Object(B, 2, 0, 4, 3)
Object(C, 3, 2, 3, 2) | Area(A, 6)
Area(B, 1)
Area(C, 1)
Adjacent(A, B)
Adjacent(A, C)
BiggerThan(A, B)
BiggerThan(A, C)
North(C, A) |

example 2 (negative)

| Object(A, 2, 0, 3, 0)
Object(B, 2, 1, 2, 1)
Object(C, 3, 2, 2, 1) | Area(A, 6)
Area(B, 1)
Area(C, 1)
Adjacent(A, C)
Inside(B, A)
BiggerThan(A, B)
BiggerThan(A, C)
North(C, A) |

example 3 (positive)

| object(A, 2, 0, 3, 0)
object(B, 2, 0, 3, 1)
object(C, 3, 1, 4, 3) | Area(A, 6)
Area(B, 4)
Area(C, 2)
Adjacent(A, C)
Inside(B, A)
BiggerThan(A, B)
BiggerThan(A, C)
East(C, A) |

example 4 (positive)

Figure 8.10   A learning set to acquire the knowledge pertaining to a particular spatial concept. The examples exhibiting the concept are labeled "positive."

knowledge by which we might produce hypotheses from the description of the examples follows:

Rule 1
>   **If**   there is no hypothesis
>   **then**   make the first positive example encountered the hypothesis.

Rule 2
>   **If**   there are statements in a positive example that contradict statements in the hypothesis
>   **then**   delete these statements, or their negation, from the hypothesis.

Rule 2.5
>   **If**   two statements contain different values for the same attributes of the same object
>   **then**   these statements *contradict* one another.

Rule 3

   If   there are statements in the hypothesis that are the same as
          statements in a negative example
  then delete these statements from the hypothesis.

Rule 4

   If   the names of some of the objects described in the hypothesis
          differ from the names of objects in a positive example **and**
          these objects have the same attributes and values **and**
          these statements are the only differences between the positive
          example and the hypothesis
  then change the names of these objects in the hypothesis into variables.

These rules assume that concepts can be described in terms of conjunctions of facts, rather than disjunctions. The identification of disjunctions poses difficulties, and further rules are required to produce disjunctive sets of statements. We should also note that these rules are heuristically based and incomplete for most learning tasks. In our example they lead to a useful result, but with a different learning set they might not be as useful.

We could employ the background knowledge to produce all possible inferred facts about each example, then employ the knowledge given in the preceding rules to produce general hypotheses. Because there are many possible ways of describing even the simple examples of Figure 8.10, we may use some kind of control heuristics to guide the production of new facts. Let us start by generalizing on the basic description (as objects), then expand the description incrementally according to the following rules.

Meta-rule 1

   If   a statement is to be deleted from the hypothesis **and**
          that statement matches the left side of a deductive rule **and**
          the statements of the right side of the rule are not within the
          hypothesis
  then fire the deductive rule **and**
          add the statements of the right side of the deductive rule to the
          hypothesis.

Meta-rule 2

   If   no learning rules are applicable **and**
          there is a match between the left side of a deductive rule and
          statements in the hypothesis **and**
          the statements of the right side of the rule are not within the
          hypothesis
  then fire the deductive rule **and**
          add the statements of the right side of the deductive rule to the
          hypothesis.

We keep this example simple by ignoring the use of the meta-rules. We also provide the learning system with a set of facts already inferred from the background knowledge. Such a strategy would result in the development of a description shown here. The first example is presented to the system. Rule 1 produces the following hypothesis:

Object(A, 2, 0, 3, 0)

Object(B, 2, 1, 1, 0)

Object(C, 2, 1, 4, 3)

Area(A, 6)

East(C, A)

Adjacent(A, C)

Adjacent(C, A)

Inside(B, A)

Area(B, 1)

East(C, B)

These are effectively the preconditions to a hypothesized interpretive rule with the consequent Positive(example).

When example 2 is encountered, this causes Rule 3 to be fired. It deletes all the facts that are also in the negative example:

Object(B, 2, 1, 1, 0)

Object(C, 2, 1, 4, 3)

East(C, A)

Inside(B, A)

Area(B, 1)

East(C, B)

Example 3 is considered. This prompts Rule 2 to delete any facts in the hypothesis that contradict those in the positive example:

Area(B, 1)

Inside(B, A)

Example 4 is considered. This prompts Rule 2 again to delete facts in the hypothesis that contradict those in the positive example:

Inside(B, A)

The fact (there is only one in this case) that describes positive examples and excludes negative examples results in the interpretive knowledge:

If   Inside(B, A)
then   Positive(example).

Note that this rule does not incorporate variables. To invoke Rule 4, we would need another positive example that included a description in which the object inside A was other than B; we would then generalize by substituting variables for the instances.

This induced interpretive rule constitutes a hypothesis. A different body of learning knowledge could have produced a different rule that is also "valid." Note that this approach favors a learning set in which there are only subtle differences. The order in which examples are encountered is also important. Winston (1975) describes this approach in learning about configurations of blocks; the appropriateness of the search strategy depends significantly on the nature of the learning examples and the general domain. Further strategies of considerably greater sophistication are discussed by Michalski (1983). The approach has been effective in fields outside design (Michalski, 1983; Michalski and Chilausky, 1981).

8.3   The Acquisition of Syntactic Knowledge

The acquisition of syntactic knowledge poses several difficulties. In general, syntactic knowledge, such as a grammar, must be learned from examples of the final product (sentences in acquiring natural language grammar, or designs in acquiring syntactic design knowledge). There are two sets of unknowns in this process: (1) the set of operators; and (2) the path followed from one operator to another in the formation of the end state. Thus, in discovering grammatical rules from design descriptions, we must determine how each of those designs can be "parsed"—that is, the sequence of actions by which they are derived—as well as the nature of those actions. The same design may be generated by different sets of actions and different sequence of actions, making the task of generalizing very difficult. In this section we review one approach to the acquisition of a design grammar, demonstrating that such a grammar can be acquired if we make certain simplifying assumptions.

A model of learning a natural language grammar is discussed by Charniak and McDermott (1985). Phrase-structure grammars are one of several formal methods of defining a space of sentences that bear some resemblance to cognitive activity. A sentence must be parsed in order to be understood; this means exploring the ways in which a string of words can be divided into categories. One of the ways in which it is thought that human beings learn language serves as an example. The abbreviated sentence

little girl drop red block

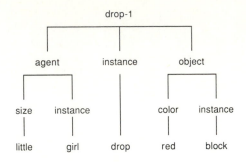

Figure 8.11 An associative network (as a parse tree) by which a simple sentence is broken down into phrase-structure rules.

can be parsed according to the network of Figure 8.11. When a young child is given simple sentences to start with, the process of acquiring a grammar is made relatively simple. Nouns are objects, and verbs are action words. The patterns of *relation*, *agent*, and *object* are categories generalized from simple sentences. The process is similar to the formation of hypotheses about interpretive rules. More complex sentences are introduced gradually so that further categories can be assimilated.

An example of a further refinement of categories is with the phrase "the red block." A child soon discovers that this phrase is grammatically acceptable, but that "red the block" is unacceptable. This suggests that "the" and "red" belong to different categories. How such a hypothesis might be formed is discussed by Berwick (1980). In human language, learning takes place over a long period of time and involves trial and error; there are powerful feedback and correction mechanisms from the environment. If the child makes an utterance from its own mistaken grammar, it will either be corrected, or the failure in communication will imply that that particular grammatical pattern (hypothesis) has failed and another one is required. The theoretical basis of these processes is explored by Anderson (1977).

It is possible that similar learning processes can be applied to the acquisition of design grammars. The approach followed here involves the following steps: (1) devise a set of general candidate actions, (2) generate alternative action sequences that produce the designs (their parse trees), and (3) inspect the parse trees to discover a set of restrictive actions that generate the same designs.

There appear to be many action sequences we could use to arrive at a single design, such as a floorplan. There are also many kinds of actions. For example, we could produce a floorplan design by beginning with a rectangle, which we then divide into rooms. An alternative approach is to begin with a single room and then add the other rooms around it one at a time. The results of the two processes may be entirely the same. We may also formulate generative knowledge that bears little relation to our intuitions about design knowledge, but that is operable and that effectively defines a space of designs. The work on shape grammar exemplifies this (Section 6.4). It is not necessary that the particular rewrite rules accord with the way we look at the composition of the design.

Thus let us turn to methods of representing generative knowledge that provide some kind of computational advantage. Their intuitive appeal as a means of representing design knowledge is a secondary concern here. One useful approach is demonstrated by a system for learning a grammar appropriate to generating floorplans. This system, developed by Mackenzie (1988), induces a grammar for generating simple floorplans from a set of existing plans. Several assumptions underlie the operation of the system. The first assumption is that a design can be described in terms of geometrically orthogonal relationships. Second, it is assumed that we need not be concerned with dimensions of spaces in describing designs. Third is the assumption that designs can be generated through actions that split spaces in a north-south or east-west direction; these spaces can be split further in a recursive fashion. Finally, we assume that this system of actions can be employed to account for the commonalities between plans. It can effectively form the basis for a set of actions that defines a space of designs.

The system commences with a given set of designs. The strategy embedded in the system is first to find an appropriate parse tree for each design. There may be many possible trees that could be produced—that is, there may be many ways in which the design could be successively split. Thus the system uses certain heuristics to find the "best" tree. One heuristic is to favor generative procedures that combine simple shapes in simple ways. Several trees may be generated for each design. The task is then to generalize on parse trees—that is, refine the sequences and partial sequences of actions to a more succinct but general set of actions. When no useful generalizations can be made, alternative parse trees are generated. The generalizing mechanism is based on the methods of Fu and Booth (1975) for inducing relational "tree grammars."

It is worth elaborating on this procedure. Figure 8.12 shows a progression of states in which successive splitting results in a spatial layout. The action sequence could be described informally as follows:

Start with house zone.

Split house zone into eating zone north of living zone.

Split eating zone into kitchen west of dining room.

Replace living zone with living room.

Zones are like syntactic categories in natural language grammars (such as "noun phrase," "verb phrase," and "article"), while spaces such as "kitchen" and "living room" are *terminal words*. A "legal design" may contain only terminal words.

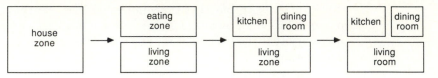

Figure 8.12   "Parsing" a design.

The objective of a grammar-induction system is to derive a set of actions that is effective in producing designs that are similar in some respects. They may belong to the same "style." Initially, the system must "parse" designs such as that shown in Figure 8.12. In this example there are not very many ways in which the design can be parsed in terms of splitting. Figure 8.13 shows a more complex example, which can be parsed in different ways—that is, it can be decomposed into several different binary configurations. There are also systems of subdivision that do not always result in rectangles, as shown in Figure 8.14. When several paths lead from an initial state to an end state, some heuristics are employed to restrict the kind of parse trees that are generated.

The heuristics used by the system embody a kind of simplicity principle: paths that involve intermediate states in the development of the design and that contain simple rectangles are favored above states with complex shapes. A further principle is that "T" shapes are favored above "L" shapes. One of the most unlikely intermediate states is considered to be that in which a stepped configuration occurs. The appropriateness of such heuristics depends on the domain. These simplicity principles may be appropriate to learning about these particular floorplans, which are essentially symmetrical, but different styles may warrant other heuristics. The method of generalization is based on an algorithm by Levine (1982). The algorithm utilizes certain parameters that can be varied to control the degree of generality of the derived grammar.

The method just outlined has been applied to the induction of a grammar for defining a space of Palladian villa plans (Mackenzie, 1988). The learning set is shown in Figure 8.15. A grammar produced by the system and appropriate to this class of designs is shown formally as rewrite rules in Figure 8.16. These

Figure 8.13   A design that can be "parsed" in several different ways (after Mackenzie, 1988).

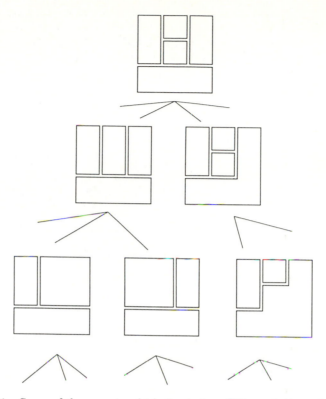

Figure 8.14 Some of the ways in which the design of Figure 8.13 can be "parsed" (after Mackenzie, 1988).

rules could be readily translated into descriptive actions, as shown previously. Some new designs generated by the grammar appear in Figure 8.17.

A further example provided by Mackenzie is the generation of floorplans of standard tract homes. The learning set is shown in Figure 8.18, while some new designs conforming to the grammar are shown in Figure 8.19. This grammar turns out to be very general, permitting such configurations as a house com-

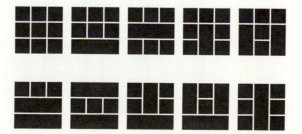

Figure 8.15 A set of Palladian-like plans (after Mackenzie, 1988).

V = {start, 1, 2, 3, 4, 5, 6, 7, 9, 12, 13, 14}
A = {(,), *east, north*, rectangle, square}
P = start → *north*(1, 1)   start → *north*(1, 6)   start → *north*(6, 1)
      start → *east*(12, 5)   start → *east*(14, 12)   start → *north*(6, 6)
      start → *east*(4, 4)   1   → rectangle   2   → square
      3   → *east*(1, 9)   4   → *east*(4, 4)   4   → *east*(1, 4)
      4   → *north*(2, 9)   5   → *north*(1, 2)   6   → *east*(3, 1)
      6   → *east*(13, 9)   6   → *east*(2, 7)   6   → *north*(1, 6)
      7   → *east*(2, 2)   9   → *north*(2, 2)   12   → *east*(5, 14)
      13   → *east*(9, 1)   14   → *north*(9, 2)
S = {start}

Figure 8.16   A grammar for producing "Palladian" floorplans (after Mackenzie, 1988).

prised only of bathrooms; for large numbers of rooms, some rooms that need natural light are sometimes placed internally. These anomalies may indicate a shortcoming in the splitting assumption. Knowledge about the successive splitting of spaces may not facilitate the capture of knowledge about proximity to the exterior of the building. To be useful, the syntactic knowledge produced by the system would have to be enhanced with other knowledge. It could form the basis of a useful generate-and-test system along the lines discussed in Chapter 6.

This approach could be extended to consider actions other than binary splitting. The system could also produce grammars that are then further refined by other learning knowledge. An entire grammar constitutes a hypothesis, the testing of which is accomplished by attempting to employ it in regenerating known designs. The hypothesis could be subject to further testing

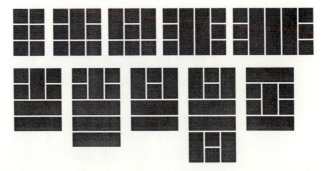

Figure 8.17   Some of the designs generated by the rules of Figure 8.16 (after Mackenzie, 1988).

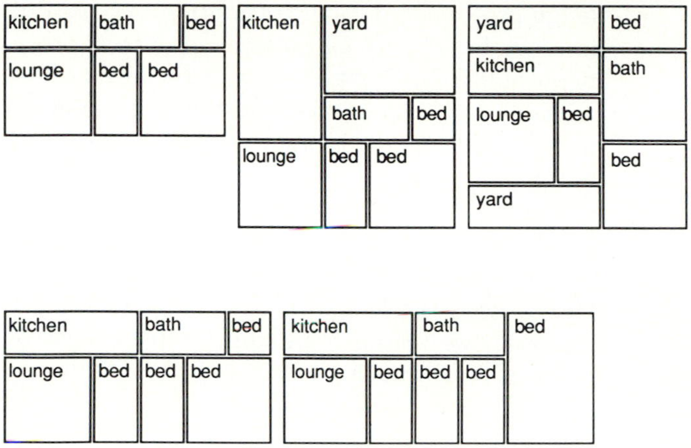

Figure 8.18 A set of tract-home layouts (after Mackenzie, 1988).

and refinement. Formulating and refining grammatical hypotheses presumably requires domain-dependent knowledge that embodies assumptions about what to look for in example designs. The representation of this kind of design knowledge is an important research area that poses considerable challenge.

8.4 Analogical Reasoning in Design Systems

So far in this chapter we have referred to the acquisition of knowledge as rules. Extracting knowledge from examples, and then discarding the examples, results in the loss of certain information. An alternative approach is to store representations of existing designs and reason with them analogically (Carbonell, 1983, 1986). This means finding similarities between the current

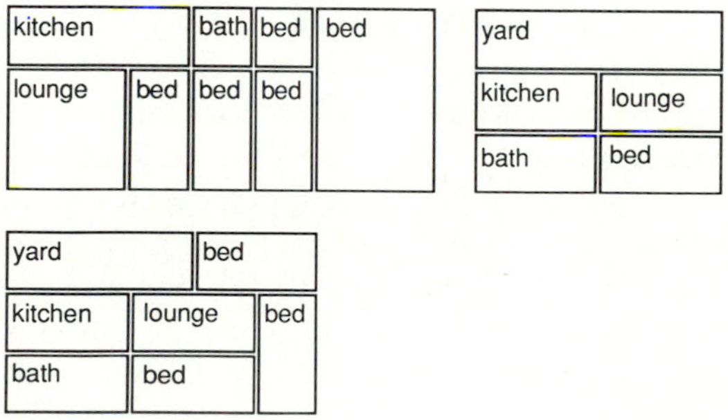

Figure 8.19 A set of designs generated by a grammar produced from Figure 8.18 (after Mackenzie, 1988).

design problem and a store of existing designs. Put simply, the idea is to make decisions by looking for patterns in data, a process considered to match human cognitive processes closely (Dreyfus and Dreyfus, 1986). The important points about this process are (1) the need to develop computer models that are more soundly based on what we understand of human cognition and (2) the ability of a system to learn and reason by memorizing past design "episodes" or "cases."

We explore two approaches to analogical reasoning here. In the first, we reason from design "episodes," attempting to match our current problem. This is accomplished with the aid of a "black box" metric. In the second approach, we reason from stored explanations of certain decisions made in previous design tasks.

8.4.1   Reasoning from Design Episodes or Cases

A simple yet useful application of analogical reasoning has been developed by Stanfill and Waltz (1986). Although not a design example, it demonstrates the basic idea. The task explored was to determine the correct pronunciation of a new word from a store of words accompanied by their correct pronunciations. The knowledge base for the system is simply the lexicon of words coupled with the string of phonemes that represents their pronunciation. When a new word is presented to the system, a carefully designed metric is brought into play to provide a measure of the difference between each of the letters in the target word and analogous letters in the database. Thus the system can determine that the "g" in "gypsum" is more closely analogous to the "g" in "gypsy" than to the "g" in "guzzle." The metric takes account of the position of each letter in the word and the proximity of letters to each other in a word. The system then produces a feasible pronunciation for the new word.

One advantage of this approach is that the system can improve its performance by simply adding a newly learned pronunciation to its lexicon. This is somewhat like adding a new word to the dictionary of an automatic spelling checker, except that the new word may help determine the pronunciation of other new words by analogy. Unfortunately there is no universal metric for all string-matching problems. The metric must be tailored to the particular problem, and the knowledge by which similarities are found is effectively a "black box."

Let us apply the idea demonstrated in the pronunciation example to an equivalently simple problem pertinent to design. We can readily extend the idea of matching strings to two dimensions. Instead of a single string we have a grid, or matrix, which might represent a tract of land. The values in the grid squares are the attributes distributed across the site, such as slope, soil type, and vegetation. The distribution of attributes represents a performance: we require a distribution of land uses across the site that accommodates the attributes of the site. The land-use distribution represents a design description.

We may store many "successful" site-attribute and land-use-distribution pairs in the knowledge base of our system. When we present a new site description to the system, by analogy with the store of existing designs it can then produce a new land-use description that is appropriate to the site distribution. Of course, we require an appropriate metric for determining the mapping of land use to site condition, bearing in mind the surrounding site features. As described, this task is similar to that of image recognition (Ballard and Brown, 1982).

We have presented a fairly superficial application of the analogical reasoning idea to design. This approach is not limited to simple spatial mappings, but could apply to any system of description.

Neural network, or PDP (parallel distributed processing), models of computation are based on mathematical models of neural processes. The PDP paradigm exploits the idea of storing past "episodes" (designs and their performances), but in a way that is claimed to be more closely related to our understanding of human information processing at the neural level (Rumelhart and McClelland, 1986). The major operational advantage is an efficient means of pattern matching that enables fast access to stored patterns. An added advantage is that access to stored patterns can be achieved even when there is only an approximate match.

Although computationally expensive, this application of analogical reasoning is becoming more feasible as computers become more powerful. It also lends itself to new kinds of computer architecture, such as those based on parallelism (McClelland and Rumelhard, 1986). It is described in more detail in Section 8.5.

8.4.2 Reasoning from Explanations of Design Decisions

Significant developments in the idea of reasoning from explanations come from Schank (1982, 1986), Mitchell, Keller, and Kedar-Cabelli (1986), and Dejong and Mooney (1986). Navinchandra (1987) developed the CYCLOPS system, a direct application of the idea to design. The example used is that of siting a house on a steeply sloping site. Building on a slope presents certain problems that are readily solved by terracing the site or placing the house on stilts. In this example, however, we assume that we are dealing with a design system that initially contains only the knowledge of a novice designer. The idea of stilts may occur to such a designer from knowledge of a different problem, that of placing huts on flood planes in a Thai village.

In reasoning from explanations in a design system, we attempt to model this kind of behavior. More than a superficial understanding is required (Navinchandra and Sriram, 1987). In order to draw analogies, we require some explanation of why things are done in a certain way.

Such a system might consider the following reasoning in relation to the design of a village hut. The following abductive rule decides that a particular

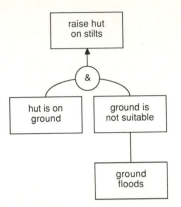

Figure 8.20   An abductive rule for deciding on the action "raise hut on stilts."

action should be implemented:

> **If**   hut is on ground **and**
>   ground is not suitable
> **then**   raise hut on stilts.

We assume we are dealing with some kind of generative design system that searches a design space defined by actions. The detection of a situation such as the preceding one (hut is on ground and ground is not suitable) is used as a means of selecting an appropriate action, "raise hut on stilts." We may also use knowledge for determining when the ground is not suitable (Figure 8.20):

> **If**   ground floods
> **then**   ground is not suitable.

Having generated a satisfactory design, we can store this "experience" along with the description of the final design. This knowledge can be represented as an explanation of why the design is as it is: the hut is on stilts because the ground is unsuitable, and the ground is unsuitable because it floods. Similar explanations may accompany other decisions.

We now turn to the design task of locating a house on a typical urban site. We wish to use something of the experience gained in the preceding problem to help solve the problem of a sloping site. We may have knowledge such as the following:

> **If**   site has good view **and**
>   structure is stable
> **then**   design is suitable.

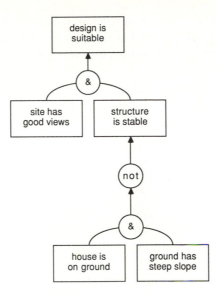

Figure 8.21 The conditions under which a design is suitable.

We assume that we can reason with such rules "abductively," as discussed in Chapter 6. To be suitable, a design must have "good views" and a "stable structure." We may also have a rule that states some conditions under which a structure is not stable (Figure 8.21):

If   house is on ground **and**
      ground has steep slope
then structure is not stable.

Suppose that the design system encounters a state in the development of the design in which it is true that

    Site has good views and site has steep slope.

It follows that the design is therefore "unsuitable." The system may then search for a design action to ameliorate the situation. We assume for the purposes of this example that there is no such action. In Chapter 6 we considered backtracking as a way of resolving an undesirable interpretation. In the approach considered here, we discuss how the system may reason analogically to determine the best action to consider at each state in the development of the design.

There are two tasks in this kind of analogical reasoning: (1) finding analogous situations and (2) finding analogous objects. Let us consider the task of finding analogous situations first. An analogical reasoning system may inspect previous design episodes, such as that encountered in the design of the Thai hut, for a situation that matches the current one. The system will look for a design that was at some stage "unsuitable" and attempt to find an action that ameliorated the problem. This should be evident by inspecting a record of explanations. That the design is "unsuitable", however, is too general a complaint. The system will then look for the cause of the unsuitability—that the "structure is not stable." The system will search for a similar situation and an action that solved the problem. It there is none it then looks for a match with the causes of that particular problem: "house is on ground" and "ground has steep slope." If it can find an action that provided a remedy to one of these problems, then that action is selected as appropriate in this situation—the fact that "house is on ground" matches "hut on ground" of the previous example. The solution was to raise the hut on stilts, and the same action can be applied to the house.

Of course, we need some mechanism to detect that huts and houses are similar. This is relatively straightforward if the system contains an explicit representation of object types. The system needs the knowledge to determine that a hut is sufficiently like a house in this particular context. As described here, the process is essentially one of finding partial matches between inference trees. The inference trees of prior design experience are reconstructed from explanations.

The common situation is that, in both designs, building on the ground is undesirable, though for different reasons. In the case of the hut, the ground floods. In the case of the house, the ground has a steep slope. The action "raise building on stilts" is appropriate in both cases. The CYCLOPS design system (Navinchandra, 1987) can exploit this kind of reasoning in producing designs.

A similar approach is developed in EDISON, an engineering design "invention" system (Dyer, Flowers, and Hodges, 1986). In this system, generic designs are organized and indexed in an "episodic memory." The system selects designs from memory and attempts to create novel devices by altering them. Some of these generic designs may be used quite differently from their intended purpose, such as using an umbrella as the basis of a new kind of door. (At a particular level of abstraction we can describe a door as functioning to block the passage of physical objects. An umbrella can also be so described.) EDISON uses a system of indexing in which these levels of abstract description are accessed easily to form links between problems and generic designs.

Similar ideas are developed by Maher and Zhao (1987) for the design of building structures "from experience" and by Murthy and Addanki (1987), who developed PROMPT, a system for selecting types (standard cross sections for structural members) and modifying them to create "novel" profiles. Discovering the right analogies for drawing on and modifying types is a very sophisticated form of reasoning. These ideas are at an early stage of development and their realization represents an important step in the development of knowledge-based design systems.

8.5 Neural Network Models of Learning in Design

8.5.1 The Neural Network Model

In neural network or parallel distributed processing (PDP) models, knowledge about design is represented as patterns and is generally stored simply as weights and threshold values in a network. Nodes in the network are called *units*. Input units, output units, and hidden units are connected by directed arcs in some configuration. *Input units* receive values from the "outside world." *Output units* transmit their output values to the world. *Hidden units* only receive and transmit values to other units. The parameters of the system are weights on the arcs, threshold values on the units, and input and output values of units. It is simplest to think of units as receiving binary inputs and producing binary outputs. The output value of a unit—1 or 0—is a function of the sum of the products of the weights of arcs directed toward the unit and the output value of the connected unit. A net value greater than the threshold generally means that the unit will "fire", producing a value of 1. Net values less than or equal to the threshold produce a 0. This kind of operation is propagated throughout the network to produce a configuration of output values in response to a given set of inputs.

Learning involves automatically adjusting these weights and thresholds to account for all input and output pairs presented to the system. Having learned these patterns, the system should reproduce the appropriate output pattern when presented with only one of the input patterns. One of the issues of interest in the context of design is what happens when the system is presented with an input pattern that it has not seen before. This is explored in the next section. As their name implies, PDP systems lend themselves to parallel computer architectures, though the whole process can be simulated with iterative computer programs using standard architectures. Parallel distributed-processing machines are considered by some to resemble closely the operations of the human brain at a micro level. As pointed out by McClelland, Rumelhart, and Hinton (1986), this leads to interesting and useful analogies between cognition and computing, as well as interesting explanations of cognitive phenomena. We summarize some of these features here. Once a PDP system has learned a set of input and output patterns, the system can be regarded as operating "holistically" in producing interpretations of patterns presented to it. Input patterns are not reduced to primitives for analysis, but are processes in totality. The process of discovering similarities and categorizing input features is essentially automatic.

PDP models are a good way of simulating the human capacity for pattern completion. Successful matches between input and output patterns can still be accomplished with imperfect input patterns. Even when there is poor or ambiguous input, the system will always produce some form of output—perhaps an amalgamation of outputs, or perhaps the output corresponding to the most closely matching stored input pattern. It could be said that the knowledge degrades gracefully at the boundaries of its domain.

Unlike rules, produced through other learning paradigms, the content of the PDP network is not generally meaningful on inspection. The learned examples, constituting a kind of knowledge base, however, are not lost but can be reconstructed from the network. An interesting parallel between the PDP method of storage and that used in the brain is that storage of information is not local. Destruction or modification of one part of the network generally results in uniform degradation of the performance of the system rather than the loss of any particular item of memory. Further parallels may be drawn with cognitive behavior at a macro level. Rumelhart and McClelland (1986) describe an interesting experiment in which it was possible to replicate the "behavior" of children learning the past tense of verbs. The same sequence of standardizing and "over-regularizing" can apparently be observed in human behavior and the behavior of the PDP model.

Without delving into psychology, we can observe a simple correspondence between design tasks and the PDP model. The task of recalling output patterns to match inputs is similar to the task of interpreting designs. A PDP system appears to perform most effectively when there are unique input and output pattern pairs, or when there are several input patterns to one output pattern. (No input pattern has more than one corresponding output pattern—there are only one-to-one or many-to-one mappings of input to output.) The input pattern can be seen as a description of a design. The output is its performance, and several designs may have the same performance. The interpretation of designs is a relatively straightforward process. The reverse operation, abduction, is less direct.

8.5.2   A Simple Model of Pattern Association

There are many models of parallel distributed processing. We describe a simple model here based on the pattern associator of Rumelhart and McClelland (1986). Although the model is not as powerful as some, the approach is simple to grasp and provides a good starting point for considering other models. In this model there are input and output units and directed arcs between every input and output unit. There are no hidden units, which, as explained later, would be used to detect features. The model can be understood by considering the simple network in Figure 8.22, in which there are three input units and one output unit. Each input unit is connected to the output unit with a directed arc of weight 1. The output unit has an associated threshold value (θ) of 0. If we present this network with a set of inputs, then we would expect a particular output value to be generated. This value is calculated by summing the product of each input unit with each weight (to produce a net input)

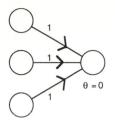

Figure 8.22   A network with weights and a threshold value (θ).

and comparing this with the threshold of the output unit. If the net input
is a greater value than the threshold, the system produces a 1 as output. If
the value of the net input is less than or equal to the threshold, the system
generates a 0 output value. As shown in Figure 8.23, the input pattern {1 0
1} produces 1 (the net sum is 2, which is greater than the threshold value 0):
the input pattern {0 1 0} also produces 1 (the net sum is 1, which is greater
than the threshold value 0).

Suppose we require the system to generate 0 in response to the input
pattern {0 1 0}. In other words, we wish to teach the system to respond to {0
1 0} with a 0 as output. By inspecting the network of Figure 8.24, it should be
apparent that we wish to decrease the influence of weights from input units
with value 1. The algorithm that can be applied is as follows.

1. Calculate the value of the output unit produced from the given input
   pattern.

2. Compare this predicted value with the teaching output presented to the
   system.

3. Inspect each of the input values in turn.

   If   the input value is 1, the predicted output is 1,
           and the teaching output is 0
   then   subtract 1 from the weight of the arc emanating from that unit
           and add 1 to the threshold of the output unit.

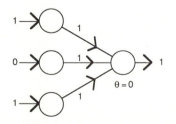

Figure 8.23   A network with inputs and the resultant output.

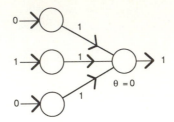

Figure 8.24   The network of Figure 8.23 with a different input.

By decreasing the weight and raising the threshold of the output unit, it becomes more likely that the system will produce a 0 as output. This results in the revised network of Figure 8.25. If we now present this system with the input pattern {0 1 0}, it produces the required output pattern of {0}.

We should check, of course, that the previous input {1 0 1} still gives the desired output of {1}. As shown in Figure 8.26, this is the case. We also need a rule for adjusting weights and thresholds in the event that we require the output to be 1 and the system predicts a 0:

> **If**   the input value is 1, the predicted output is 0,
>   and the teaching output is 1
> **then**   add 1 to the weight of the arc emanating from that unit and
>   subtract 1 from the threshold of the output unit.

Of course, there is no change to weights and thresholds if the teaching and predicted outputs are the same. The resultant network (Figure 8.27) provides the required match for both input-output patterns.

Thus there are two operations in the use of a PDP system: (1) when the system is taught various input and output patterns—the *teaching* operation— and (2) when we present the system with the input only and it generates an output—the *simulation* or *matching* operation. The teaching operation involves cycling through the set of given input-output patterns. The computation time obviously increases with the number of examples presented.

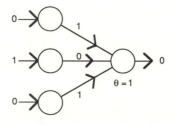

Figure 8.25   The network of Figure 8.24 with weights and threshold value adjusted to produce a different output.

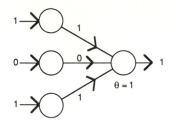

Figure 8.26   The network of Figure 8.25 producing the same behavior shown in Figure 8.23.

Simulation is a simpler operation, in which computation time is independent of the number of teaching examples.

In the learning operation it is usual to cycle through the algorithm several times because the adjustment of weights and thresholds may cause the network to "forget" a pattern it has just learned. When there is no overlap between the input patterns—that is, when the input patterns are linearly independent, as in this example—the algorithm is guaranteed to find a system of weights and thresholds to store all input and output pattern pairs. When any input node has the same value (1 or 0) in more than one teaching pattern (the most usual case), no such assurances can be given. We now describe how stochastic methods can be used to improve the performance of the system in the most general case.

Rumelhart and McClelland (1986) refine this model further. The difference between the net input and the threshold to an output unit actually carries some useful information. The difference can be seen not only as a gauge of whether the output should be 0 or 1, but it also gives some indication of the strength of "belief" in the output value. It can be seen as a measure of the probability that the output unit will "fire." In simulation we could therefore produce a "shaded" output to indicate non-binary measures of certainty in the output. There is a convenient function for distributing the probability from 0 to 1, calculated from the difference between the net inputs and the output threshold. Figure 8.28 shows the *logistic function* for mapping the difference between the input and the threshold to a unit (*Diff*) onto a probability value

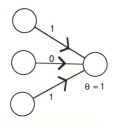

Figure 8.27   The network after it has learned the two input and output patterns.

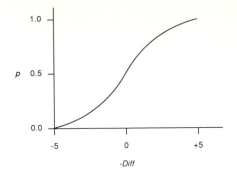

Figure 8.28   The logistic function for mapping the difference between the input and the threshold (*Diff*) onto a probability value (*p*). The slope of the curve varies according to a constant T. Very small values of T produce a stepped function.

(p) depending on a constant T, which determines the slope of the curve. The function is

$$p = \frac{1}{1 + e^{-Diff/T}}$$

When similarities exist in the input patterns—that is, where different input patterns have certain units in common—the probabilistic values of the output units can reflect the overlapping nature of the corresponding outputs.

The operations of a PDP system based on this model are interesting to observe. A peculiar problem that a system may encounter is that during successive cycles the system may change certain weights in response to a particular input-output pair in one cycle, only to undo the effect in another cycle in response to another input-output pair. The system can become "locked" into a set of weights that is not optimal and that is influenced by the order in which input-output pairs are presented to the system. This can be overcome with one further refinement to the model.

A common technique is to introduce controlled randomness into the learning process. As just described, when the PDP system learns, it adjusts weights and thresholds on the basis of the predicted value of an output node and the desired output. Using the logistic function, we can calculate the likelihood of the predicted value being a 1 or 0 and generate a 1 or 0 at random according to this probability. Surprisingly, this introduction of randomness over a large number of learning cycles produces a more accurate result. Over many learning cycles, a smoother series of weights and thresholds is produced. The process allows the system to learn more slowly. The analogy is often made between controlled randomness and thermodynamic principles. The constant factor T in the preceding logistic function can be seen as temperature. The higher the temperature of the system, the more random the predicted values at output units and the greater the means of escaping from a set of weights and thresholds that is sub-optimal. With this stochastic approach the difference between net values and thresholds becomes a better measure of the degree of certainty attached to the output.

There are many possible variations to this model. One enhancement is to introduce hidden units into the network, which in essence increases the number of parameters by which patterns can be stored. With large numbers of hidden units, there is more chance that the system will generalize on features rather than locationally bound binary patterns. A simple bench test of the ability of a system to learn features is the "shifter problem" (Hinton and Sejnowski, 1986). A PDP system is "taught" an input pattern in which a 1 is contained within a string of 0's. The output pattern contains a 1 surrounded by 0's but located one position to the right of the corresponding position in the input pattern. Several such patterns are given. If we then present the system with a new input pattern, when there is a 1 in a new position it should produce an output pattern with the 1 shifted to the right. The system has then learned to generalize on this feature of always transposing a 1 to the right. The simple pattern associator we described is unable to perform this learning feat. A system that makes use of hidden units is required.

With hidden units, the task of finding a set of weights and thresholds to account for input and output pairs is more difficult, involving the solution of a complex optimization problem. Probabilistic approaches are used here, too, as exemplified in the Boltzmann machine (Hinton and Sejnowski, 1986). We will not explore these models further, but rather demonstrate the application of the simpler pattern-associator approach.

8.5.3   Pattern Completion

The binary nature of input-output patterns just demonstrated may appear somewhat limiting. If we configure these patterns into two-dimensional arrays, however, the application to visual perception and computer-aided design becomes apparent. Here we explore the use of two kinds of patterns: spatial patterns across a grid and patterns of relationships on a matrix. In design tasks, we are usually interested in mappings in both directions between descriptions and performances. Thus we blur the distinction between input and output patterns. We also explore the use of a simple feedback mechanism to decide between closely matching patterns.

In the preceding discussion we assumed that there is a set of input and output patterns to be learned. If we make the input and output patterns exactly the same, and present the system with a partial input pattern, we would expect the pattern to be completed in the output pattern. User interaction is simplified if we regard each unit as both input and output. From the point of view of the system, which units we designate design as descriptions and which indicate performances then become arbitrary. This means that the tasks of

Figure 8.29   A set of learned patterns.

interpretation and abduction can be reduced to that of *pattern completion*, an idea that is also exploited in applications to visual perception.

Pattern completion in a graphical domain is demonstrated with the example in Figure 8.29. Nine patterns are learned. (In this example, as in those that follow, there were 100 learning cycles, and the temperature was set at 5 degrees Celsius). Following the learning phase, we present a partially complete pattern to the system (Figure 8.30). The system produces the complete pattern of Figure 8.31. The "noise" in the image is due to the fact that the partial pattern of Figure 8.30 matches more than one of the patterns of Figure 8.29, though clearly one is more dominant than the rest. A certain amount of noise is also attributable to inaccuracies inherent in the stochastic process. (Subsequently we discuss how the noise can be reduced considerably.)

It is interesting to observe what happens when the system is given an ambiguous pattern during simulation—that is, when the partial pattern contains fragments of several different learned patterns. In this case the output is a combination of the closest matching patterns; a kind of "fuzzy union." This result is not always informative. We can reinforce the closest match by

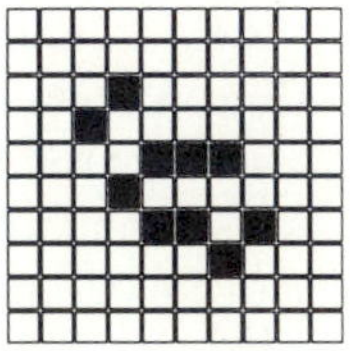

Figure 8.30   A partial pattern for matching against the "memory" of Figure 8.29.

Figure 8.31 Completion of the pattern of Figure 8.30.

feeding the output pattern back into the system, not as grey values but as the closest binary values. The result of cycling though this feedback process for a particular input pattern six times is shown in Figure 8.32. This operation reduces the noise due to overlapping and finds the image that matches the partial pattern most closely. (We could say that this process is analogous to an ill-formed idea undergoing some kind of refinement. It also acts as a kind of reinforced learning.) As we demonstrate later in this section, this process of iterative feedback provides a mechanism for handling some of the difficulty of abduction.

There are limited set-theoretic aspects to the performance of a PDP system. These can be exploited readily if we abandon the iterative feedback idea temporarily. In Figure 8.33 the units are arranged into two configurations: an association matrix on the left and a gridded building site on the right. Figure 8.34 shows the distribution of various attributes across the site. These attributes are presented to the system during the learning phase. The interesting feature about the patterns on the left is that each attribute is represented

Figure 8.32 The results of iterative feedback. The original pattern is at the top left corner. After successive iterations, the system "homes in" on the closest matching pattern.

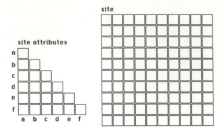

Figure 8.33   A system for organizing site information.

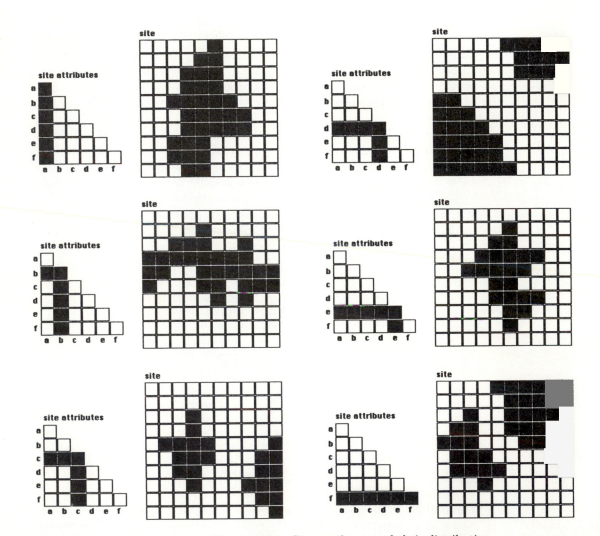

Figure 8.34   Site attributes and their distribution.

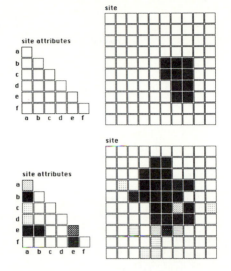

Figure 8.35   A partial pattern and the resultant match.

by the same number of units in the matrix and that there is only one unique unit to each attribute (on the diagonal of the matrix).

If we present an arbitrary spatial configuration of units to the system, it should produce a pattern on the association matrix that tells us what proportion of each attribute is represented. Figure 8.35 shows an application of this idea. The system is presented with a partial pattern in the form of a building plan. In "completing" the pattern the system fills in the site-attributes matrix. We can see most of that the building occupies parts of the site with attributes a and e, while c and d are not well represented. The completed site drawing gives an indication of the envelope within which the building can be moved before the site attributes change appreciably. Figure 8.36 shows another example, in which attributes a and b are well represented but f is poorly represented. This is probably as far as we can take this application. There are superior numerical methods of accomplishing these kinds of operations, and the results here are not entirely predictable. Of course, a larger site grid and a greater number of learning cycles produce an even better result.

In certain applications, set union and intersection are less useful than in this example. An interesting design application is matching floorplan layouts to their performances. Because of the gross nature of pattern completion, it is useful to think in terms of matching performances and types, specifically building-plan forms. Figure 8.37 shows how the units can be configured to represent a set of plan forms and their attributes (performances). The attributes are a combination of high-level plan descriptors and the opinions of one designer as to the suitability of the various forms for certain uses and siting conditions. We show several plan types and their attributes. For the

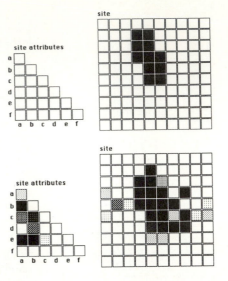

Figure 8.36   A partial pattern and the resultant match.

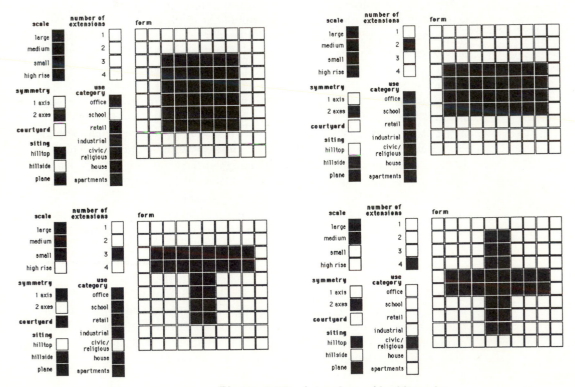

Figure 8.37   A typology of building plans.

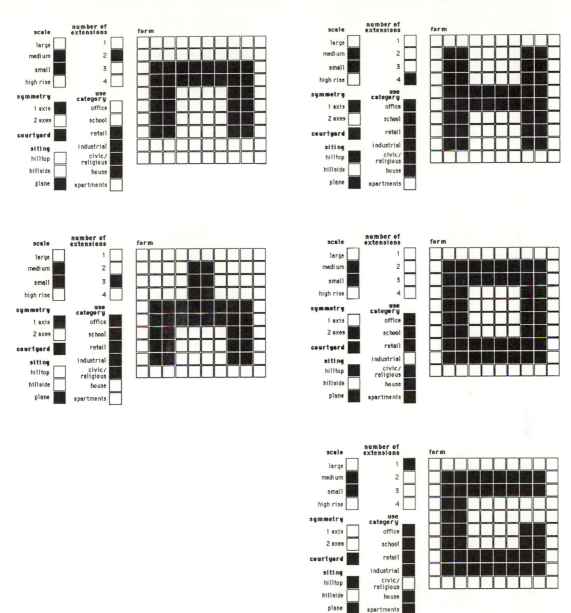

Figure 8.37 Continued.

system to be really useful, we should teach it all possible orthogonal plan forms—a large number, but certainly enumerable. If we then present the system with any one of the plans, it will generate the attributes. It is interesting that the plan "drawing" need not be perfectly accurate in order to find the right attribute match. Figure 8.38 shows a progression from the inaccurate, partial pattern presented to the system and the result after several feedback iterations.

The abductive operation presents the system with a set of attributes that then generates a plan form. If we exploited the set-theoretic properties of the system just demonstrated, we would produce loosely drawn unions of the closest matching plan types. There are three ways of treating this information. First, it can be seen as a loose kind of synthesis for producing new types (assuming that not all types are stored). Unfortunately the combination of plans rarely produces attributes that are simply derived from the union or intersection of the attributes of each of the plans. Second, the overlapping patterns may serve to indicate the closest matching patterns from which the operator of the system can make a choice. Third, we can use the iterative feedback operation to produce the closest matching plan. Figure 8.39 demonstrates this approach.

In Figure 8.39 we show the stages in the iterative feedback process by which the system "homes in" on the closest matching plan form. The iterative process need not be visible to the operator, of course, but it is informative here in showing what the system is doing. We can make three important observations about this operation, in which we provide a partial pattern of only a small number of performances during simulation. The operation is abductive and we are attempting to find a matching form when there are in fact several forms. The most prominent forms appear shadowed after the first iteration in Figure 8.39. Also, we are attempting to complete a pattern in which we supply the system with only a very small pattern fragment, considerably increasing the noise in the result. Additionally, the partial pattern may closely match with several plan forms. On successive iterations the system produces a plan shape that closely matches the combination of these plans. The resultant form may have very different performances from those specified in the original partial pattern. This difficulty can be overcome by introducing some kind of weighting factor into the representation so that one unit of performance equals several units on the form grid. For some applications, of course, the combination of plans may be very useful.

In the system described here, generalizations are made on the basis of absolute positions on a grid. There is no mechanism for detecting features that exist independently of absolute positions. For example, the system would be unable to induce an implicit rule that simple forms have the largest number of possible uses, or that an upside-down T form has exactly the same properties as a T that is right-side up.

In essence, the PDP approach described here is concerned with a simple form of analogical reasoning with a store of design performance and de-

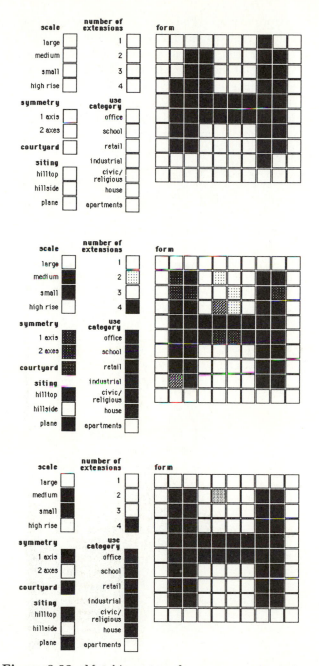

Figure 8.38 Matching a type from an approximate image.

scription pairs. The approach has been to find the closest match between the current task and those stored in the PDP network. There are, however, conventional database search techniques for achieving similar ends. The advantage of the approach here is that a suitable match can generally be found even when performance requirements are poorly defined. The ill-definedness of goals is an important characteristic of design, and it is clear that PDP models can be of assistance.

A major implementational advantage of the approach is that the time taken to retrieve patterns from a PDP network does not increase with the number of examples stored. When stored patterns overlap in certain ways, there is a degradation in the quality of what can be retrieved, just as when people are faced with such a task. There would appear to be a limit to the subtlety of the information that can be stored for a given number of units. Some of the limitations of the simple pattern-associator model described here

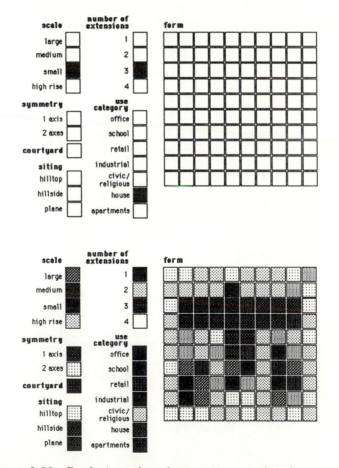

Figure 8.39   Producing a form from a given set of performances.

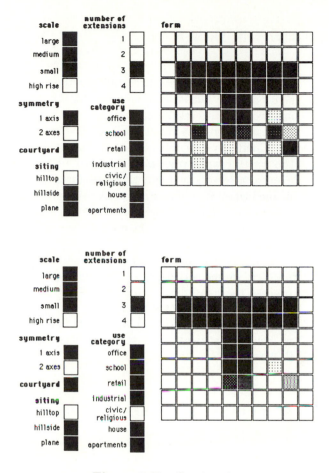

Figure 8.39 Continued.

can be overcome with the introduction of more complex models that make use of hidden units.

In this discussion we have considered only the acquisition of interpretive knowledge—that which maps decisions to performances. More attention must be directed to the other aspect of design knowledge—the application to generative knowledge. Design appears to be concerned with the changing of ideas under the influence of actions. Design involves much more than just pattern matching, though the PDP approach may have a lot to say about the mechanism by which patterns within partial design descriptions trigger change and development.

Much of what we have just discussed will prove useful in the next section, in which we discuss creativity in knowledge-based design systems.

8.6   Creativity in Design

In this book we do not presuppose any particular role for computers in design practice. The issue of creativity is important in whatever way computers are brought into service in design practice—as "impartial tools," "intelligent assistants," or "automated-design" systems.

Creativity can mean many different things. Chomsky (1975) identified the use of rewrite rules as a kind of "rule-governed creativity." We generally expect, however, that creativity requires more than the application of well-defined rules. It is important to identify different models of creativity to relate the concept to knowledge-based systems. We discuss two different views of creativity before settling on one that we regard as useful for knowledge-based design.

The term creativity may be discussed in relation to both product and process. In the former case, we mean that an artifact is the product of a process, the creativity of which is assessed by the quality of the product. In the latter case, we mean that a process is designated as creative, irrespective of the quality of the product. Both approaches are discussed in this section, although we believe that the latter is more applicable to a discussion of knowledge-based design.

8.6.1   Creativity According to Product

There are three ways in which an artifact can be said to be the product of a creative process: in relation to the degree to which it is innovative, in relation to its value, and according to the richness of interpretations that can be placed on it.

Innovation

If an artifact is different in some way from the products of other designers (or design systems), we may say that an artifact is the product of a designer who is particularly creative. We may qualify this by saying that the designer is creative if something is produced that never existed before. If we have a way to measure the originality of artifacts, then we can compare one with another, and determine which designer is more creative than others. In this view creativity is dependent on historical context. To "invent" the steam engine in 1989 would not be regarded as very creative—nor for that matter, would the rediscovery of the general theory of relativity, even if it were derived by someone whose mathematical training did not extend beyond high-school level.

A novel artifact may also serve as a prototype, effectively defining a class of things. Hence a certain invention—such as the barrel vault, the rotary engine, or the spring-loaded clothespin—is regarded as the product of a highly creative mind (or group of minds) in that it provides the forerunner to a whole body of variations.

Value

Another view of creativity is that it is related to the value of an artifact in some way. We may describe a designer as creative if something is produced that is generally regarded as of high value, by whatever criteria we may wish to employ, such as beauty, simplicity, sophistication, suitability for the intended purpose, or marketability.

Richness of Interpretation

As well as exhibiting qualities of high value, an artifact may be multi-valued. Certain artistic endeavors thrive on ambiguity (Lansdown, 1985; Cohen, Cohen, and Nii, 1985). An artifact may be regarded as the product of a creative process if it can be interpreted in many different ways: "there is always something new to discover!" This may be manifested as the ability of a design to assume many roles such that it can serve several purposes, including those for which it was not intended originally. Richness can also be evident in terms of the social and other messages that the design may convey.

8.6.2 Creativity According to Process

A further explanation of creativity is to see it as an attribute of an information process. This view of creativity as concerned with the quality of a design, measurable in terms of both its product and its process, is important because of its mapping with the way we are often accustomed to talking about creativity in design. This aspect of creativity is considered under the headings of entropy, efficiency, and richness.

Entropy

A design system may be regarded as creative if it produces a complex artifact description from information that contains the "seeds" of the design to only a very small degree at the outset, and if the initial state exhibits a low level of information content. Hence the "re-invention" of the steam engine by someone with no knowledge of mechanisms and of physics may be regarded as more creative than a similar process carried out by an engineer. The process by which a schoolboy comes up with the general theory of relativity might rank highly in terms of creativity because his information would be at a fairly rudimentary state. The fact that the discovery had already been made by someone else would not significantly affect our view of his creative ability. (We may think, however, that he had wasted his time.)

A computer system that selects from a database of building designs might therefore be regarded as less creative than a system that generates descriptions by combining building elements, and then selecting from among them. In the former case, the designs are already in existence. In the latter, the designs must be both generated and evaluated. There is also a link between the idea of creativity and that of generality. The more general the descriptions available to a system, the greater the capacity of the system to be creative. This view of creativity is closely connected to Shannon and Weaver's definition of *entropy* (Shannon and Weaver, 1949). Entropy can be measured in terms of the amount of organization, or structure, exhibited between the components of a system. A high entropy implies little organization. The higher the entropy of a design system, the greater its creative potential.

Efficiency and richness, two other aspects of process, qualify the issue of entropy as a criterion of creativity.

Efficiency

Efficiency has a bearing on the degree to which we may describe a system as creative. If a rule-based generative system exhibits a high entropy because of the unstructured nature of its processes, it may be regarded as more creative than a highly structured system. If designs are not realized in a finite time, however, we would probably not regard the system as creative.

Richness

If a design system is capable of producing only one design for a given context in which other designs could serve equally well, then the system may be regarded as less creative than one that can generate a range of designs. Richness in terms of process may also be exhibited if a system is able to accommodate a wide range of task environments. This also reflects the adaptability of a system.

8.7   Creativity in Knowledge-Based Design Systems

In this section we develop the definition of creativity based on process, relating creativity to knowledge. We identify those aspects of the design process that provide a key to realizing creative design systems. These are drawn from an examination of strategies employed by human beings during their creative acts.

8.7.1   Design Spaces that Change

It is commonly held that creative individuals are those who are able to break out of conventional modes of thinking (Altshuller, 1984). This view was popularized by deBono (1967) in his use of the term "lateral thinking." Various views on the subject of creativity, based on psychological models, are provided by Vernon (1970), Storr (1972) and Koestler (1976). In terms of knowledge-based systems, we can say that creativity means going beyond the bounds of our knowledge, extending the space of design possibilities beyond that dictated by our knowledge (Gero, Maher, and Zhao, 1988b).

Because design knowledge is not usually well-articulated, a search through a space of designs is therefore especially difficult—like trying to find one's way across an ever-changing platform of logs floating down a river. The shape of the design space appears to be determined as much by the search process itself as by external factors. The search results in modifications to the character and boundaries of the space. The definition of design spaces is therefore dynamic, and the design process involves not only search within such spaces but the acquisition and modification of the knowledge that defines those spaces.

We have suggested that learning plays a major role in creative design. Design systems acquire knowledge from other designs, as well as from their own products. How to model the dynamic aspect of design knowledge poses difficulties. It may involve exploring a space of design possibilities while holding on tenaciously to a set of knowledge as embodying certain constraints, then relaxing those constraints in order to search further afield. It may mean modifying assumptions as the design is progressing—in other words, reasoning non-monotonically (Sections 4.7 and 6.2.6). It would appear to require reasoning under uncertainty as suggested by plausible inference (Section 6.2.3) and fuzzy reasoning (Section 6.2.4). Creative design systems must embrace and integrate a range of reasoning approaches.

8.7.2   Human Creativity and Knowledge-Based Design Systems

We can examine various strategies employed by people during their creative acts and posit analogous computational processes (Maher, Zhao, and Gero, 1988). There have been numerous publications on human creativity—for example, Mansfield and Busse (1981), Talor (1975), Amabile (1983), Gruber, Terrell, and Wertheimer (1962), Osborn (1957), deBono (1973), Sandler (1985), and Buhl (1960). The subject has drawn the attention of psychologists, educators, philosophers, sociologists, architects, engineers, and others, all of whom perform research from different perspectives. Designers, however, may be the most practical researchers on the subject of creativity. They are mainly interested in the methods that can be used to design more creatively. Understanding creativity in terms of its psychological aspects will help us to construct useful problem-solving mechanisms because of the strong connection between psychology and artificial intelligence.

Some people believe creativity to be at least to some extent a gift with which a creative person is born rather than a learned ability. Some people believe creativity to be similar to intuition, without any structures or patterns, and unexplainable. Psychologists have observed, however, a regular demonstration of certain personalities and traits possessed by creative people (Mansfield and Busse, 1981; MacKinnon, 1970). Some of these traits are not much different from those of less creative people but seem to be necessary conditions for creativity, while others are demonstrated more strongly by creative people. These include the following.

Above-average intelligence. These studies show that creative professionals usually have substantially above-average intelligence. Intelligence gives a person the ability to organize information in a useful way and to apply this knowledge to solve problems. Obviously, a creative person should have good problem-solving abilities, though a highly intelligent person is not necessarily creative.

Extensive training. For any professional to be successful in a field, he or she should have very good knowledge about the field, in both depth and breadth. Such knowledge may be gained through formal and informal education.

These two conditions are necessary for creativity, but many people who are not creative or less creative may have the same qualities. Therefore there must be some additional qualities that particularly promote creativity, including the following.

Openness to experience. This is a quality opposed to that of defensiveness. People who are open to experience do not refuse to see or hear anything not in compliance with their own experiences or concepts, nor do they let incoming stimuli be distorted by preconceptions and bias. Instead, creative people are willing to engage in adventure and experimentation, and to seek new facts, interesting phenomena, and fresh ideas, no matter how odd they may look at first glance. This openness means both a greater chance of being exposed to novel ideas and a flexible way of thinking that increases this chance.

Autonomy. Creative people, although open to experience and different opinions, are not necessarily influenced by them, even if the opinions are given by authorities or experts. Anything will be accepted if judged valid. This is an essential trait for creative people, since in a complex field there is so much conflict and confusion that if a person can not maintain his or her autonomy, the person would never have a basis on which a creative work can be built.

Ability to play with ideas. Creative people have a good ability to manipulate concepts and ideas. They can change, distort, relate, combine. and dismantle them in many novel ways. Such playing with ideas sometimes may result in discoveries, inventions, and novel ideas. It is not an arbitrary and blind manipulation, however, but has some certain patterns that are determined by the thinking techniques or creative problem-solving approaches used. Creativity also appears to involve the maintenance of multiple views and communication among the abstractions represented by those views.

Aesthetic sensibility. Creative people are sensitive to the beauty of patterns, structures, ideas, and theories. They want to find not only ways to achieve their goals, but also the best ways—sometimes the simplest and most efficient mechanism to implement a desired notion, sometimes a unified theory that can explain many different phenomena.

In addition to these well-recognized characteristics of creative people, there are some others such as commitment to work and the desire to be original (Mansfield and Busse, 1981). Commitment and diligence appear to be essential for success in serious and difficult tasks. Works of genius may take years or even decades to accomplish, and may require much personal sacrifice such as lack of short-term fame or material comfort. There are certainly strong motives behind such commitments. The motive can be the desire to be creative, to be successful, to be recognized, or simply enjoyment and satisfaction. This last motive appears to be strongest of all: it is hard to imagine that a person who does not enjoy his or her work would achieve any significant accomplishment.

All these personal characteristics are frequently observed in creative people, which is not to say that they are completely missing in less creative people. The latter may have the two necessary qualities of above-average intelligence and extensive training, but miss some of the other qualities (or these other qualities may not be strong enough). Every person, however, has the potential to be creative (Schank and Childers, 1988), and the more creative characteristics the person possesses, the better the chance that he or she can be creative.

So far we have discussed creativity and the particular personal characteristics possessed by creative people. Now let us compare people and computers based on their potential for creativity. This comparison does not imply that if a computer program can possess characteristics similar to those of creative people, it will have the basis for being similarly creative. A creative work of art is infused with the artist's feelings. If a computer is to create a work of art, the mechanism must be quite different since we cannot describe an artist's complex feelings in a computer language. The comparison, however, can give us ideas about what some essentials may be needed. Only after such an exploration can we decide which directions to pursue.

The two basic conditions for professional creativity are above-average intelligence and extensive training. Although computers can perform certain specific tasks better than average people, as in the case of expert systems, no computer can do many of the simple tasks that a person with average intelligence can. Such a comparison suggests that for computers to carry out tasks of average complexity that people with average intelligence can perform, they need not have "above-average intelligence." Such a comparison, however, is not necessarily appropriate. Just as we can make computers capable of some quite complex tasks without any general intelligence—for example, expert systems—we can probably make them demonstrate some potential for creativity as well. The need to demonstrate a particular creative potential by a knowledge-based design system may be provided by certain advanced problem-solving techniques as well as by adequately encoded knowledge.

What is possible, however, is to draw an analogy between the extensive education and training of people and a knowledge base and experience base in a computer system. The knowledge base contains theories, as well as general and special methods and techniques useful in a particular domain. The experience base records the uses of the knowledge—that is, what knowledge is applied in what situations, and the results. As with people, however, an extensive knowledge base or experience base is not sufficient for creativity. Many knowledge-based design systems do not produce very creative designs (see, for example, Maher, 1984; Rosenman, Coyne, and Gero, 1987).

Relaxing constraints in a computer system may be seen as corresponding to the openness of creative people. By relaxing constraints, a knowledge-based design system can produce more solutions. Also, by delaying decision taking or evaluation, many solutions that might otherwise be determined inferior based on incomplete information may be retained further development. Navinchandra (1987) reports an attempt to build an innovative design system by relaxing constraints.

The ability of people to play with ideas may be simulated by certain computerized problem-solving techniques. We can program a computer to produce numerous variations of a concept, provided this concept can be expressed in a manipulatible form. We can also program a computer to search in its knowledge base and draw any relations between two concepts. Interesting variations and analogies of different concepts are probably the primary sources of creative products.

Aesthetic sensibility is extremely difficult to encode in computer programs. Aesthetic sensibility is useful for a designer when evaluating an idea or a product for good form. Knowledge-based systems that evaluate a design using non-numerical criteria—for example, structural and architectural compatibility—tend to quantify the criteria as a ranking based on recognized patterns of features (Jain and Maher, 1988). Certain evaluations that involve personal feelings, such as in architectural design or works of art, may be extremely difficult or impossible for computers to perform.

One very important difference between people and computers lies in their learning ability. People learn subconsciously and apparently continuously, while computers currently have weak learning abilities. Since creation is a process of producing new products that never existed before in that form—that is, they move from the unknown to known—learning is involved. Creative people must be very good at learning because they can acquire new ideas that do not occur to other people. For a computer to be able progressively to improve itself and produce novel solutions, it should have some minimum learning ability. AM (Lenat, 1982), although not a design system, is an example of a program that can learn new concepts.

One advantage of computers over people is that computers do not require personal motives to drive them. They can be programmed to work constantly. Therefore, if creativity can be obtained by applying the right knowledge and techniques persistently, it may be a good task for computers.

8.7.3 Modeling Human Strategies for Creativity

In this section we discuss techniques for creative thinking, which along with other thinking techniques have been described by deBono (1973), Buhl (1960), Osborn (1957), and Sandler (1985). Sometimes these techniques are classified or labeled in slightly different ways. Perhaps the most promising approach for developing creative knowledge-based design systems is to adapt for computers the well-known creative thinking techniques that people find useful. The techniques presented can be or have shown the potential to be used in computer systems. We describe six such techniques and briefly discuss the relation of each one to existing or possible artificial intelligence techniques.

Trial and Error

Trial and error, a conventional method, is applied to a wide variety of problems, resulting in both routine and creative solutions. The process of trial and error involves generating a potential solution and then testing the solution. This process is repeated until a solution passes the test. When a solution fails to pass the test, this is because of errors that make it invalid. The errors can be recognized and be useful in finding a new potential solution in the next attempt that is free of these errors. A computer version close to trial and error is "generate and test." Generate and test has two important components: a mechanism that can automatically generate solutions for a given problem, and an evaluation method that can be used to test whether a solution is valid. The process is similar to that of trial and error, in which a cycle of generating a solution and testing it continues until a valid solution is obtained. There is a significant difference, however, between a person using the trial-and-error method and the computer application of generate and test. That is, people can learn well from their errors while computers cannot, or learn only poorly. Recently, there have been interesting developments in artificial intelligence in incremental learning that begin to address this issue (Burstein, 1988).

Brainstorming

Brainstorming is a technique suitable for group thinking. For a particularly difficult problem, people with good knowledge of the problem who have different backgrounds and different technical expertise participate in a brainstorming session, during which they are encouraged to think freely and present their ideas without any restrictions or criticism. The advantage of working as a group rather than individually is that each person in the group is exposed to other people's ideas, which may stimulate creative thinking. SOAR (Laird, Newell, and Rosenbloom, 1987) manifests some characteristics of brainstorming by providing a mechanism for defining distinct knowledge sources. When a problem is specified, all relevant knowledge from the separate sources is invoked and put into the working memory. Brainstorming, however, is not simply throwing in available ideas, but a process during which new ideas emerge and seemingly irrelevant ideas become related—that is, creation is involved. Therefore, for a computer to be able to simulate a brainstorming process, it needs not only diverse knowledge about problems, possibly represented as multiple knowledge sources, but also the ability to generate new relations among the existing knowledge elements and new concepts.

Mutation

Mutation is the deliberate action of changing features or attributes of an object or concept in an unconventional manner. By "unconventional" manner we mean that the change is not restricted by the usual rules and constraints. The purpose of mutation is to find new properties, functions, and meanings of an existing object or concept by modifying it. Mutation can be performed in many different ways. Some variations of this technique are multiplication, division, addition, and subtraction performed on the features, functions, structures, and mechanisms of concepts. Mutation is one process to produce prototype adaptation, described in Section 2.6.2. Computerization of mutation is relatively easy if the representation is appropriate. AM (Lenat, 1982; Lenat and Brown, 1983) is a successful example of applying simple mutation in a computer program. Because we cannot yet incorporate all appropriate knowledge in a knowledge base, it is difficult to obtain meaningful mutation operations. This often makes mutation seem somewhat arbitrary. Without appropriate control, its nature is to some extent opportunistic and success cannot be guaranteed. However, knowledge-based design systems based on mutation look promising as a process to produce potentially creative design (McLaughlin and Gero, 1988).

Analogy

Analogy is a particularly useful approach for solving an unfamiliar new problem without adequate or directly applicable knowledge. By analogy, relations between the new problem and some past experience or knowledge about a particular design can be found. This experience or knowledge can be placed in the new situation so that the new problem can be better understood or a new plan for solving the problem can be generated. Analogy is not only an effective problem-solving approach, but also an effective learning vehicle (see Section 8.4). Because analogy does not require the two episodes or concepts compared to bear a close resemblance, it leaves room for imagination and thus creativity. In order to use analogy effectively, however, considerable experience and knowledge related to the problems are needed, and powerful methods for relating past experience and new problems are essential. Currently, the analogical approach in computer systems is an active research area in artificial intelligence and its application (Prieditis, 1988). Based on the nature of the knowledge transfer from old experience to the new problem, the analogical reasoning approach can be classified into two categories: transformational analogy and derivational analogy (Carbonell, 1983, 1986). Transformational analogy adapts the solutions to the past problems for the new problem. Derivational analogy applies the past problem-solving processes or methods to solve the new problem.

Inversion

Inversion, also called reversion, is a technique for creative thinking by reversing the cause and effect of a phenomenon. A typical example is the movement of a conductor in a magnetic field. When the conductor is moved in the magnetic field transversely, a current is generated in the conductor. Conversely, if a current passes through a conductor, a magnetic field is produced around the conductor. To apply inversion in knowledge-based design systems, designs must be described in terms of causes and effects. Moreover, in order to be able to infer a reverse relation, more complicated knowledge is required about relationships among the causes and effects. The techniques for reasoning qualitatively, reasoning based on first principles, and physical models are in a primary research stage. PROMPT (Murthy and Addanki, 1987) is an attempt to develop a tool for innovative design of physical systems that reasons using first principles.

Systematic Approach

This approach is applied by moving from the end to the beginning of a problem (Sandler, 1985). If an appropriate final goal can be formulated, a solution may be obtained by finding all the possible ways of satisfying this goal. This is a recursive process, in which finding possible ways of satisfying the final goal results in several sub-goals, and all the possible ways of satisfying the sub-goal induce new sub-sub-goals. It is important that the new goal only reflect the final goal, not introducing unnecessary restrictions that may reduce the chance of producing creative solutions. The computer implementation requires causal reasoning techniques. At present there are some versions of systematic approach such as sub-goaling and synthesis (for example, Howe et al., 1986). The implementations, however, allow only for a predefined goal decomposition and predefined groupings of elements for synthesis, thereby possibly precluding creative design solutions.

8.8   Summary

In this chapter we have briefly considered how learning can be modeled in knowledge-based design systems. We discussed learning as an isolated phenomenon, in which it is possible to acquire both interpretive and generative knowledge. The two approaches we considered are learning by the successive discrimination of examples on the basis of their attributes and learning through knowledge about generalization. We also considered analogical reasoning, in which knowledge is stored in the form of prior design "experiences," episodes, and explanations. We argued that learning, analogical reasoning, and multiple abstractions of the design process provide keys to understanding and appropriating creativity in knowledge-based design systems.

We examined creativity from a human perspective, describing computer implementations of various processes that appear to be involved in human creativity.

Clearly, to address these key issues, we must consider their integration into systems. Throughout this book we have mainly discussed issues in isolation. This serves a useful didactic purpose, but also indicates the state of the art. The integration of ideas poses an enormous challenge, of greater order than the individual ideas. We expect the development of theories and models pertinent to knowledge-based design to extend well beyond what has been described in these pages. We hope that this book will stimulate further research and development in that direction.

8.9   Further Reading

Two major texts on machine learning that collect contributions that largely define the field are Michalski, Carbonell, and Mitchell (1983 and 1986). A good introductory text on machine learning is Forsyth and Rada (1986). Reasoning from episodes or case-based reasoning was the subject of a recent workshop (Kolodner, 1988).

Very little definitive information has been published on knowledge-based creative design. Interesting ideas can be found in Lansdown (1987), Schank

(1986), and Schank and Childers (1988). Conference proceedings that contain relevant material are Fox and Navinchandra (1988) and Gero and Oksala (1989).

Exercises

8.1 What is the role of learning in knowledge-based design systems?

8.2 What is the essential difference between the two learning approaches introduced in Section 8.2? To what situations are each best suited? What is their application to learning in design?

8.3 What are the difficulties in acquiring syntactic design knowledge? How can these difficulties be overcome in design systems?

8.4 List the different ways in which designers appear to reason analogically. Describe attempts that have been made to automate analogical reasoning. What are the difficulties in modeling this kind of reasoning?

8.5 What are the requirements of a knowledge-based design system that is to behave creatively? What is the role of learning in such a system?

8.6 Discuss the issue of integration in knowledge-based design systems. Can integration be achieved? What are some of the theoretical issues involved?

The Conventions of Logic

First-order predicate calculus is a system of logic that provides a formal language for representing knowledge and making processes of deduction explicit. In this book we have adopted the conventions of this language to explain certain features about representation and reasoning. Some of the terminology was introduced informally in Section 2.1. Here we review the subject in slightly more detail.

Formal logic is essentially concerned with the form or syntax of logical statements, and with the determination of truth by the manipulation of formulas. *Propositional calculus* is the simplest mathematical formalism of logic. Statements (also referred to as axioms, clauses, or sentences) consist of propositions and sentential connectives. Propositions are simply statements that can be considered either true or false. *Sentential connectives* combine propositions and are represented by various conventions, some of which are listed here.

$\wedge$	and
$\vee$	or
$\sim$	not
$\Rightarrow$	implies
$\Leftarrow$	if
$\Leftrightarrow$	if and only if
$\equiv$	equals
()	parentheses
[ ]	brackets

Throughout the book we frequently used words rather than symbols to represent sentential connectives. Also, we adopted the *if-then* construct, as well as the use of the symbols $\Rightarrow$ and $\Leftarrow$. We should note in passing that the following statements are logically equivalent:

If X **then** Y.

Y **if** X.

The preceding symbols enable us to construct sentences such as

$$(X \Rightarrow (Y \wedge Z)) \equiv ((X \Rightarrow Y) \wedge (X \Rightarrow Z)).$$

This statement can be interpreted as follows: X implies Y and Z is equivalent to the statement that X implies Y and X implies Z, where X, Y, and Z are sentential constants for which any logical sentence can be substituted. These constants can be replaced by groups of words that carry some meaning, such that a sentence such as the following is produced: Cloudy skies imply rain and thunder is the same as saying that cloudy skies imply rain and cloudy skies imply thunder. Such tautologies become more useful when they make it possible to deduce other statements of value.

General rules of inference, by which such sentences can be manipulated, enable new sentences to be deduced from previously given sentences. The most important rule is *modus ponens*, which states that if two sentences of the form X and X implies Y are true then it can be inferred that the sentence Y is true:

$$(X \wedge (X \Rightarrow Y)) \Rightarrow Y$$

Predicate calculus is an extension of propositional calculus. Using predicate calculus, sentences about objects or sentences that describe relationships between objects can be constructed and manipulated. Predicate calculus also makes it possible to generalize about objects and relationships by means of variables. Statements about individual objects, or about relationships between objects, can be represented in the general predicate form

$$X(a_1, a_2, ..., a_i, ..., a_n),$$

where X is the head of a predicate that is a word (or atom) that carries some descriptive meaning, and a_i is an argument. The entire predicate has a value of either true or false. Examples of predicates and their values follow:

LessThan(3, 5)	true
IsRed(FireEngine)	true
Supports(Column, Beam)	true
GreaterThan(3, 5)	false

Arguments can be variables, in which case the true or false values of the predicates will not be known until the variables are instantiated. (The convention here is that arguments beginning with a lowercase character are variables. Arguments beginning within uppercase characters are constants. The head of a predicate cannot be a variable.) An advantage of using predicates of this form is that they can assume whatever interpretation one wishes to place on them, yet the predicate and its arguments lend themselves to rigorous symbolic manipulation.

A further qualification of logic statements can enhance their expressive power. The device of making explicit the nature of the variables employed in statements by means of quantification allows us to handle the expression of such tautologies as every person has a mother. The logical statement

$$\text{Person}(x) \Rightarrow \text{Mother}(y)$$

might be interpreted erroneously as any person has any mother. It must be made clear that the statement refers to all people, but only a particular mother for each person. Quantifiers make clear whether statements are universally true for all values of a variable, or whether they are true for a particular instance of a variable. Quantifiers are defined as follows:

The universal quantifier: $\forall x[p]$,
where p is a statement containing the variable x and is true for all x.

The existential quantifier: $\exists x[p]$,
where p is a statement containing the variable x which is true for some instance of x.

The preceding sentence describing the person/mother relationship can therefore be written as

$$\forall x \exists y[\text{Human}(x) \rightarrow \text{Mother}(y)]$$

As discussed next, it is possible to write logic statements in a form that dispenses with the need to make these distinction about variables.

First-order logic introduces functions or operators into the formalism. Functions resemble predicates, but they are more powerful in that they can return values for variable arguments. This is one of the major operations within logic programming. Values are known as substitution instances; the process of discovering substitution instances for two expressions that will make them equivalent is known as *unification*. There is generally no distinction between functions and predicates in logic programming. Let us return to an example given in Section 2.2.1. An example of unification is that in which we determine the values of the variables that will make the following statement true:

$$\text{Object}(\text{Building}(b), \text{NoOfStoreys}(s), \text{Location}(loc), \text{Owner}(\text{Govt})) \equiv$$
$$\text{Object}(\text{Building}(\text{Office}), \text{NoOfStoreys}(6), \text{Location}(\text{City}(\text{Cbd}, \text{South})), q)$$

In order for this expression to be true, the following substitution instances must be made:

$$
\begin{aligned}
b &= \text{Office} \\
s &= 6 \\
loc &= \text{City(Cbd, South)} \\
q &= \text{Owner(Govt)}
\end{aligned}
$$

A procedural algorithm for unification was originally described by Robinson (1965). Such an algorithm is also described by Nilsson (1982).

As a further refinement on the use of logic, we introduce the convention of *Horn clauses*. There is a certain redundancy built into the syntactic conventions of the logic language so far described. The same set of statements can be expressed in many ways, and it is possible to rewrite any set of statements in logic in a form known as *clausal form*. This form eliminates the need for the implies and the existential and universal quantifiers. The major sentential connectives retained are "and" and "if". All variables are assumed to be universally quantified. A further refinement on clausal form is the use of Horn clauses (named after their most influential exponent). These take the following form:

$$ B \Leftarrow A_1, ..., A_i, ..., A_n $$

The *head* of the clause, B, is true if the *body* of the clause consisting of the terms $A_1, ..., A_n$ is true. It can be demonstrated that little is lost of the expressive power of logic with this form, and considerable advantages are gained when it comes to automating proof procedures (Kowalski, 1979). It is possible to restate any declaration in standard logic in this form.

Conversion to clausal form involves substitutions based on resolution rules that are employed in theorem proving. (Although these rules are often called *rules of inference*, that term can also refer to the Horn clauses themselves.) The following rules are given a full explanation by Nilsson (1981).

$$
\begin{aligned}
\sim(\sim X) &\equiv X \\
(X \Rightarrow Y) &\equiv \sim X \vee Y
\end{aligned}
$$

de Morgan's laws:
$$
\begin{aligned}
\sim(X \wedge Y) &\equiv \sim X \vee \sim Y \\
\sim(X \vee Y) &\equiv \sim X \wedge \sim Y
\end{aligned}
$$

Distributive laws:
$$
\begin{aligned}
X \wedge (Y1 \vee Y2) &\equiv (X \wedge Y1) \vee (X \wedge Y2) \\
X \vee (Y1 \wedge Y2) &\equiv (X \vee Y1) \wedge (X \vee Y2)
\end{aligned}
$$

Commutative laws:

$X \wedge Y$	$\equiv$	$Y \wedge X$
$X \vee Y$	$\equiv$	$Y \vee X$

Associative laws:

$(X \wedge Y1) \wedge Y2$	$\equiv$	$X \wedge (Y1 \wedge Y2)$
$(X \vee Y1) \vee Y2$	$\equiv$	$X \vee (Y1 \vee Y2)$

Contrapositive law:

$X \Rightarrow Y$	$\equiv$	$\sim Y \Rightarrow \sim X$

Quantifier laws:

$\sim \exists x P$	$\equiv$	$\forall x [\sim P]$
$\forall x P$	$\equiv$	$\exists x [\sim P]$
$\forall x [P \wedge Q]$	$\equiv$	$\forall x P \wedge \forall x Q$
$\exists x [P \vee Q]$	$\equiv$	$\exists x P \vee \exists x Q$

(P and Q are statements containing the variable x.)

With these rules, it is possible to convert a statement in standard form to other forms, including Horn clause form.

Example Frame Systems in LISP and PROLOG

B.1   A Simple Frame System in LISP

The following functions carry out the basic operations of a frame system. Function `fget` returns information given the frame, slot, and facet:

```
(defun fget (frame slot facet)
    (mapcar 'car
        (cdr (assoc facet
            (cdr (assoc slot
                (cdr (get frame 'frame)))))))))
```

Function `fput` places information given the frame, slot, and facet:

```
(defun fput (frame slot facet value)
    (cond ((member value (fget frame slot facet)) nil)
    ((fassoc value
        (fassoc facet
            (fassoc slot
                (fgetframe frame)))) value)))
```

Function `fremove` removes a particular value given the frame, slot, and facet:

```
(defun fremove (frame slot facet value)
    (prog (slots facets values target)
        (setq slots (fgetframe frame))
        (setq facets (assoc slot (cdr slots)))
        (setq values (assoc facet (cdr facets)))
        (setq target (assoc value (cdr values)))
        (delete target values)
        (cond ((null (cdr values)) (delete values facets)))
        (cond ((null (cdr facets)) (delete facets slots)))
        (return (not (null target)))))
```

Function `fgetframe` returns the frame representation given an object:

```
(defun fgetframe (frame)
    (cond ((get frame 'frame))
          (t (putprop frame (list frame) 'frame))))
```

Function `fgetclasses` finds all other frames that are related to a given frame:

```
(defun fgetclasses (frame)
    (prog (queue progeny classes)
    (setq queue (cons frame (fget frame 'is_a 'value)))
    tryagain
    (cond ((null queue) (return (reverse classes)))
       ((not (member (car queue) classes))
          (setq classes (cons (car queue) classes))))
          (setq progeny (append (fget (car queue)
                            'ako 'value)
             (car (fget (car queue) 'memberOf 'value))))
             (setq queue (cdr queue))
             (setq queue (append queue progeny))
             (go tryagain)))
```

Function `fget_z` finds a value by exploiting defaults; `if_needed` demons through the `ako` path in zig-zag style:

```
(defun fget_z (frame slot)
    (prog (classes result)
    (setq classes (fgetclasses frame))
```

```
      loop
      (cond ((null classes) (return nil))
            ((setq  result
                (or (fget (car classes) slot 'value)
                    (mapcan '(lambda (e) (apply e nil))
                  (fget (car classes) slot 'if_needed))
                    (fget (car classes) slot 'default)))
                (return result))
            (t (setq classes (cdr classes))))
      (go loop)))))

(defun fassoc (key a_list)
    (cond ((assoc key (cdr a_list)))
        (t (cadr (rplacd (last a_list) (list (list key)))))))
```

Function `show_frame` displays the contents of a frame:

```
(defun show_frame (frame)
   (prog (slots facets values)
   (cond ((null (setq slots (cdr (fgetframe
                        frame)))) (return nil)))
   (terpr)
   (princ '|Frame : |)
   (printl frame)
   (terpr)
   (mapc 'print_slots slots)
   (return t)))
```

Function `print_slots` displays all the slots of a frame:

```
(defun print_slots (s)
   (prog (slot facets)
   (setq slot (car s) facets (cdr s))
   (tab 3)
   (princ slot)
   (princ '| : |)
   (terpr)
   (mapc 'p_facet facets)
   (terpr)))
```

Function `print_facet` displays the information associated with different facets:

```
(defun print_facet (s)
   (prog (facet values)
   (setq facet (car s) values (apply 'append (cdr s)))
```

```
(tab 12)
(princ facet)
(princ '| : |)
(cond ((onep (length values)) (printl (car values)))
        (t (printl values)))))
```

Example Frames in LISP

```
(defprop building (building
    (ako (value (structure)))
    (number_of_stories (default (10)))
    (color (default (white)))) frame)

(defprop myhouse (myhouse
    (ako (value (building)))
    (number_of_rooms   (value (7)) (default (10)))
    (color (value (red)))) frame)
```

An Example Session

This example uses the two frames just defined. User input is shown in bold.

```
=> (show_frame 'myhouse)
Frame : myhouse
        ako :
            value : building
        number_of_rooms :
            value : 7
            default : 10
        color :
            value : red


=> (fget 'myhouse 'color 'value)
(red)


=> (fput 'myhouse 'number_of_stories 'value 2)
2


 => (show_frame 'myhouse)
Frame : myhouse
        ako :
            value : building
        number_of_rooms :
            value : 7
            default : 10
```

```
                color :
                     value : red
                number_of_stories :
                     value : 2
```

=> **(show_frame 'building)**
```
Frame : building
           ako :
                value : structure
           number_of_stories :
                default : 10
           color :
                default : white
```

=> **(fput 'building 'shape 'default 'rectangle)**
```
rectangle
```

=> **(show_frame 'building)**
```
Frame : building
           ako :
                value : structure
           number_of_stories :
                default : 10
           color :
                default : white
           shape :
                default : rectangle
```

=> **(fget_z 'myhouse 'shape)**
```
(rectangle)
```

B.2   A Simple Frame System in PROLOG

The following predicates carry out the basic operations of a frame system. Predicate `fget` returns information given the frame, slot, and facet:

```
fget(Frame, Slot, Facet, Value) :-
    X =..[Frame, Slot, Facet, Value],
    X, !.
fget(_, _, _, unknown).
```

Predicate `fput` places information given the frame, slot, and facet:

```
fput(Frame, Slot, Facet, Value) :-
    fget(Frame, Slot, Facet, Value), !.
fput(Frame, Slot, Facet, Value) :-
    X =.. [Frame, Slot, Facet, Value],
    assertz(X).
```

Predicate `fremove` removes a particular value given the frame, slot, and facet:

```
fremove(Frame, Slot, Facet, Value) :-
    X =.. [Frame, Slot, Facet, Value],
    retract(X), !.
```

Predicate `fgetclasses` finds all frames that are related to a given frame:

```
fgetclasses(Frame, Classes) :-
    ini_list(Frame, L), getclass(L, Classes).

ini_list(Frame,[Frame|[X]]) :-
    fget(Frame,is_a,value, X), X \== unknown, !.
ini_list(Frame,[Frame]).

getclass([],[]).
getclass([H|T],[H|C]) :-
    fget(H, ako, value, X), make_list(X, X1),
    fget(H, memberOf, value, Y), make_list(Y, Y1),
    append(X1, Y1, Z), append(Z, T, V),
    getclass(V, C).

make_list(unknown,[]) :- !.
make_list(X, [X]) :- (number(X) ; atom(X)),
!. make_list(X, X) :- is_list(X), !.
```

Predicate `fget_z` finds a value by exploiting defaults; `if_needed` demons through `ako` path in zig-zag style.

```
fget_z(Frame, Slot, X) :-
    setq(frame, Frame), setq(slot, Slot),
    fgetclasses(Frame, Classes),
aux_fget_z(Classes, Slot, X).

aux_fget_z([],_,unknown) :- !.
aux_fget_z([H|T], Slot, X) :-
    fget(H, Slot, value, X), X \== unknown, !.
```

```
aux_fget_z([H|T], Slot, X) :-
    fget(H, Slot, if_needed, Y), Y \== unknown, !,
    Z =.. [Y, X],call(Z).
aux_fget_z([H|T], Slot, X) :-
    fget(H, Slot, default, X), X \== unknown, !.
aux_fget_z([_|T], Slot, X) :-
     aux_fget_z(T, Slot, X).
```

Predicate `show_frame` displays the contents of a frame:

```
show_frame(Frame) :-
    get_slots(Frame, Slots),
    Slots \== [], !,
    nl, write('frame :'), write(F), nl,
    print_slots(Frame, Slots).

show_frame(Frame) :-
    write(Frame), write(' frame does not exist'),
    nl, nl.
```

Predicate `print_slots` displays all the slots of a frame:

```
print_slots(Frame, []).
print_slots(Frame, [H|T]) :-
    print_slot(Frame, H),
    print_slots(Frame, T).
```

Predicate `print_slot` displays a particular slot of a frame:

```
print_slot(Frame, Slot) :-
    tab(5), write(Slot), write(' : '),nl,
    X =.. [Frame, Slot, Facet, Value],  X, tab(9),
    write(Facet),
    write(' : '),write(Value), nl,
    fail.

print_slot(_, _) :- !.
```

Example frames in PROLOG

```
building(ako, value, structure).
building(number_of_stories, default, 10).
building(color, default, white).

myhouse(ako, value, building).
myhouse(number_of_rooms, value, 7).
myhouse(number_of_rooms, default, 10).
myhouse(color, value, red).
```

An Example Session

This example uses the two frames just defined. User input is shown in bold.

```
| ?- show_frame(myhouse).
frame : myhouse
        ako :
                value : building
        number_of_rooms :
                value : 7
                default : 10
                color :
                value : red

yes


  | ?- fget(myhouse, color, value, X).
X = red


  | ?- fput(myhouse, number_of_stories, value, 2).
yes


  | ?- show_frame(myhouse).
frame : myhouse
        ako :
                value : building
        number_of_rooms :
                value : 7
                default : 10
        color :
                value : red
        number_of_stories :
                value : 2
yes


  | ?- show_frame(building).
frame : building
        ako :
                value : structure
        number_of_stories :
                default : 10
        color :
                default : white
yes


  | ?- fput(building, shape, default, rectangle).
yes
```

```
 | ?- show_frame(building).
frame : building
        ako :
             value : structure
        number_of_stories :
             default : 10
        color :
             default : white
        shape :
             default : rectangle
yes

 | ?- fget_z(myhouse, shape, X).
X = rectangle
```

A Simple Rule Interpreter in LISP and PROLOG

C.1   A Simple Rule Interpreter in LISP

The following functions carry out the basic operations in a rule interpreter.

Forward-Chaining Procedure

Function `infer` carries out a forward-chaining procedure:

```
(defun infer (facts)
    (setq Facts facts)
    (infer_all))
```

Function `infer_all` continually scans the rules and fires the appropriate rules. It terminates when no more rules become available for firing:

```
(defun infer_all ()
    (prog (progress)
        loop
        (cond ((fire_rule) (setq progress t)
            (t (return progress)))
        (go loop)))
```

533

Function `fire_rule` selects a rule in which all the preconditions are satisfied and the consequent is not proved yet, and fires that rule:

```
(defun fire_rule ()
    (prog (rule_list)
          (setq rule_list rules)
          loop
          (cond ((null rule_list) (return nil))
                (and (not (member (conseq (car rule_list)) Facts)
                     (subset (antece (car rule_list)) Facts))
                (update_facts (conseq (car rule_list)))
                (return t))
                (T (setq rule_list (cdr rule_list))))
          (go loop)))
```

Function `subset` returns `T` if all the elements in the first list are in the second list:

```
(defun subset (list1 list2)
    (cond ((null list1) 'T)
          ((member (car list1) list2)
               (subset (cdr list1) list2))
          (T 'nil)))
```

Function `update_fact` updates the fact base with new inferred facts:

```
(defun update_fact(fact)
    (print fact) (terpri)
    (setq Facts (append1 Facts fact)))
```

Function `antece` returns the antecedent part of a rule:

```
(defun antece (rule)
    (cadr rule))
```

Function `conseq` returns the consequent part of a rule:

```
(defun conseq (rule)
    (cadddr rule))
```

Backward-Chaining Procedure

Function `prove` carries out the backward-chaining procedure:

```
(defun prove (prop)
    (prog ()
        (cond ((prove_rule prop rules)
            (print prop) (terpri) (return t))
            (t (ask_user prop)))))
```

Function `prove_rule` finds a rule in which the given proposition is the consequent and all of its antecedents are satisfied. If such rule is found, then it returns t, otherwise, it returns nil:

```
(defun prove_rule (prop rule_list)
    (prog (rule)
    loop
    (setq rule (car rule_list))
    (cond ((null rule_list) (return nil))
          ((and (equal prop (conseq rule))
                (prove_all (antece rule)) (return t)))
    (setq rule_list (cdr rule_list))
    (go loop)))
```

Function `prove_all` proves a list of propositions:

```
(defun prove_all (props)
   (prog ()
   loop
   (cond ((null props) t)
         ((not (prove (car props))) (return nil))
   (setq props (cdr props))
   (go loop)))
```

Function `ask_user` proposes a query to the user:

```
(defun ask_user (prop)
    (and (print prop)
    (cond ((equal 'yes (read)) t))))
```

Example Rules in LISP

```
(setq rules '(

    (if (shape_is_cylinder extends_floor_to_ceiling)
     then column)

    (if (plan_is_circle elevation_is_rectangle)
    then shape_is_cylinder)))
```

An Example Session with LISP

This example uses the two rules above. User input is shown in bold.

=> (infer '(plan_is_circle elevation_is_rectangle

```
     extends_floor_to_ceiling))
shape_is_cylinder
column
t

=> (prove 'column)
plan_is_circle? yes
elevation_is_rectangle? yes
extends_floor_to_ceiling? yes
column
t
=>
```

C.2   A Simple Rule Interpreter in PROLOG

These operators define the arity of and, if and then.

```
:-op(780, xfy, and).
:-op(950, fx, if).
:-op(900, xfx, then).
```

Forward-Chaining Procedure

Predicate infer carries out the forward chaining procedure:

```
infer(A) :- assert_all(A), infer_all.
```

Predicate infer_all selects a rule in which all the preconditions are satisfied and the consequent is not proved yet, and fires that rule. This procedure is repeated until no rule becomes available:

```
infer_all :- (if C then D), not(D), prove_all(C),
    write(D), nl, assert(D), infer_all.
infer_all.
```

Predicate assert_all puts all the given facts into memory:

```
assert_all(A and B) :- assert(A), assert_all(B).
assert_all(A) :- assert(A).
```

Predicate match checks whether all the preconditions of a rule are satisfied:

```
not(A) :- A, !, fail.
not(A).
```

Backward-Chaining Procedure

Predicate prove carries out the backward-chaining procedure:

```
prove(A) :- A.
prove(A) :- (if B then A), prove_all(B), write(A), nl, assert(A).
prove(A) :- ask_user(A).
```

Predicate prove_all proves a list of propositions:

```
prove_all(A and B) :- prove(A), prove_all(B).
prove_all(A) :- prove(A).
```

Predicate ask_user proposes a query to the user:

```
ask_user(A) :- write(A), write('?'), read(Yes), Yes == 'yes',
  assert(A).
```

Example Rules in PROLOG

```
if shape_is_cylinder and extends_floor_to_ceiling
then column.

if plan_is_circle and elevation_is_rectangle
then shape_is_cylinder.
```

An Example Session with PROLOG

This example uses the two preceding rules. User input is shown in bold.

```
?- infer(plan_is_circle and elevation_is_rectangle
  and extends_floor_to_ceiling).
shape_is_cylinder
column
yes

?- prove(column).
plan_is_circle? yes.
elevation_is_rectangle? yes.
extends_floor_to_ceiling? yes.
column
yes.
```

References

Abelson, R. P., and Black, J. B. (1986). Introduction to knowledge structures, *in* J. A. Galambos, R. P. Abelson, and J. B. Black (eds.), *Knowledge Structures*, Erlbaum, Hillsdale, New Jersey, pp. 1–18.

Aikins, J. S. (1979). Prototypes and production rules: an approach to knowledge representation for hypothesis formation, *IJCAI-79*, Tokyo, Japan, pp. 1–3.

Akin, O. (1978). How do architects design? *in* J.-C. Latombe (ed.), *Artificial Intelligence and Pattern Recognition in Computer Aided Design*, North-Holland, Amsterdam, pp. 65–119.

Akin, O. (1979). *Models of Architectural Knowledge: An Information Processing View of Architectural Design*, Ph.D. Thesis, Department of Architecture, Carnegie Mellon University, Pittsburgh.

Akin, O. (1986). *Psychology of Architectural Design*, Pion, London.

Akiner, T. (1985). TOPOLOGY1: *A System that Reasons about Objects and Spaces in Buildings*, Ph.D. Thesis, Department of Architectural Science, Sydney University, Sydney.

Alexander, C. (1964). *Notes on the Synthesis of Form*, Harvard University Press, Cambridge, Massachusetts.

Allwood, R. J., and Stewart, D. J. (1984). *Interim Report on Expert System Shells Evaluation for Construction Industry Applications*, Loughborough University of Technology, Loughborough, England.

Altshuller, G. S. (1984). *Creativity as an Exact Science*, Gordon and Breach, New York.

Amabile, T. M. (1983). *The Social Psychology of Creativity*, Springer-Verlag, New York.

Anderson, J. R. (1977). Induction of augmented transition networks, *Cognitive Science*, **1**(2):125–157.

Arbab, F., and Wing, J. M. (1985). Geometric reasoning: a new paradigm for processing geometrical information, *in* H. Yoshikawa and E. Warman (eds.), *Design Theory for CAD*, North-Holland, Amsterdam, pp. 145–159.

Archer, B. L. (1969). The structure of the design process, *in* G. Broadbent and A. Ward (eds.), *Design Methods in Architecture*, Lund Humphries, London, pp. 76–102.

Archer, B. L. (1970). An overview of the structure of the design process, *in* G. T. Moore (ed.), *Emerging Methods in Environmental Design and Planning*, MIT Press, Cambridge, Massachusetts, pp. 285–307.

Asimow, W. (1962). *Introduction to Design*, Prentice-Hall, Englewood Cliffs, New Jersey.

Australian Model Uniform Building Code (1982). AUBRCC Directorate, P.O. Box 111, Dickson, ACT 2602, Australia.

Bachant, J., and McDermott, J. (1984). R1 revisited: four years in the trenches, *AI Magazine*, **5**(3):21–32.

Balachandran, M. (1988). *A Model for Knowledge-Based Design Optimization*, Ph.D. Thesis, Department of Architectural Science, Sydney University, Sydney.

Balachandran, M., and Gero, J. S. (1987a). Dimensioning of architectural plans under conflicting objectives, *Environment and Planning B*, **14**:29–37.

Balachandran, M., and Gero, J. S. (1987b). A knowledge-based approach to mathematical design modelling and optimization, *Engineering Optimization*, **12**(12):99–115.

Balachandran, M., and Gero, J. S. (1988). A model for knowledge-based graphical interfaces, *in* J. S. Gero and R. Stanton (eds.), *Artificial Intelligence Developments and Applications*, North-Holland, Amsterdam, pp. 147–163.

Ballard, D. H., and Brown, C. M. (1982). *Computer Vision*, Prentice-Hall, Englewood Cliffs, New Jersey.

Balzer, R., Erman, L. D., London, P. E., and Williams, C. (1980). HEARSAY-III: A domain-independent framework for expert systems, *Proceedings of the First Annual National Conference on AI (AAAI)*, Stanford University, Stanford, California, pp. 108–110.

Barr, A., and Feigenbaum, E. A. (eds.) (1981). *The Handbook of Artificial Intelligence—Volume 1*, William Kaufmann, Los Altos, California (republished by various other publishers).

Bartlett, F. C. (1932). *Remembering*, Cambridge University Press, Cambridge.

Begg, V. (1984). *Developing Expert CAD Systems*, Kogan Page, London.

Berwick, R. C. (1980). Computational analogues of constraints on grammars: a model of syntactic acquisition, *Proceedings of the 18th Conference of*

the Association for Computational Linguistics, Association for Computational Linguistics, pp. 49–53.

Bijl, A. (1987). An approach to design theory, *in* H. Yoshikawa and E. A. Warman (eds.), *Design Theory in CAD*, North-Holland, Amsterdam, pp. 3–25.

Bø, K., and Lillehagen, F. (eds.) (1983). *CAD Systems Framework*, North-Holland, Amsterdam.

Bobrow, D. G., and Collins, A. M. (eds.) (1975). *Representation and Understanding*, Academic Press, New York.

Bobrow, D. G., and Stefik, M. (1981). *The Loops Manual*, Technical Report KB-VLSI-81-13, Knowledge Systems Area, Xerox Palo Alto Research Center, Palo Alto, California.

Bobrow, D. G., and Winograd, T. (1977a). Experience with KRL-0, one cycle of a knowledge representation language, *IJCAI-77*, Cambridge, Massachusetts, pp. 213–222.

Bobrow, D. G., and Winograd, T. (1977b). An overview of KRL, a knowledge representation language, *Cognitive Science*, 1:3–46.

Brachman, R. J., and Levesque, H. J. (eds.) (1985). *Readings in Knowledge Representation*, Morgan Kaufmann, Los Altos, California.

Broadbent, G. (1981). *Design in Architecture*, Wiley, London.

Brown, D. C., and Chandrasekaran, B. (1985). Expert systems for a class of mechanical design activity, *in* J. S. Gero (ed.), *Knowledge Engineering in Computer-Aided Design*, North-Holland, Amsterdam, pp. 259–282.

Brownston, L., Farrell, R., Kant, E., and Martin, M. (1985). *Programming Expert Systems in OPS5: An Introduction to Rule-Based Programming*, Addison-Wesley, Reading, Massachusetts.

Buchanan, B. G. (1982). New research on expert systems, *in* J. E. Hayes and D. Michie (eds.), *Machine Intelligence 10*, Wiley, London, pp. 269–299.

Buchanan, B. G., and Feigenbaum, E. A. (1978). DENDRAL and Meta-DENDRAL: their applications dimension, *Artificial Intelligence*, 11(1):5–24.

Buchanan, B. G., and Shortliffe, E. H. (eds.) (1984). *Rule-Based Expert Systems*, Addison-Wesley, Reading, Massachusetts.

Buhl, H. R. (1960). *Creative Engineering Design*, Iowa State University Press, Ames.

Buis, M., Hamer, J., Hosking, J. G., and Mugridge, W. B. (1987). An expert advisory system for fire safety code, *in* J. R. Quinlan (ed.), *Applications of Expert Systems*, Addison-Wesley, Sydney, New South Wales, Australia, pp. 85–101.

Burstein, M. (1988). Incremental learning from multiple analogies, *in* A. Prieditis (ed.), *Analogica*, Pitman/Morgan Kaufmann, Los Altos, California, pp. 37–62.

Carbonell, J. (1983). Learning by analogy: formulating and generalizing plans from past experience, *in* R. Michalski, J. Carbonell, and T. Mitchell (eds.), *Machine Learning*, Tioga, Palo Alto, California, pp. 137–161.

Carbonell, J. (1986). Derivational analogy: a theory of reconstructive problem solving and expertise acquisition, *in* R. Michalski, J. Carbonell, and T. Mitchell (eds.), *Machine Learning—Volume II*, Morgan Kaufmann, Los Altos, California, pp. 371–392.

Carbs Ltd., (1985). EAGLE 3D *Modelling System Command Reference Manual*, Cadcom Proprietary Ltd, Greenwich, New South Wales, Australia.

Chalmers, A. F. (1982). *What is this Thing Called Science?* University of Queensland Press, St. Lucia, Queensland, Australia.

Charniak, E., and McDermott, D. (1985). *Introduction to Artificial Intelligence*, Addison-Wesley, Reading, Massachusetts.

Chomsky, N. (1957). *Syntactic Structures*, Mouton, The Hague.

Chomsky, N. (1963). Formal properties of grammars, *in* R. Luce, D. Bush, and E. Galanter (eds.), *Handbook of Mathematical Psychology*, Wiley, New York, pp. 323–418.

Chomsky, N. (1971). Deep structure, surface structure and semantic interpretation, *in* D. Steinberg and L. Jakobovits (eds.), *Semantics*, Cambridge University Press, Cambridge, pp. 183–216.

Chomsky, N. (1975). *The Logical Structure of Linguistic Theory*, Plenum Press, New York.

Clark, K. L., and McCabe, F. G. (1982). PROLOG: a language for implementing expert systems, *in* J. E. Hayes, D. Michie, and Y.-H. Pao (eds.), *Machine Intelligence 10*, Wiley, London.

Clocksin, W. F., and Mellish, C. S. (1981). *Programming in Prolog*, Springer-Verlag, Berlin.

Cluster Titles Committee (1979). *Model Cluster Code*, Department of Local Government, Melbourne, Australia.

Cohen, D. I. A. (1986). *An Introduction to Computer Theory*, Wiley, New York.

Cohen, H., Cohen, B., and Nii, P. (1985). *The First Artificial Intelligence Coloring Book*, Kaufmann, Oxford.

Cohon, J. L. (1978). *Multiobjective Programming and Planning*, Academic Press, New York.

Coyne, R. D. (1986). *A Logic Model of Design Synthesis*, Ph.D. Thesis, Department of Architectural Science, University of Sydney, New South Wales, Australia.

Coyne, R. D. (1988). *Logic Models of Design*, Pitman, London.

Cross, N., Naughton, J., and Walkers, D. (1981). Design method and scientific method, *Design Studies*, **2**(4):195–201.

Dahl, O.-J., Dijkstra, E. W., and Hoare, C. A. R. (1972). *Structured Programming*, Academic Press, London.

Davis, R., and King, J. (1977). An overview of production systems, *in* E. W. Elcock and D. Michie (eds.), *Machine Intelligence 8*, Ellis Horwood, Chichester, England, pp. 300–332.

deBono, E. (1967). *The Use of Lateral Thinking*, Penguin, Harmondsworth, Middlesex, England.

deBono, E. (1973). *Lateral Thinking: Creativity Step by Step*, Harper and Row, New York.

de Kleer, J. (1986). An assumption-based truth maintenance system, *Artificial Intelligence*, **28**(2):127–162.

Dejong, G. and Mooney, R. (1986). Explanation-based learning: an alternative view, *Machine Learning*, **1**(2):145–176.

Dietterich, T. G., and Buchanan, B. G. (1981). The role of the critic in learning systems, *Technical Report* No. HHP-81-19, Department of Computer Science, Stanford University, Stanford, California.

Dietterich, T. G, and Ullman, D. G. (1987). FORLOG: a logic-based architecture for design, *in* J. S. Gero (ed.), *Expert Systems in Computer-Aided Design*, North-Holland, Amsterdam, pp. 1–17.

Downing, F., and Flemming, U. (1981). The bungalows of Buffalo, *Environment and Planning B*, **8**:269–290.

Doyle, J. (1981). A truth maintenance system, *in* B. L. Webber and N. J. Nilsson (eds.), *Readings in Artificial Intelligence*, Tioga, Palo Alto, California, pp. 496–516.

Dreyfus, H., and Dreyfus, S. (1984). Mindless machines, *The Sciences*, November/December, pp. 18–22.

Dreyfus, H., and Dreyfus, S. (1986). Why computers may never think like people, *Technology Review*, January, pp. 41–61.

Duda, R. O., Hart, P. E., and Nilsson, N. J. (1976). Subjective Bayesian methods for rule-based inference systems, *AFIPS Conference Proceedings*, Artificial Intelligence Center, Technical Note 124, **45**:1075–1082.

Duffy, F., Cave, C., and Worthington, J. (eds.), (1976). *Planning Office Space*, Architectural Press, London.

Dyer, M. G., Flowers, M., and Hodges, J. (1986). EDISON: an engineering design invention system operating naively, *Artificial Intelligence in Engineering*, **1**(1):36–44.

Dym, C. L. (ed.) (1985). *Applications of Knowledge-Based Systems to Engineering Analysis and Design*, American Society of Mechanical Engineers, New York.

Earl, C. F. (1977). A note on the generation of rectangular dissections, *Environment and Planning B*, **4**:241–246.

Earl, C. F. (1985). Creating design worlds, *Research Paper*, Centre for Configurational Studies, Open University, Milton Keynes, Buckinghamshire, England.

Eco, U. (1984). *Semiotics and the Philosophy of Language*, Macmillan, London.

Engelmore, R. and Morgan, T. (eds.) (1988). *Blackboard Systems*, Addison-Wesley, Wokingham, England.

Erman, L. D., Hayes-Roth, F., Lesser, V. R., and Reddy, D. R. (1980). The Hearsay-II speech understanding system: integrating knowledge to resolve uncertainty, *Computing Surveys*, **12**(2):213–253.

Erman, L. D., London, P. E., and Fickas, S. F. (1981). The design and an example use of HEARSAY-III, *IJCAI-81*, Vancouver, British Columbia, Canada, pp. 409–415.

Feibleman, J. (1970). *An Introduction to the Philosophy of Charles S. Peirce*, MIT Press, Cambridge, Massachusetts.

Fenves, S. J., and Baker, N. C. (1987). Spatial and functional representation language for structural design, *in* J. S. Gero (ed.), *Expert Systems in Computer-Aided Design*, North-Holland, Amsterdam, pp. 511–530.

Fenves, S. J., and Liadis, S. A. (1976). A data structure for computer-aided design of buildings, *APEC Journal*, Fall, pp. 14–18.

Fenves, S. J., and Wright, R. N. (1977). The representation and uses of design specifications, *NBS Technical Note 940*, The National Bureau of Standards, U.S. Department of Commerce.

Fikes, R., and Kehler, T. (1985). The role of frame-based representation in reasoning, *Communication of the ACM*, **28**(9):904–920.

Fikes, R., and Nilsson, N. J. (1971). STRIPS: A new approach to the application of theorem proving to problem solving, *Artificial Intelligence*, **2**(3–4):189–208.

Flemming, U. (1978). Wall representations of rectangular dissections and their use in automated space allocations, *Environment and Planning B*, **5**:215–232.

Flemming, U., Coyne, R., Glavin, T., and Rychener, M. (1986). A generative expert system for the design of building layouts, *in* D. Sriram and R. Adey (eds.), *Applications of Artificial Intelligence in Engineering Problems— Volume II*, Springer-Verlag, Berlin, pp. 811–822.

Forgy, C. L. (1981). *OPS5 User's Manual*, Technical Report, Department of Computer Science, Carnegie Mellon University, Pittsburgh.

Forsyth, R., and Rada, R. (1986). *Machine Learning: Applications in Expert Systems and Information Retrieval*, Ellis Horwood, Chichester, England.

Forwood, B. S. (1971). Computer simulated lift design: analysis for office buildings, *Architectural Science Review*, **14**(2):41–47.

Fox, M., and Navinchandra, D. (eds.) (1988). *Artificial Intelligence in Design: Proceedings of the AAAI Workshop*, American Association for Artificial Intelligence, Menlo Park, California.

Fu, K. S., and Booth, K. L. (1975). Grammatical inference: introduction and survey, parts 1 and 2, *IEEE Transactions Sys. Man. Cybern*, **SMC-5**(1):95–111, (2):409–423.

Garvey, T. D., Lowrance, J. D., and Fischler, M. A. (1981). An inference technique for integrating knowledge from disparate sources, *IJCAI-81*, Vancouver, British Columbia, Canada, pp. 319–325.

Gaschnig, J. (1982). Application of the PROSPECTOR system to geological exploration problems, *in* J. E. Hayes, D. Michie, and Y.-H. Pao (eds.), *Machine Intelligence 10*, Wiley, London, pp. 301–323.

Georgeff, M. P., and Lansky, A. L. (eds.) (1987). *Reasoning about Actions and Plans*, Morgan Kaufmann, Los Altos, California.

Genesereth, M. R., and Nilsson, N. J. (1987). *Logical Foundations of Artificial Intelligence*, Morgan Kaufmann, Los Altos, California.

Gero, J. S. (1977). Note on 'Synthesis and Optimization of Small Rectangular Floor Plans' of Mitchell, Steadman and Liggett, *Environment and Planning B*, **4**:81–88.

Gero, J. S. (ed.) (1985a). *Knowledge Engineering in Computer-Aided Design*, North-Holland, Amsterdam.

Gero, J. S. (ed.) (1985b). *Optimization in Computer-Aided Design*, North-Holland, Amsterdam.

Gero, J. S. (ed.) (1987a). *Expert Systems in Computer-Aided Design*, North-Holland, Amsterdam.

Gero, J. S. (1987b). *Prototypes*: a new schema for knowledge–based design, *Working Paper*, Architectural Computing Unit, Department of Architectural Science, University of Sydney, Sydney.

Gero, J. S. (ed.) (1988a). *Artificial Intelligence in Engineering: Design*, Elsevier/CMP, Amsterdam.

Gero, J. S. (ed.) (1988b). *Artificial Intelligence in Engineering: Diagnosis and Learning*, Elsevier/CMP, Amsterdam.

Gero, J. S. (ed.) (1988c). *Artificial Intelligence in Engineering: Robotics and Processes*, Elsevier/CMP, Amsterdam.

Gero, J. S., Akiner, V. T., and Radford, A.D. (1983). What's what and what's where—knowledge engineering in the representation of buildings by computer, *PARC83*, Online, Middlesex, England, pp. 205–215.

Gero, J. S., Maher, M. L., and Zhang, W. (1988a). Chunking structural design knowledge as prototypes, *in* J. S. Gero (ed.), *Artificial Intelligence in Engineering: Design*, CMP/Elsevier, Amsterdam, pp. 3–21.

Gero, J. S., Maher, M. L., and Zhao, F. (1988b). A model for knowledge-based creative design, *in* M. Fox and D. Navinchandra (eds.), *Workshop on Artificial Intelligence in Design*, American Association of Artificial Intelligence, Menlo Park, California, unnumbered.

Gero, J. S., and Oksala, T. (eds.) (1989). *Knowledge-Based Systems in Architecture*, Acta Polytechnica Scandinavica, Helsinki.

Gips, J., and Stiny, G. (1980). Production systems and grammars: a uniform characterization, *Environment and Planning B*, **7**:399–408.

Goldberg, A., and Kay, A. (1976). *Smalltalk-72*, Instruction Manual SSL-76-6, Xerox Palo Alto Research Center, Palo Alto, California.

Goodman, N., and Quine, W. V. (1947). Steps toward a constructive nominalism, *The Journal of Symbolic Logic*, **12**:105–122.

Gregory, S. (ed.) (1966). *The Design Method*, Butterworth, London.

Gruber, H. E., Terrell, G., and Wertheimer, M. (1962). *Contemporary Approaches to Creative Thinking*, Prentice-Hall, Englewood Cliffs, New Jersey.

Hammond, P. (1982). Logic programming for expert systems, *Technical Report*, DOC 82/4, Imperial College of Science and Technology, London.

Hammond, P., and Sergot, M. (1983). *A PROLOG Shell for Logic Based Expert Systems*, Imperial College of Science and Technology, London.

Harmon, P., and King, D. (1985). *Expert Systems: Artificial Intelligence in Business*, Wiley, New York.

Hay, R. D. (ed.) (1984). *DZ4226: 1984 Draft New Zealand Standard Design for Fire Safety*, Standards Association of New Zealand, Wellington.

Hayes-Roth, B., Garvey, A., Johnson, M. V., and Hewett, M. A. (1986). Application of the BB1 blackboard control architecture to arrangement assembly tasks, *Artificial Intelligence in Engineering*, **1**(2):85–94.

Hayes-Roth, B., and Hayes-Roth, F. (1979). A cognitive model of planning, *Cognitive Science*, **3**(4):275–309.

Hayes-Roth, F., and Lesser, V. R. (1977). Focus of attention in the Hearsay-II System, *IJCAI-77*, Cambridge, Massachusetts, pp. 27–35.

Hayes-Roth, F., Waterman, D. A., and Lenat, D. B. (eds.) (1983). *Building Expert Systems*, Addison-Wesley, Reading, Massachusetts.

Head, H. (1926). *Aphasia and Kindred Speech Disorders*, Cambridge University Press, Cambridge.

Heath, T. (1984). *Method in Architecture*, Wiley, Chichester, England.

Hewitt, C. (1972). Description and theoretical analysis of PLANNER, a language for providing theorems and manipulating models in a robot, *Research Report*, TR-258, Artificial Intelligence Laboratory, MIT, Cambridge, Massachusetts.

Hillier, B., Musgrove, J., and O'Sullivan, O. (1972). Knowledge and design, *in* W. J. Mitchell (ed.), *Environmental Design: Research and Practice*, University of California, Los Angeles, pp. 29.3.1–29.3.14.

Hinton, G. E., and Sejnowski, T. J. (1986). Learning and relearning in Boltzmann machines, *in* D. E. Rumelhart and J. L. McClelland (eds.), *Parallel Distributed Processing: Explorations in the Microstructure of Cognition—Volume 1, Foundations*, MIT Press, Cambridge, Massachusetts, 282–314.

Holden, T. (1987). *Knowledge Based CAD and Microelectronics*, North Holland, Amsterdam.

Hopgood, F. R. A., and Duce, D. A. (1978). A production system approach to interactive graphic program design, *Report*, Atlas Computing Design, Rutherford Appleton Laboratory, Didcot, Oxon, England.

Howe, A. E., Cohen, P. R., Dixon, J. R., and Simmons, M. K. (1986). DOMINIC: a domain-independent program for mechanical engineering design, *Artificial Intelligence in Engineering*, **1**(1):23–28.

Hunt, E. B., Martin, J., and Stone, P. J. (1966). *Experiments in Induction*, Academic Press, New York.

Hutchinson, P. (1985). *An Expert System for the Selection of Earth Retaining Structures*, M.Bdg.Sc. Thesis, Department of Architectural Science, University of Sydney.

Hutchinson, P. J., Rosenman, M. A., and Gero, J. S. (1987). RETWALL: an expert system for the selection and preliminary design of earth retaining structures, *Knowledge-Based Systems*, **1**(1):11–23.

Jain, D., and Maher, M. L. (1988). Combining expert systems and CAD techniques, *in* J. S. Gero and R. Stanton (eds.), *Artificial Intelligence: Developments and Applications*, North-Holland, Amsterdam, pp. 65–81.

Jones, C., and Thornley, D. (eds.) (1963). *Conference on Design Methods*, Pergamon, Oxford.

Jones, J. C. (1970). *Design Methods*, Wiley, New York.

Kant, I. (1781). *Critique of Pure Reason*, Anchor Books, Garden City, New York (reprint).

Kaplan, S., and Kaplan, R. (1982). *Cognition and Environment: Functioning in an Uncertain World*, Praeger, New York.

Kolodner, J. (ed.) (1988). *Case-Based Reasoning Workshop*, DARPA/ISTO, Morgan Kaufmann, San Mateo, California.

Koning, H., and Eizenberg, J. (1981). The language of the prairie: Frank Lloyd Wright's prairie houses, *Environment and Planning B*, **8**:295–323.

Kostem, C. L., and Maher, M. L. (eds.) (1986). *Expert Systems in Civil Engineering*, American Society of Civil Engineers, New York.

Kowalski, R. A. (1979). *Logic for Problem Solving*, North-Holland, Amsterdam.

Koestler, A. (1976). *The Act of Creation*, Hutchinson, London.

Krishnamurti, R. (1980). The arithmetic of shapes, *Environment and Planning B*, **7**:463–484.

Krishnamurti, R. (1981). The construction of shapes, *Environment and Planning B*, **8**:5–40.

Krishnamurti, R., and Giraud, C. (1984). Towards a shape editor, *Research Paper*, EdCAAD, University of Edinburgh, Edinburgh.

Kuhn, T. S. (1970). *The Structure of Scientific Revolutions*, University of Chicago Press, Chicago.

Laird, J. E., Newell, A., and Rosenbloom, P. S. (1987). SOAR: an architecture for general intelligence, *Artificial Intelligence*, **33**(1):1–64.

Langley, P., and Carbonell, J. S. (1984). Approaches to machine learning, *Journal of the American Society for Information Science*, **35**(5):306–316.

Lansdown, J. (1985). Computing in the creative professions, *Research Paper*, Department of Design Research, Royal College of Art, London.

Lansdown, J. (1987). The creative aspects of CAD: a possible approach, *Design Studies*, **8**(2):76–81.

Latombe, J.-C. (ed.) (1978). *Artificial Intelligence and Pattern Recognition in Computer-Aided Design*, North-Holland, Amsterdam.

Lawson, B. R. (1980). *How Architects Think*, Architecture Press, London.

Lemmon, E. J. (1971). *Beginning Logic*, Nelson, London.

Lenat, D. B. (1982). An artificial intelligence approach to discovery in mathematics as heuristic search, *in* R. Davis and D. B. Lenat, *Knowledge-Based Systems in Artificial Intelligence*, McGraw-Hill, New York.

Lenat, D. B., and Brown, J. S. (1983). Why AM and Eurisko appear to work, *Proceedings AAAI-83*, pp. 236–240.

Le Texier, J. Y., Doman, A., Marksjo, B. S., and Sharpe, R. (1985). Use of

Prolog with graphics including CAD, *AUSGRAPH85*, Australasian Computer Graphics Association, Brisbane, pp. 12–16.

Levine, B. (1982). The use of tree derivatives and a sample support parameter for inferring tree systems, *IEEE Trans. Pattern Anal. Machine Intelligence*, **PAMI-4**:25–34.

Lopez, L. A., Elam, S. L., and Christopherson, T. (1985). SICAD: a prototype implementation system for CAD, *Proceedings of the ASCE Third Conference on Computing Civil Engineering*, American Society of Civil Engineers, New York, pp. 84–94.

Luckman, J. (1967). An approach to the management of design, *Operational Research Quarterly*, **18**(4):345–358.

Lynch, K. (1960). *The Image of the City*, MIT Press, Cambridge, Massachusetts.

Mackenzie, C. (1988). Inducing relational grammars from design interpretations, *in* J. S. Gero and R. Stanton (eds.), *Artificial Intelligence Developments and Applications*, North-Holland, Amsterdam, pp. 315–328.

MacKinnon, D. W. (1970). Creativity: a multi-faceted phenomenon, *in* J. D. Roslansky (ed.), *Creativity: A Discussion at the Nobel Conference*, North-Holland, Amsterdam.

Maher, M. L. (1984). HI-RISE*: An Expert System for the Preliminary Structural Design of High Rise Buildings*, Ph.D. Thesis, Department of Civil Engineering, Carnegie Mellon University, Pittsburgh.

Maher, M. L. (ed.) (1987). *Expert Systems for Civil Engineers: Technology and Application*, American Society of Civil Engineers, New York.

Maher, M. L., and Fenves, S. J. (1985). HI-RISE: an expert system for the preliminary structural design of high rise buildings, *in* J. S. Gero (ed.), *Knowledge Engineering in Computer-Aided Design*, North-Holland, Amsterdam, pp. 125–164.

Maher, M. L., and Zhao, F. (1987). Using experience to plan the synthesis of new designs, *in* J. S. Gero (ed.), *Expert Systems in Computer-Aided Design*, North-Holland, Amsterdam, pp. 349–369.

Maher, M. L., Zhao, F., and Gero, J. S. (1988). Creativity in humans and computers, *in* J. S. Gero and T. Oksala (eds.), *Preprints Knowledge-Based Design in Architecture*, Helsinki University of Technology, Helsinki, pp. 29–44.

Manago, C., and Gero, J. S. (1986). Some aspects of the communication between expert systems and a CAD model, *Working Paper*, Architectural Computing Unit, Department of Architectural Science, University of Sydney, Sydney.

Mansfield, R. S., and Busse, T. V. (1981). *The Psychology of Creativity and Discovery*, Nelson-Hall, Chicago.

March, L. (1976). The logic of design and the question of value, *in* L. March (ed.), *The Architecture of Form*, Cambridge University Press, Cambridge, pp. 1–40.

March, L., and Stiny, G. (1985). Spatial systems in architecture and design: some history and logic, *Environment and Planning B*, **12**:31–53.

Markus, T. A., Whyman, P. , Morgan, J., Whitton, D., Mavert, J., Carter, D., and Fleming, J. (1972). *Building Performance*, Applied Science, London.

Mathlab Group (1977). *MACSYMA Reference Manual*, Computer Science Laboratory, Massachusetts Institute of Technology, Cambridge.

McCarthy, J. (1968). Programs with commonsense, *in* M. Minsky (ed.), *Semantic Information Processing*, MIT Press, Cambridge, Massachusetts.

McClelland, J. L., and Rumelhart, D. E. (eds.) (1986). *Parallel Distributed Processing: Explorations in the Microstructure of Cognition*, MIT Press, Cambridge, Massachusetts.

McClelland, J. L., Rumelhart, D. E., and Hinton, G. E. (1986). The appeal of parallel distributed processing, *in* D. E. Rumelhart and J. L. McClelland (eds.), *Parallel Distributed Processing: Explorations in the Microstructure of Cognition—Volume 1, Foundations*, MIT Press, Cambridge, Massachusetts, pp. 3–44.

McDermott, J. (1982). R1: A rule-based configurer of computer systems, *Artificial Intelligence*, **19**(1):39–88.

McDermott, J., and Doyle, J. (1980). Non-monotonic logic 1, *Artificial Intelligence*, **13**(1):41–72.

McLaughlin, S., and Gero, J. S. (1987). Acquiring expert knowledge from characterized designs, *Artificial Intelligence for Engineering Design, Analysis, and Manufacturing*, **1**(2):73–87.

McLaughlin, S., and Gero, J. S. (1988). Requirements of a reasoning system that supports creative and innovative design activity, *Working Paper*, Design Computing Unit, Department of Architectural Science, University of Sydney, Sydney.

Medewar, P. B. (1964). Is the scientific paper a fraud?, *in* D. Edge (ed.), *Experiment: A Series of Scientific Case Histories*, BBC Publications, London, pp. 7–12.

Mesarovic, M. D., Macko, D., and Takahara, Y. (1970). *The Theory of Hierarchical Multilevel Systems*, Academic Press, New York.

Michalski, R. S. (1983). A theory and methodology of inductive learning, *Artificial Intelligence*, **20**:111–161.

Michalski, R. S., Carbonell, J. G., and Mitchell, T. M. (eds.) (1983). *Machine Learning: An Artificial Intelligence Approach*, Tioga, Palo Alto, California.

Michalski, R. S., Carbonell, J. G., and Mitchell, T. M. (eds.) (1986). *Machine Learning: An Artificial Intelligence Approach—Volume II*, Morgan Kaufmann, Palo Alto, California.

Michalski, R. S., and Chilausky, R. L. (1981). Knowledge acquisition by encoding expert rules versus computer induction from examples: a case study involvng soybean pathology, *in* E. H. Mamdani and B. R. Gaines (eds.), *Fuzzy Reasoning and Its Applications*, Academic Press, London.

Michie, D., Muggleton, S., Riese, C., and Zubrick, S. (1984). Rule Master: a second-generation knowledge engineering facility, *The First Conference on Artificial Intelligence Applications*, IEEE Computer Society Press, Washington, D.C., pp. 591–597.

Minsky, M. (1975). A framework for representing knowledge, *in* P. H. Winston (ed.), *The Psychology of Computer Vision*, McGraw-Hill, New York, pp. 211–277.

Mitchell, J. R., and Radford, A. D. (1986). Adding knowledge to computer-aided detailing, *AUSGRAPH-86*, Australasian Computer Graphics Association, Sydney, New South Wales, Australia, pp. 31–35.

Mitchell, T. M., Keller, R., and Kedar-Cabelli, S. (1986). Explanation-based generalizations: a unifying view, *Machine Learning*, **1**(1):47–80.

Mitchell, W. J. (1977). *Computer-Aided Architectural Design*, Van Nostrand Reinhold, New York.

Mitchell, W. J., Steadman, J. P., and Liggett, R. S. (1976). Synthesis and optimization of small rectangular floor plans, *Environment and Planning B*, **3**:37–70.

Mittal, S., Dym, C. L., and Morjaria, M. (1986). PRIDE: an expert system for the design of paper handling systems, *in* C.L. Dym (ed.), *Applications of Knowledge-Based Systems to Engineering Analysis and Design*, American Society of Mechanical Engineers, New York, pp. 99–115.

Mizoguchi, F. (1983). Prolog based expert system, *New Generation Computing*, **1**(1):99–104.

Moneo, R. (1978). On typology, *Oppositions*, **13**:23–45.

Moore, G. T. (ed.) (1970). *Emerging Methods in Environmental Design and Planning*, MIT Press, Cambridge, Massachusetts.

Murthy, S. S., and Addanki, S. (1987). PROMPT: an innovative design tool, *in* J. S. Gero (ed.), *Expert Systems in Computer-Aided Design*, North-Holland, Amsterdam, pp. 323–341.

Navinchandra, D. (1987). *Exploring Innovative Designs by Relaxing Criteria and Reasoning from Precedent Knowledge*, Ph.D. Thesis, Department of Civil Engineering, MIT, Cambridge, Massachusetts.

Navinchandra, D., and Sriram, D. (1987). Anology-based engineering problem solving: an overview, *in* D. Sriram and R. A. Adey (eds.), *Artificial Intelligence in Engineering: Tools and Techniques*, CMP, Southampton, England, pp. 273–285.

Newell, A. (1973). Artificial intelligence and the concepts of mind, *in* R. C. Schank and K. M. Colby (eds.), *Computer Models of Thought and Language*, W. H. Freeman, San Francisco.

Newell, A., and Simon, H. A. (1972). *Human Problem Solving*, Prentice-Hall, Englewood Cliffs, New Jersey.

Nii, H. P., and Aiello, N. (1979). AGE (Attempt to Generalize): a knowledge-based program for building knowledge-based programs, *IJCAI-6*, Tokyo, Japan, pp. 645–655.

Nilsson, N. J. (1982). *Principles of Artificial Intelligence*, Springer-Verlag, Berlin.

Ogden, C. K., and Richards, I. A. (1923). *The Meaning of Meaning*, Kegan Paul, London.

O'Neill, J. L. (1987). Plausible reasoning, *Australian Computer Journal*, **19** (1):2–15.

Osborn, A. F. (1957). *Applied Imagination*, Scribner, New York.

Osherson, D. N., and Smith, E. E. (1981). On the adequacy of prototype theory as a theory of concepts, *Cognition*, **9**(1):35–58.

Oxman, R., and Gero, J. S. (1987). Using an expert system for design synthesis and design diagnosis, *Expert Systems*, **4**(1):4–15.

Oxman, R., and Gero, J. S. (1988). Designing by prototype refinement in architecture, *in* J. S. Gero (ed.), *Artificial Intelligence in Engineering: Design*, CMP/Elsevier, Amsterdam, pp. 395–412.

Pahl, G., and Beitz, W. (1984). *Engineering Design*, Springer-Verlag, New York.

Pereira, L. M. (1982). Logic control with logic, *Proceedings of the First International Logic Programming Conference*, Marseille, pp. 9–18.

Pham, D. T. (ed.) (1988). *Expert Systems in Engineering*, IFS Publications/ Springer-Verlag, Berlin.

Popper, K. (1972). *Conjectures and Refutations, the Growth of Scientific Knowledge*, Routledge and Kegan Paul, London.

Post, E. (1943). Formal reductions of the general combinatorial decision problems, *American Journal of Mathematics*, **65**:197–268.

Preziosi, D. (1979). *Architecture, Language and Meaning*, Mouton, The Hague, Holland.

Prieditis, A. (ed.) (1988). *Analogica*, Pitman/Morgan Kaufmann, Los Altos, California.

Quillian, M. R. (1968). Semantic memory, *in* M. Minsky (ed.), *Semantic Information Processing*, MIT Press, Cambridge, Massachusetts, pp. 216–270.

Quine, W. V., and Ullian, J. S. (1970). *The Web of Belief*, Random House, New York.

Quinlan, J. R. (1979). Discovering rules by induction from a large collection of examples, *in* D. Michie (ed.), *Expert Systems in the Micro-Electronic Age*, Edinburgh University Press, Edinburgh, pp. 168–201.

Quinlan, J. R. (1982). *Consistency and Plausible Inference*, Report, Rand Corporation, Santa Monica, California.

Quinlan, J. R. (1983). Inferno: a cautious approach to uncertain inference, *The Computer Journal*, **26**(3):255–269.

Quinlan, J. R. (1986). Induction of decision trees, *Machine Learning*, **1**:81–106.

Quinlan, J. R., Horn, K. A., Compton, P. J., and Lazarus, L. (1987). Inductive knowledge acquisition: a case study, *in* J. R. Quinlan (ed.), *Applications of Expert Systems*, Addison-Wesley, Sydney, pp. 157–173.

Radford, A. D., and Gero, J. S. (1988). *Design by Optimization in Architecture, Building and Construction*, Van Nostrand Reinhold, New York.

Robinson, J. A. (1965). A machine-oriented logic based on the resolution principle, *Journal of the ACM*, **12**:23–41.

Robinson, J. A. (1983). Logic programming—past, present and future, *New Generation Computing*, **1**:107–124.

Rosenman, M. A., Coyne, R. D., and Gero, J. S. (1987). Expert systems for design applications, *in* J. R. Quinlan (ed.), *Applications of Expert Systems*, Addison-Wesley, Sydney, pp. 66–84.

Rosenman, M. A., and Gero, J. S. (1985). Design codes as expert systems, *Computer-Aided Design*, **17**(9):399–409.

Rosenman, M. A., Manago, C., and Gero, J. S. (1986). A model-based expert system shell, *1AAI'86*, First Australasian Artificial Intelligence Congress, Melbourne, Victoria, Australia, pp. c:1:1–15.

Rowe, P. G. (1987). *Design Thinking*, MIT Press, Cambridge, Massachusetts.

Rumelhart, D. E., and McClelland, J. L. (1986). On learning the past tense of English verbs, *in* D. E. Rumelhart and J. L. McClelland (eds.), *Parallel Distributed Processing: Explorations in the Microstructure of Cognition— Volume 2, Psychological and Biological Models*, MIT Press, Cambridge, Massachusetts, pp. 216–271.

Rychener, M. D. (1985). Expert systems for engineering design, *Expert Systems*, **2**(1):30–44.

Rychener, M. (ed.) (1988). *Expert Systems for Engineering Design*, Academic Press, New York.

Sacerdoti, E. D. (1974). Planning in a hierarchy of abstraction spaces, *Artificial Intelligence*, **5**:115–135.

Sacerdoti, E. D. (1977). *A Structure for Plans and Behavior*, Elsevier, New York.

Sandler, B. (1985). *Creative Machine Design*, Paragon House, New York.

Schank, R. C. (1982). *Dynamic Memory: A Theory of Reminding and Learning in Computers and People*, Cambridge University Press, New York.

Schank, R. C. (1986). *Explanation Patterns: Understanding Mechanically and Creatively*, Lawrence Erlbaum, Hillsdale, New Jersey.

Schank, R. C., and Abelson, R. (1975). Scripts, plans and knowledge, *IJCAI-75*, Tbilisi, Georgia, USSR, pp. 151–157.

Schank, R. C., and Abelson, R. (1977). *Scripts, Plans, Goals and Understanding*, Lawrence Erlbaum, Hillsdale, New Jersey.

Schank, R., and Childers, P. (1988). *The Creative Attitude*, Macmillan, New York.

Shafer, G. (1976). *A Mathematical Theory of Evidence*, Princeton University Press, Princeton, New Jersey.

Shannon, C., and Weaver, W. (1949). *The Mathematical Theory of Communication*, University of Illinois Press, Urbana, Illinois.

Shortliffe, E. H., and Buchanan, B. G. (1975). A model of inexact reasoning in medicine, *Mathematical Biosciences*, **23**:351–379.

Simon, H. A. (1969). *The Sciences of the Artificial*, MIT Press, Cambridge, Massachusetts.

Simon, H. A. (1973). The structure of ill-structured problems, *Artificial Intelligence*, **4**:181–201.

Simon, H. A. (1983). Search and reasoning in problem solving, *Artificial Intelligence*, **21**:7–29.

Smith, A. (ed.) (1986). *Knowledge Engineering and Computer Modelling in CAD*, Butterworths, London.

Sowa, J. F. (1984). *Conceptual Structures*, Addison-Wesley, Reading, Massachusetts.

Spillers, W. R. (1972). On the use of examples in adaptive systems, *in* H. W. Robinson and D. E. Knight (eds.), *Cybernetics, Artificial Intelligence, and Ecology*, Spartan Books, New York.

Sriram, D., and Adey, R. (eds.) (1986a). *Applications of Artificial Intelligence in Engineering Problems—Volume I*, Springer-Verlag, Berlin.

Sriram, D., and Adey, R. (eds.) (1986b). *Applications of Artificial Intelligence in Engineering Problems—Volume II*, Springer-Verlag, Berlin.

Sriram, D., and Adey, R. (eds.) (1987a). *Artificial Intelligence in Engineering: Tools and Techniques*, Computational Mechanics Publications, Southampton, England.

Sriram, D., and Adey, R. (eds.) (1987b). *Knowledge-Based Expert Systems in Engineering: Planning and Design*, Computational Mechanics Publications, Southampton, England.

Sriram, D., and Adey, R. (eds.) (1987c). *Knowledge-Based Expert Systems in Engineering: Classification, Education and Control*, Computational Mechanics Publications, Southampton, England.

Stabler, P. (1986). Object-oriented programming in Prolog, *AI Expert*, **1**(10): 46–57.

Stanfill, C., and Waltz, D. (1986). Toward memory-based reasoning, *Communications of the ACM*, **29**(12):1213–1228.

Steadman, J. P. (1983). *Architectural Morphology*, Pion, London.

Stefik, M. (1981a). Planning with constraints (MOLGEN: Part 1), *Artificial Intelligence*, **16**:111–140.

Stefik, M. (1981b). Planning with constraints (MOLGEN: Part 2), *Artificial Intelligence*, **16**:141–169.

Stefik, M., and Bobrow, D. G. (1981). *The Loops Manual*, Technical Report KB.VLSI.81.13, Xerox Palo Alto Research Center, Palo Alto, California.

Stefik, M., and Bobrow, D. G. (1986). Objected-oriented programming—themes and variations, *AI Magazine*, **6**(4):40–62.

Stiny, G. (1980). Introduction to shape and shape grammars, *Environment and Planning B*, **7**:343–351.

Stiny, G. (1982). Shapes are individuals, *Environment and Planning B*, **9**:359–367.

Stiny, G., and Gips, J. (1978). *Algorithmic Aesthetics*, University of California Press.

Stiny, G., and Mitchell, W. J. (1978). The Palladian grammar, *Environment and Planning B*, **5**:5–18.

Storr, A. (1972). *The Dynamics of Creation*, Secker and Warburg, London.

Sussman, G. J. (1975). *A Computer Model of Skill Acquisition*, Elsevier, New York.

Swinson, P. S. G. (1983). Prolog: a prelude to a new generation of CAAD, *Research Paper*, EdCAAD, University of Edinburgh, Edinburgh.

Szalapaj, P., and Bijl, A. (1985). Knowing where to draw the line, *in* J. S. Gero (ed.), *Knowledge Engineering in Computer-Aided Design*, North-Holland, Amsterdam, pp. 147–165.

Talor, I. A. (1975). A retrospective view of creative investigation, *in* I. A. Talor and J. W. Getzels (eds.), *Perspective in Creativity*, Aldine, Chicago.

Tate, A. (1975). Interacting goals and their use, *IJCAI-75*, Tbilisi, Georgia, USSR, pp. 215–218.

ten Hagen, P., and Tomiyama, T. (eds.) (1987). *Intelligent CAD Systems I*, Springer-Verlag, Berlin.

Thomson, J. V. (1987). A water penetration expert system using PROLOG with graphics, *in* J. R. Quinlan (ed.), *Applications of Expert Systems*, Addison-Wesley, Sydney, pp. 48–65.

Tommelein, I. D., Johnson, M. V., Hayes-Roth, B., and Levitt, R. E. (1987). SIGHTPLAN: A blackboard expert system for construction site layout, *in* J. S. Gero (ed.), *Expert Systems in Computer-Aided Design*, North-Holland, Amsterdam, pp. 153–167.

Vernon, P. E. (1970). *Creativity*, Penguin, Harmondsworth, Middlesex, England.

Waldinger, R. (1977). Achieving several goals simultaneously, *in* E. W. Elcock and D. Michie (eds.), *Machine Intelligence 8*, Halstead/Wiley, New York.

Warren, D. H. D. (1974). WARPLAN: A system for generating plans, *Memo 76*, Department of Computational Logic, School of Artificial Intelligence, University of Edinburgh, Edinburgh.

Waterman, D. (1986). *A Guide to Expert Systems*, Addison-Wesley, Reading.

Waterman, D., and Hayes-Roth, F. (eds.) (1978). *Pattern-Directed Inference Systems*, Academic Press, New York.

Wilensky, R. (1983). *Planning and Understanding: A Computational Approach to Human Reasoning*, Addison-Wesley, Reading, Massachusetts.

Winston, P. H. (1970). *Learning Structural Descriptions from Examples*, Ph.D. Thesis, MIT, Cambridge, Massachusetts.

Winston, P. H. (ed.) (1975). *The Psychology of Computer Vision*, McGraw-Hill, New York.

Winston, P. H. (1984). *Artificial Intelligence*, 2nd ed., Addison-Wesley, Reading, Massachusetts.

Wittgenstein, L. (1922). *Tractatus Logico-Philosophicus*, Routledge and Kegan Paul, London.

Woodbury, R. and Oppenheim, I. (1988). Geometric reasoning: concepts and demonstration, *in* J. S. Gero (ed.), *Artificial Intelligence in Engineering: Robotics and Processes*, CMP/Elsevier, Amsterdam, pp. 61–75.

Yoshikawa, H., and Warman, E. A. (eds.) (1987). *Design Theory for CAD*, North-Holland, Amsterdam.

Zadeh, L. A. (1975). Fuzzy logic and approximate reasoning, *Synth'ese*, **30**: 407–428.

Zadeh, L. A. (1983). The role of fuzzy logic in the management of uncertainty in expert systems, *Fuzzy Sets and Systems*, **11**(3):199–228.

Journals

Artificial Intelligence in Engineering, published by Computational Mechanics Inc., Southampton, England, and Billerica, Massachusetts.

Artificial Intelligence for Engineering Design, Analysis and Manufacturing, published by Academic Press, London and New York.

AI Magazine, published by the American Association for Artificial Intelligence, Menlo Park, California.

Artificial Intelligence, published by North-Holland, Amsterdam.

Engineering Applications of Artificial Intelligence, published by Pineridge Periodicals, Swansea, UK.

Knowledge-Based Systems, published by Butterworths, Guildford, England.

Index